Lecture Notes in Mathematics

[illegible] Dold, F. Takens and B. Teissier

Editorial [illegible]

[illegible] the [illegible] of monographs

[illegible]

[illegible] should [illegible] in [illegible] [illegible] [illegible] several

chapters.

A [illegible] should [illegible] that [illegible] incomplete or [illegible] close to [illegible]

[illegible]

3. Manuscripts should [illegible]

[illegible] should contain at least 100 pages of [illegible] text, and

[illegible]

[illegible] of contents;

[illegible] introduction, with adequate motivation and [illegible]

[illegible] remarks, should be accessible [illegible]

[illegible]

[illegible] this is [illegible]

Lecture Notes in Mathematics

Edited by A. Dold, F. Takens and B. Teissier

Editorial Policy
for the publication of monographs

1. Lecture Notes aim to report new developments in all areas of mathematics – quickly, informally and at a high level. Monograph manuscripts should be reasonably self-contained and rounded off. Thus they may, and often will, present not only results of the author but also related work by other people. They may be based on specialized lecture courses. Furthermore, the manuscripts should provide sufficient motivation, examples and applications. This clearly distinguishes Lecture Notes from journal articles or technical reports which normally are very concise. Articles intended for a journal but too long to be accepted by most journals, usually do not have this "lecture notes" character. For similar reasons it is unusual for doctoral theses to be accepted for the Lecture Notes series.

2. Manuscripts should be submitted (preferably in duplicate) either to one of the series editors or to Springer-Verlag, Heidelberg. In general, manuscripts will be sent out to 2 external referees for evaluation. If a decision cannot yet be reached on the basis of the first 2 reports, further referees may be contacted: the author will be informed of this. A final decision to publish can be made only on the basis of the complete manuscript, however a refereeing process leading to a preliminary decision can be based on a pre-final or incomplete manuscript. The strict minimum amount of material that will be considered should include a detailed outline describing the planned contents of each chapter, a bibliography and several sample chapters.
Authors should be aware that incomplete or insufficiently close to final manuscripts almost always result in longer refereeing times and nevertheless unclear referees' recommendations, making further refereeing of a final draft necessary.
Authors should also be aware that parallel submission of their manuscript to another publisher while under consideration for LNM will in general lead to immediate rejection.

3. Manuscripts should in general be submitted in English.
Final manuscripts should contain at least 100 pages of mathematical text and should include
- a table of contents;
- an informative introduction, with adequate motivation and perhaps some historical remarks: it should be accessible to a reader not intimately familiar with the topic treated;
- a subject index: as a rule this is genuinely helpful for the reader.

Lecture Notes in Mathematics 1732

Springer
Berlin
Heidelberg
New York
Barcelona
Hong Kong
London
Milan
Paris
Singapore
Tokyo

Karsten Keller

Invariant Factors, Julia Equivalences and the (Abstract) Mandelbrot Set

Springer

Author

Karsten Keller
Institute of Mathematics and Computer Science
Ernst-Moritz-Arndt University
Jahnstr. 15a
17487 Greifswald, Germany

E-mail: keller@mail.uni-greifswald.de

Cataloging-in-Publication Data applied for

Die Deutsche Bibliothek - CIP-Einheitsaufnahme

Keller, Karsten:
Invariant factors, Julia equivalences and the (abstract) Mandelbrot set / Karsten Keller. - Berlin ; Heidelberg ; New York ; Barcelona ; Hong Kong ; London ; Milan ; Paris ; Singapore ; Tokyo : Springer, 2000
(Lecture notes in mathematics ; 1732)
ISBN 3-540-67434-9

Mathematics Subject Classification (2000): 30D05, 54H20, 37B10

ISSN 0075-8434
ISBN 3-540-67434-9 Springer-Verlag Berlin Heidelberg New York

Typesetting: Camera-ready TEX output by the author
Printed on acid-free paper SPIN: 10724999 41/3143/du 543210

To Anke, Dörthe, Svea and Wiete

Preface

This monograph is primarily concerned with the combinatorial structure of quadratic dynamics. It provides an enhanced and elaborated version of my habilitation thesis at Greifswald University, submitted at the end of 1995 and having its root in a series of papers in collaboration with my teacher and colleague Christoph Bandt. I want to thank first of all him for all his ideas and his immense share of work now hidden in this monograph.

I also want to express my deep gratitude to Dierk Schleicher and Christopher Penrose for fruitful discussions and for sharing many of their ideas. Both have discussed the combinatorics of complex quadratic dynamics by using approaches similar to that presented here.

I should thank Adrien Douady and John Hubbard for opening the interesting research field of quadratic dynamics and William P. Thurston for providing the basis of Chapters 2 and 3. Further, I want to underline that Chapter 4 has greatly benefited from the work of John Milnor and Dierk Schleicher.

It is often not easy to trace back results and ideas to their rightful owners, particularly if they are known as 'folklore' but nobody has ever written them down. I apologize for any omission in this respect.

I am grateful to Rebecca Heier and to my colleagues Christoph Bandt, Petra Gummelt, Ekkehard Krause and Andreas Seib, who read the whole manuscript or parts of it, and eliminated a lot of language errors and misprints. Last, but not least, I thank my family for their unlimited patience.

Greifswald, January 2000 *Karsten Keller*

Contents

1. Introduction

1.1 Quadratic iteration and Julia equivalences

1.1.1 Some introductory remarks

A little bit of history. When DOUADY gave a survey on Julia sets and the Mandelbrot set in [121], he started with the words

> *... I must say that, in 1980, whenever I told my friends that I was just starting with* J. H. HUBBARD *a study of polynomials of degree 2 in one complex variable (and more specifically those of the form* $z \mapsto z^2 + c$*), they would all stare at me and ask: Do you expect to find anything new?*
>
> A. Douady in: PEITGEN, RICHTER [121]

Although the book appeared in 1986, quadratic polynomials are still in the focus of mathematical research. More precisely, many people are interested in the iteration of the family of quadratic maps p_c given by

$$p_c(z) = z^2 + c$$

and acting on the complex plane $\mathbb{C}$, where the parameter c runs through $\mathbb{C}$. The central question is how the structure of an orbit

$$\{z, p_c(z), p_c(p_c(z)), p_c(p_c(p_c(z))), \ldots\}$$

of a point z changes when z and the parameter c vary. Manifold facets of this question were illustrated by fascinating computer experiments; for example, see PEITGEN and RICHTER's beautiful book [121]. The published mathematical literature on quadratic iteration however is quite diffuse, which was one reason for writing this book.

The study of holomorphic maps under iteration was already started at the end of the last century. It goes back to CAYLEY [26], who investigated Newton's method for complex mappings, and to SCHRÖDER [153], KOENIGS [87], LEAU [94] and BÖTKHER [17], who analyzed the local behavior of holomorphic maps at fixed points in general.

The first who studied the global behavior of iterated rational maps were FATOU [46, 47, 48, 49] and JULIA [70] in the early twenties. They observed that by the action of such maps the complex sphere divides into a stable and an instable set, today named the Fatou set and the Julia set, respectively.

For the maps p_c the unstable set consists of those points z where the orbits change from being unbounded to being bounded: By the action of p_c, the complex plane splits into the *filled-in Julia set* K_c, consisting of all points with bounded orbits, and its complement. The common boundary of K_c and $\mathbb{C} \setminus K_c$ is called the *Julia set* of p_c and is denoted by J_c.

FATOU [46, 47, 48, 49] and JULIA started to classify the components of the stable set for rational maps. Then, with the exception of some substantial progress by CREMER [30], SIEGEL [156] and a few other mathematicians, there followed a period of dormancy.

The second period of interest in the subject began in 1980, and it is not surprising that the revival of complex iteration theory fell in the time of modern computer graphics: what was found by FATOU and JULIA could now be seen on a computer display, and the pictures obtained were so fascinating that they forced the further development of the mathematical theory. In particular, the *Mandelbrot set*

$$M = \{c \in \mathbb{C} \mid J_c \text{ is connected}\}$$

became extremely popular, also in the non-mathematical world.

SULLIVAN [159, 163] introduced quasi-conformal techniques into the field of holomorphic dynamics, and his proof of non-wandering domains allowed completion of the classification of components in the stable set. DOUADY and HUBBARD, the pioneers in modern complex quadratic dynamics, started a deep exploration of the Mandelbrot set in their famous papers [43].

In the meantime the investigations were extended to higher degree polynomials, to entire functions and to several variables, but there remained open problems concerning the quadratic case. In particular, the celebrated problem of local connectivity of the Mandelbrot set is still open, although substantial progress was made by YOCCOZ (see [66]) and LYUBICH [101, 99, 100].

What we want to discuss. The main part of the work presented is devoted to complex quadratic dynamics, but the work is only partially about quadratic maps in the complex plane. Our aim is to point out that much of complex quadratic dynamics can be understood from the pure combinatorial point of view, and we will try to convince the reader that 'quadratic combinatorics' has a right to exist along with the theory in the complex plane.

The conceptual kernel of our work is a self-contained abstract counterpart of connected quadratic Julia sets and of the Mandelbrot set, presented as Chapters 2 and 3. It is built on the base of THURSTON's quadratic invariant laminations given in his unpublished but widely circulated paper [167] and of symbolic descriptions of the angle-doubling map.

However, this whole text were extremely dry if this kernel would not be embedded into the 'concrete' complex plane theory. Therefore, in Chapter 1 we will outline the few facts from quadratic iteration being necessary for motivating our further discussion. In particular, we will display some central well-known and some partially new statements on the structure of the Mandelbrot set and of quadratic iteration.

Chapter 4 provides rigorous proofs of these and some further statements by linking the theory obtained in Chapters 2 and 3 to the complex plane. In some parts we will refer to the survey [114] by MILNOR and the monographs [25] by CARLESON and GAMELIN, [160] by STEINMETZ, and [108] by MCMULLEN, and we will use some special statements and ideas in GOLDBERG and MILNOR [58], MILNOR [117], PETERSEN [129], and SCHLEICHER [150].

In our abstract approach to quadratic Julia sets and to the Mandelbrot set, we share many ideas with C. PENROSE [123, 124] and with LAU and SCHLEICHER [92, 148], who developed an abstract theory into different directions: PENROSE's work is mainly based on quotients of the shift, and his level of abstraction goes far beyond that of our work. LAU and SCHLEICHER emphasize the Mandelbrot set and the geometric point of view. Their concept of internal addresses allowing a compact description of the Mandelbrot set is involved in our work.

Quadratic iteration is a special topic in the theory of discrete dynamical systems. Nevertheless, we will not need much about this theory. For some general background on an adequate level the reader is referred to DEVANEY's book [36] on chaotic dynamical systems. In order to facilitate the reader's orientation, we have labelled some statements in our text as follows.

This is one of the statements mentioned in the Introduction (Chapter 1) and recommended to look at for getting a broad overview.

---

and

This is another main statement.

Also note that Chapter 4.1.1 interpreting some of the results of Chapter 2 in the dynamical plane does not need any statement of Chapter 3.

1.1.2 The framework: Topological dynamics and symbolics

The basic topological dynamical system. Our first view to quadratic iteration is devoted to the simplest case $p_0(z) = z^2$. This case is the starting point for presenting some basic concepts and ideas.

Clearly, the filled-in Julia set of p_0 coincides with the closed unit disk, and the Julia set is equal to the unit circle. We identify the latter with $\mathbb{T} = \mathbb{R}/\mathbb{Z} = [0,1[$ via $\beta \longleftrightarrow e^{2\pi\beta i}; \beta \in [0,1[$.

With this identification, the restriction of p_0 onto the unit circle is no more than the *angle-doubling map* defined by $h(\beta) = 2\beta \bmod 1$ for all $\beta \in \mathbb{T}$. Each point on $\mathbb{T}$ has two preimages with respect to h. If one - say β - is fixed, then the other one is given by β', where $'$ denotes the rotation by 180^0, defined by $\beta' = (\beta + \frac{1}{2}) \bmod 1$ for all $\beta \in \mathbb{T}$. Clearly, $'$ provides the restriction of the map $- : z \mapsto -z$ to the unit circle. The topological dynamical system

$$(\mathbb{T}, h, ') \tag{1.1}$$

will be in the center of our interest. Before saying more about this system, let us give the concepts from topological dynamics we will use.

Definition 1.1. (Basic concepts from topological dynamics)
1. By a topological dynamical system $(X, (f_i)_{i\in I})$ *we understand a topological space X equipped with a family $(f_i)_{i\in I}$ of continuous maps. (Instead of $(X, (f_i)_{i=1}^n)$, we also write $(X, f_1, f_2, \ldots, f_n)$.)*

2. Two topological dynamical systems $(X, (f_i)_{i\in I})$ and $(Y, (g_i)_{i\in I})$ are said to be conjugate *(*semi-conjugate*) if there exists a homeomorphism (continuous map) ϕ from X onto Y satisfying $\phi \circ f_i = g_i \circ \phi$ for all $i \in I$. Then the map ϕ is called a* conjugacy *(*semi-conjugacy*).*

Invariance of equivalence relations and Julia equivalences. Next look at the topological dynamical system $(J_{-2}, p_{-2}, -)$. It is well known that J_{-2} coincides with the real interval $[-2, 2]$. At first glance, the following is surprising: The map $\beta \mapsto 2\cos 2\pi\beta$ forms a semi-conjugacy from the topological dynamical system $(\mathbb{T}, h, ')$ onto $(J_{-2}, p_{-2}, -)$.

So the interval 'arises' from $\mathbb{T}$ by identifying each β between 0 and $\frac{1}{2}$ with $1 - \beta$, and the identification is compatible with a given dynamical structure, namely with the maps h and $'$.

Generally, compatibility of identification with some dynamical structure will be treated in the framework of invariant equivalence relations:

Definition 1.2. (Identification compatible with dynamics)
1. For an equivalence relation $\equiv$ on a set X denote the transitive hull $\{x \mid x \equiv a \text{ for some } a \in A\}$ of $A \subseteq X$ by $[A]_\equiv$, and let $X/\!\equiv\, = \{[x]_\equiv \mid x \in X\}$ be the set of all equivalence classes. Let $\phi_\equiv$ be the natural projection from X onto $X/\!\equiv$, defined by $\phi_\equiv(x) = [x]_\equiv$ for all $x \in X$. If X is a topological space, we equip $X/\!\equiv$ with the quotient topology. (A set is open if its preimage with respect to $\phi_\equiv$ is.)

2. Let f be a continuous map acting on a topological space X. An equivalence relation $\equiv$ on X being closed as a subset of the product space $X \times X$ *is said to be* f-invariant *and* completely f-invariant *if f maps each equivalence* *class into and onto an equivalence class, respectively. The* factor map $f/\equiv$ *is defined by* $(f/\equiv)([x]_\equiv) = [f(x)]_\equiv$ *for* $x \in X$.

3. If $(X, (f_i)_{i\in I})$ *is a topological dynamical system, then* $\equiv$ *is said to be* **(completely) invariant** *(on* $(X, (f_i)_{i\in I})$*) if* $\equiv$ *is (completely)* f_i*-invariant for all* $i \in I$. *In this case the factor space* $X/\equiv$ *is called a* (completely) invariant *factor of* X *(with respect to* $(f_i)_{i\in I}$*) and the topological dynamical system* $(X, (f_i)_{i\in I})/\equiv \;:=\; (X/\equiv, (f_i/\equiv)_{i\in I})$ *a* (completely) invariant *factor of* $(X, (f_i)_{i\in I})$.

Remark. f-invariance of $\equiv$ guarantees the existence of the factor map $f/\equiv$ in 3., and one easily sees that this map is continuous.

There are various reasons for studying invariant equivalence relations. One is to look for simple 'dynamical' factorizations of simple topological spaces in order to get good descriptions of more complicated ones. For example, the idea to investigate 'rigid' topological structure by discovering 'hidden dynamical structure' has influenced the research in continuum theory (see Appendix A.3).

However, we are mainly interested in the description of complicated dynamics by the identification of simpler one. Some conceptional background for our discussion is given in Appendix A, which we will refer to only a few times. On the whole, the text should be understandable without the appendix.

Let us come back to the system $(J_{-2}, p_{-2}, -)$, and let $\approx$ be the equivalence relation on $\mathbb{T}$ which is defined by $\beta_1 \approx \beta_2$ iff $\beta_1 = \beta_2$ or $\beta_1 + \beta_2 = 1$. Clearly, $\approx$ is completely invariant on (1.1), and the topological dynamical systems $(J_{-2}, p_{-2}, -)$ and $(\mathbb{T}, h, ')/\approx$ are conjugate.

Generally, connected Julia sets J_c are strongly related to special completely invariant equivalence relations on (1.1), which we will call ***Julia equivalences*** and which we will develop on the base of THURSTON's theory of quadratic invariant laminations (see [167]) in combination with symbolic descriptions of the angle-doubling map. Julia equivalences are our frame for discussing the combinatorics of quadratic dynamics. Note that $\approx$ as constructed above is a Julia equivalence.

A generalization of binary expansion. In order to explain what kind of symbolic descriptions we will use, recall a well known statement from number theory: for each $\beta \in [0,1]$ there exists a 0-1-sequence $\mathbf{b}(\beta) = b_1 b_2 b_3 \ldots$ with $\beta = \sum_{i=1}^{\infty} \frac{b_i}{2^i}$, namely the *binary expansion* of β.

Two sequences of the form $\mathbf{w}0\overline{1}$ and $\mathbf{w}1\overline{0}$ for some 0-1-word $\mathbf{w}$ represent the same real number, and multiplication of a binary number by 2 means no more than shifting it by one digit to the left. In particular, this provides an interesting representation of our system (1.1):

On the set $\{0,1\}^{\mathbb{N}}$ of all 0-1-sequences with product topology consider the equivalence relation $\equiv_0$ which for each 0-1-word $\mathbf{w}$ identifies $\mathbf{w}0\overline{1}$ with $\mathbf{w}1\overline{0}$ and beyond this $\overline{0}$ with $\overline{1}$. (The latter is necessary since 0 and 1 coincide in

$\mathbb{T} = \mathbb{R}/\mathbb{Z}$.) Clearly, h corresponds to the shift-map σ which assigns to each 0-1-sequence $s_1s_2s_3\ldots$ the sequence $s_2s_3s_4\ldots$ and $'$ to the map $^{\perp}$ changing the first symbol of a 0-1-sequence. (1.1) is conjugate to $(\{0,1\}^{\mathbb{N}}, \sigma, ^{\perp})/\equiv_0$.

To get a unique binary expansion, we want to exclude the representation by a sequence $\mathbf{w}0\overline{1}$ for $\mathbf{w} \in \{0,1\}^*$. Then binary coding of points $\beta \in \mathbb{T}$ can easily be described within dynamics: $\alpha = 0$ has the two preimages $\frac{\alpha}{2} = 0$ and $\frac{\alpha+1}{2} = \frac{1}{2}$ dividing $\mathbb{T}$ into the semi-circles $[0, \frac{1}{2}[$ and $[\frac{1}{2}, 1[$. The n-th digit b_n of $\mathbf{b}(\beta)$ is 0 if $h^{n-1}(\beta)$ lies in the first semi-circle and 1 otherwise.

Now take a division of $\mathbb{T}$ induced by another $\alpha \in \mathbb{T}$. Consider the two disjoint open semi-circles $]\frac{\alpha}{2}, \frac{\alpha+1}{2}[$ and $]\frac{\alpha+1}{2}, \frac{\alpha}{2}[$, where intervals are taken in counter-clockwise direction, and generalize binary expansion as follows.

Definition 1.3. (Itinerary and kneading sequence)
For given points $\alpha, \beta \in \mathbb{T}$, the sequence

$$I^{\alpha}(\beta) = s_1s_2s_3\ldots \quad \textit{with} \quad s_i = \begin{cases} 0 & \textit{for} \;\; h^{i-1}(\beta) \in \,]\frac{\alpha}{2}, \frac{\alpha+1}{2}[\\ 1 & \textit{for} \;\; h^{i-1}(\beta) \in \,]\frac{\alpha+1}{2}, \frac{\alpha}{2}[\\ * & \textit{for} \;\; h^{i-1}(\beta) \in \{\frac{\alpha}{2}, \frac{\alpha+1}{2}\} \end{cases} \tag{1.2}$$

is said to be the itinerary *of the point β with respect to α. Further, the sequence $\hat{\alpha} = I^{\alpha}(\alpha)$ is called the* kneading sequence *of the point α.*

There is a slight difference between binary expansion and the concept of an itinerary in the case $\alpha = 0$. The symbol $*$ does not distinguish between the points $0, \frac{1}{2} \in \mathbb{T}$. It stands for both symbols 0 and 1, and this 'wild card' role will be substantial for general $\alpha \in \mathbb{T}$.

Itineraries will be used to describe the identification induced by Julia equivalences. To illustrate this, let $\alpha = \frac{1}{2}$ and consider the equivalence relation $\approx$ given above. If $\beta \approx \tilde{\beta}$ for different $\beta, \tilde{\beta} \in \mathbb{T}$, hence $\beta + \tilde{\beta} = 1$, then also $h(\beta) + h(\tilde{\beta}) = 1$. So by induction one shows $I^{\alpha}(\beta) = I^{\alpha}(\tilde{\beta})$.

Conversely, for different $\beta, \tilde{\beta}$ with $\beta + \tilde{\beta} \neq 1$ let $b_1b_2b_3\ldots$ and $\widetilde{b_1}\widetilde{b_2}\widetilde{b_3}\ldots$ be the (different) binary expansions. Fix some $n \in \mathbb{N}$ with $b_n = \widetilde{b_n}$ and $b_{n+1} \neq \widetilde{b_{n+1}}$, or if this is impossible with $b_n \neq \widetilde{b_n}, b_{n+1} = \widetilde{b_{n+1}}$ but $b_{n+2}b_{n+3}\ldots \neq \widetilde{b_{n+2}}\widetilde{b_{n+3}}\ldots$. Since binary expansion in $[\frac{1}{4}, \frac{3}{4}[$ starts with 01 or 10 and in $[\frac{3}{4}\frac{1}{4}[$ with 00 or 11, the itineraries $I^{\alpha}(\beta)$ and $I^{\alpha}(\tilde{\beta})$ are different.

So we have $\beta \approx \tilde{\beta}$ iff $I^{\alpha}(\beta) = I^{\alpha}(\tilde{\beta})$, and $\alpha = \frac{1}{2}$ determines the equivalence relation $\approx$. For each $\alpha \in \mathbb{T}$ we will get a Julia equivalence and a description similar to that for $\alpha = \frac{1}{2}$: roughly speaking, equal itineraries imply equivalence, where the symbol $*$ plays its 'wild card' role.

An immediate consequence of the symbolic descriptions is that many (abstract) Julia sets can be considered as shift-invariant factors of the space of all 0-1-sequences (see Appendix A.2 and Chapter 4.2.2).

Remark. The concepts of an itinerary and a kneading sequence have been introduced by MILNOR and THURSTON for the investigation of real quadratic polynomials (see [118]), and their use in the context of complex quadratic polynomials (in a more or less explicit form) is due to different authors (compare [2, 6, 24, 67]).

1.1.3 Prerequisites from quadratic iteration

Let us start with some simple notions and agreements concerning iteration in general.

Definition 1.4. (Orbits, iterates, and (pre)periodic points)

1. For a map f acting on a set X, the set $\{f^0(x) = x, f(x), f^2(x) = f(f(x)), \ldots\}$ is called the orbit *of a point x, and its elements the* iterates *of x. (By definition, each point belongs to its orbit.) Analogously, all points y with $f^n(y) = x$ for some $n \in \mathbb{N}_0$ are said to be the* backward iterates *of x and the set of them is called the* backward orbit *of x.*

2. x (and its orbit) is periodic *of* period *$m = PER(x)$ if $f^m(x) = x$ for some $m \in \mathbb{N}$ and $f^n(x) \neq x$ for $n = 1, 2, \ldots, m-1$.* (We emphasize that there will be some slightly different usage of the notation *PER* according to Definition 3.9.) *In particular, a* fixed point *is a periodic point of period one.*

3. By a preperiodic *point x we understand a non-periodic point x with finite orbit. Its* preperiod *is the minimum number n for which $f^n(x)$ is periodic, and its* period *is that of $f^n(x)$.*

Now consider the quadratic polynomials p_c on the complex plane $\mathbb{C}$, or their extensions by $p_c(\infty) = \infty$ to the Riemann sphere $\overline{\mathbb{C}} = \mathbb{C} \cup \{\infty\}$. Note that the restriction to the family of these quadratic polynomials yields each possible quadratic dynamic behavior since each quadratic polynomial is conjugate to one of the form p_c by a linear map.

Let us provide some prerequisites being necessary for the following. The list of the well-known statements will be continued at the beginning of Chapter 4. The standard reference is [43], and further we refer to [12, 25, 160, 114, 97, 15, 58].

I. p_0, and p_c near ∞. Recall that the filled-in Julia set of the simple quadratic polynomial p_0 is the unit disk. On its complement consider the rays $\mathbf{R}_0^\beta$ consisting of all points in $\mathbb{C} \setminus K_c$ whose argument is equal to $2\pi\beta$ for some given $\beta \in \mathbb{T}$. Then $p_0(\mathbf{R}_0^\beta) = \mathbf{R}_0^{h(\beta)}$ and $-\mathbf{R}_0^\beta = \mathbf{R}_0^{\beta'}$.

The crucial point for understanding the combinatorial structure of the Julia set corresponding to a general quadratic polynomial p_c given, is that p_c is 'near' to p_0 in a neighborhood of the infinity point ∞: the more a point $z \in \mathbb{C}$ approaches ∞, the more the influence of the term c in the formula $z^2 + c$ decreases. In fact, it was shown by BÖTKHER [17] at the beginning of this century that between sufficiently small neighborhoods of ∞ there exists a unique conformal isomorphism Φ_c with $\Phi_c(\infty) = \infty$ and $\lim_{z \to \infty} \Phi_c(z)/z = 1$

which conjugates p_c and p_0, i.e. satisfies $\Phi_c p_c = p_0 \Phi_c$ (compare also [114]). It is called the *Bötkher map.*

II. Connectivity of the Julia set, and the postcritical orbit. One can try to extend the domain where Φ_c is defined as a conformal conjugacy along the backward iterates of points near ∞ step by step and as far as possible. Of course, the extension must be unique, and the following is valid (see DOUADY and HUBBARD [43] and compare [58]):

- THE CONNECTED CASE: If $c \in K_c$, then Φ_c is defined on the whole complement of K_c and maps onto the complement of the closed unit disk. In particular, J_c is connected.
- THE DISCONNECTED CASE: If $c \notin K_c$, then Φ_c is defined on an open connected set containing c. Thus there exists an $\alpha \in \mathbb{T}$ with $\Phi_c(c) \in \mathbf{R}_0^\alpha$. The map Φ_c cannot be defined completely at the backward iterates of c since the map p_c is not invertible at c, but p_0 is invertible at $\Phi_c(c)$. However, the definition domain of Φ_c can be chosen as an open set whose image contains all rays $\mathbf{R}_0^\beta$ with $h^n(\beta) \neq \alpha$ for all $n = 1, 2, 3, \ldots$. Moreover, J_c is equal to K_c, and (J_c, p_c) is topologically conjugate to the shift space $(\{0,1\}^{\mathbb{N}}, \sigma)$.

Note that if $\Phi_{c_0}(z)$ is defined for some parameter c_0 and some $z_0 \in \mathbb{C} \setminus K_c$, then for c in some sufficiently small neighborhood of c_0 the map $c \mapsto z = \Phi_c^{-1}(\Phi_{c_0}(z_0))$ is continuous.

The above statements indicate that the structure of the whole map p_c is strongly determined by the properties of the *postcritical orbit*, the orbit of the *postcritical point* c. The reason for this is that c is the only point in $\mathbb{C}$ at which p_c is not invertible. Its unique preimage, the *critical point* 0, is the only finite point where the derivative vanishes.

III. Types of periodic orbits. If $z \in \mathbb{C}$ is a periodic point of period m, then by the chain rule the value $(p_c^m)'(z)$ of the derivative of p_c^m at z satisfies

$$(p_c^m)'(z) = 2^m \prod_{i=0}^{m-1} p_c^i(z). \tag{1.3}$$

So $(p_c^m)'$ yields the same number for all points in the orbit of z, which is called the *multiplier* of z and of the orbit determined by z.

A periodic orbit and the points contained in it are said to be *attractive* (*repelling, indifferent*) if the corresponding multiplier has absolute value greater than 1 (less than 1, equal to 1). Moreover, depending on whether the argument of the multiplier (relative to 2π) is rational or irrational, one speaks of a *parabolic* or an *irrationally indifferent* periodic orbit (point).

Among the attractive periodic orbits and points we consider the *superattractive* ones - those with multiplier 0. The following facts will be crucial:

1. Each quadratic polynomial possesses at most one non-repelling periodic

orbit in $\mathbb{C}$. This is a special case of a result of DOUADY for polynomials (see [25], Theorem 1.2). Note that ∞ is a superattractive fixed point (with respect to the local coordinates $z \mapsto \frac{1}{z}$).

2. Each attractive or parabolic periodic orbit and each irrationally indifferent periodic orbit contained in J_c belongs to the closure of the postcritical orbit. In particular, in the attractive case, the periodic orbit attracts the postcritical one: if the orbit is superattractive, by formula (1.3) it coincides with the postcritical orbit, otherwise the attractive orbit is equal to the set of accumulation points of the postcritical orbit (see also Chapter 4.1.1).

3. In the case $c \in K_c$, it holds $c \notin J_c$ iff p_c has an attractive or parabolic periodic orbit.

4. J_c is the closure of the set of repelling periodic points.

IV. Dynamic rays and their landing. Let us give the concepts which link complex quadratic iteration to the abstract approach we will develop.

Definition 1.5. (Dynamic rays)
Let $\beta \in \mathbb{T}$ and for $c \notin K_c$ assume that $\Phi_c(c) \notin \mathbf{R}_0^{h^n(\beta)}$ for all $n \in \mathbb{N}$. Then $\mathbf{R}_c^\beta := \Phi_c^{-1}(\mathbf{R}_0^\beta)$ is well-defined and called a dynamic ray. *$\mathbf{R}_c^\beta$ is said to* land *at a point z of J_c if $\lim_{r\to 1} \Phi_c^{-1}(re^{2\pi\beta i})$ exists and is equal to z, where z is called a* landing point *of $\mathbf{R}_c^\beta$ and β an* external angle *of z.*

Dynamic rays need not land, and it is not easy to decide whether they do. However, the following statements provide the minimal generally known set of landing dynamic rays (compare [43]):

- THE CONNECTED CASE: If $c \in K_c$, then each dynamic ray $\mathbf{R}_c^\beta$ for rational $\beta \in \mathbb{T}$ lands at a point of J_c. The landing point is a repelling or parabolic periodic point if β is periodic, and it is preperiodic else. Conversely, each repelling or parabolic periodic point is the landing point of at least one but finitely many dynamic rays $\mathbf{R}_c^\beta$ with periodic $\beta \in \mathbb{T}$.
- THE DISCONNECTED CASE: If $c \notin K_c$, then each dynamic ray $\mathbf{R}_c^\beta$ with $\Phi_c(c) \notin \mathbf{R}_0^{h^n(\beta)}$ for all $n \in \mathbb{N}$ lands at a point of J_c, and if β is rational, the landing point is repelling.

The first (non-trivial) statement above is due to DOUADY (For a proof see MILNOR [114] and PETERSEN [129], and for the second statement compare also ATELA [2], GOLDBERG and MILNOR [58].)

Let $\beta \in \mathbb{T}$ and, if J_c is disconnected, assume that $\Phi_c(c) \notin \mathbf{R}_0^{h^n(\beta)}$ for all $n \in \mathbb{N}$. Then, since Φ_c conjugates p_c and p_0, we have

$$p_c(\mathbf{R}_c^\beta) = \mathbf{R}_c^{h(\beta)}. \tag{1.4}$$

From $\mathbf{R}_c^\beta \cup -\mathbf{R}_c^\beta = p_c^{-1}(\mathbf{R}_c^{h(\beta)}) = \mathbf{R}_c^\beta \cup \mathbf{R}_c^{\beta'}, \mathbf{R}_c^{\beta'} \cup -\mathbf{R}_c^{\beta'} = p_c^{-1}(\mathbf{R}_c^{h(\beta)}) = -\mathbf{R}_c^\beta \cup -\mathbf{R}_c^{\beta'}$ and the fact that $\mathbf{R}_c^\beta$ and $\mathbf{R}_c^{\beta'}$ are disjoint, it follows

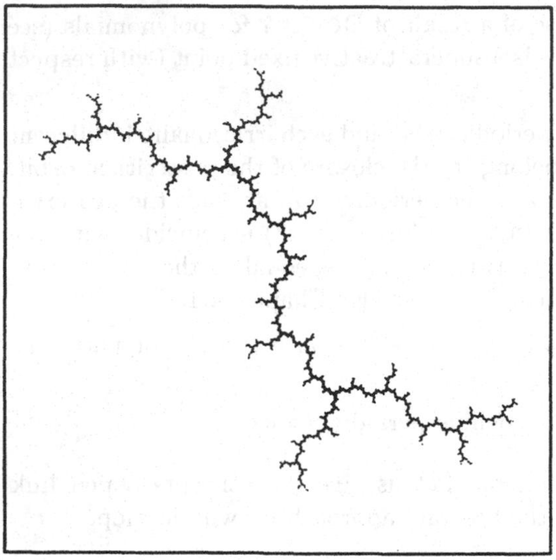

Fig. 1.1. concrete Julia set J_c for $c = i$

$$\mathbf{R}_c^{\beta'} = -\mathbf{R}_c^{\beta}. \tag{1.5}$$

The case which we will be especially interested in is that J_c (or equivalently K_c) is locally connected.

Definition 1.6. (Local connectivity)
A topological space X is said to be locally connected at a point x *if each neighborhood of $x \in X$ contains a connected neighborhood. It is called* locally connected *if it is locally connected at each of its elements.*

By Caratheodory's theorem local connectivity of J_c is equivalent to the statement that the inverse of Φ_c extends continuously to the unit circle. In particular, it implies that the dynamic rays land in a continuous manner.

1.1.4 From quadratic Julia sets to Julia equivalences

In this subsection J_c is assumed to be locally connected. Thus, by the above considerations, each dynamic ray lands at a point in J_c and each point of J_c possesses one or more external angles. So the points of J_c induce a decomposition of $\mathbb{T}$ into classes of external angles. We denote the corresponding equivalence relation by $\approx$ and the landing point of some $\mathbf{R}_c^{\beta}$ by z_c^{β}.

The map $\beta \mapsto z_c^{\beta}$ from the compact space $\mathbb{T}$ onto the (compact) Hausdorff space J_c is continuous, and (by the considerations in Appendix A.1, III.) J_c can be considered as a factor space of $\mathbb{T}$ with respect to $\approx$.

By (1.4) and (1.5) it is obvious that $z_c^{\beta_1} = z_c^{\beta_2}$ implies $z_c^{h(\beta_1)} = z_c^{h(\beta_2)}$ and $z_c^{\beta_1'} = z_c^{\beta_2'}$, meaning that $\approx$ is invariant on $(\mathbb{T}, h, ')$. Moreover, $\approx$ is completely

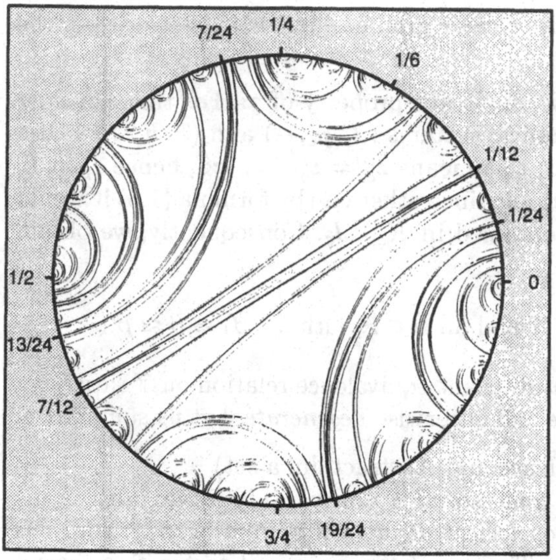

Fig. 1.2. Corresponding Julia equivalence

invariant with respect to $'$. In order to also show complete h-invariance, we need another property of $\approx$.

If two different dynamic rays $\mathbf{R}_c^{\beta_1}$ and $\mathbf{R}_c^{\beta_2}$ have a common landing point z, then the curve given by $\mathbf{R}_c^{\beta_1} \cup \{z\} \cup \mathbf{R}_c^{\beta_2}$ divides the complex plane into two open parts. Two other dynamic rays with a common landing point different from z must be contained in the same part. This implies the following property of $\approx$:

If two chords with endpoints $\beta_1, \beta_2 \in \mathbb{T}$ and $\beta_3, \beta_4 \in \mathbb{T}$ have non-empty intersection and if $\beta_1 \approx \beta_2$ and $\beta_3 \approx \beta_4$, then $\beta_1 \approx \beta_3$.

We call an equivalence relation on $\mathbb{T}$ satisfying this property ***planar***.

Now we are able to show complete h-invariance of $\approx$. If for some $\beta \in \mathbb{T}$ the set $h([\beta]_\approx)$ were not a complete equivalence class, one could fix some $\alpha \notin h([\beta]_\approx)$ satisfying $\alpha \approx h(\beta)$. Since for the two preimages $\gamma = \frac{\alpha}{2}$ and γ' of α neither $\beta \approx \gamma$ nor $\beta \approx \gamma'$ would be possible, also $\beta' \approx \gamma'$ or $\beta' \approx \gamma$ would be impossible. Each point in J_c has at most two preimages with respect to p_c. This would imply $\beta \approx \beta'$ and $\gamma \approx \gamma'$. Thus by the planarity of $\approx$ we would have $\gamma \approx \beta$, in contradiction to our assumption.

We have shown that $\approx$ forms a completely invariant equivalence relation on $(\mathbb{T}, h, ')$, and $(J_c, p_c, -)$ and $(\mathbb{T}, h, ')/\approx$ are conjugate.

Figures 1.1 and 1.2 illustrate how a Julia set can be regarded as a topological factor of the circle. The first one shows J_c for $c = i$ and the second one presents the corresponding identification on $\mathbb{T}$: two points connected by a sequence of chords are $\approx$-equivalent. (We draw 'hyperbolic' chords instead of 'straight' ones, which provides clearer graphics, but is irrelevant from the combinatorial point of view.)

We want to point out a last important property of $\approx$. For this, consider the conjugate topological dynamical systems $(J_c, p_c, -)$ and $(\mathbb{T}, h, ')/\approx$. If a point $\beta \in \mathbb{T}$ with $\beta \approx \beta'$ exists, this means $z_c^\beta = z_c^{\beta'} = -z_c^\beta$, hence $z_\beta = 0$. The critical point cannot be periodic since otherwise by formula (1.3) it would be superattractive, hence be contained in $K_c \setminus J_c$. Consequently, we obtain the following property of $\approx$:

$$\text{There do not exist a } \beta \in \mathbb{T} \text{ and an } n \in \mathbb{N} \text{ with } h^n(\beta) \approx \beta \approx \beta'.$$

Generally, we want to call an h-invariant equivalence relation on $\mathbb{T}$ satisfying this property ***non-degenerate*** and otherwise *degenerate*. Let us summarize:

Definition 1.7. (Julia equivalence and Abstract Julia set)
By a **Julia equivalence** *we understand a* **completely invariant equivalence relation** $\approx$ *on the topological dynamical system* $(\mathbb{T}, h, ')$ *which is* **planar** *and* **non-degenerate**. *The completely invariant factor* $\mathbb{T}/\approx$ *corresponding to a Julia equivalence* $\approx$ *is called* **Abstract Julia set**.

Following THURSTON's ideas, for each $\alpha \in \mathbb{T}$ we will construct a special Julia equivalence $\approx^\alpha$, the Julia equivalence mentioned in Chapter 1.1.2 which is associated to the generalized binary expansion induced by the division of the circle into the parts $]\frac{\alpha}{2}, \frac{\alpha+1}{2}[$ and $]\frac{\alpha+1}{2}, \frac{\alpha}{2}[$. The precise symbolic descriptions of $\approx^\alpha$ will be given by Theorems 2.47, 2.50 and 2.53.

In the other direction, it will turn out that the structure of each Julia equivalence $\approx$ is hidden in special points $\alpha \in \mathbb{T}$, which we call *main points* of $\approx$ (see Definition 2.58), and the investigations in Chapter 2 will culminate in the following result:

Theorem 1.8. (The set of all Julia equivalences)
Each point $\alpha \in \mathbb{T}$ *forms the main point of a unique Julia equivalence* $\approx^\alpha$. *In the other direction, if* $\approx$ *is a planar and completely invariant equivalence relation on* $(\mathbb{T}, h, ')$, *then the following statements are equivalent:*

(i) Each $\approx$*-equivalence class is finite.*

(ii) There exists an $\alpha \in \mathbb{T}$ *with* $\approx \, = \, \approx^\alpha$.

(iii) $\approx$ *is a Julia equivalence.*

In particular, $\approx \, = \, \approx^\alpha$ *for each main point* α *of a Julia equivalence* $\approx$.

The concept of a Julia equivalence was deduced from a locally connected Julia set. However, the reader will see that much of the combinatorial structure of a connected but not locally connected Julia set is contained in some Julia equivalence.

This is one reason why we will develop a self-contained 'theory' of Julia equivalences in Chapters 2 and 3. The advantages are obvious: on the one hand, one does not need to care about local connectivity, and on the other hand, the approach is (real) one-dimensional, which allows application of the powerful concepts 'itinerary' and 'kneading sequence'.

1.1.5 Renormalization of quadratic polynomials (I)

The class of quadratic polynomials $p_c; c \in \mathbb{C}$ is too small to understand all phenomena of quadratic iteration, and also from the viewpoint of our topological approach it is useful to extend this class in a natural manner.

Often, on a part of the complex plane, a holomorphic map behaves like a polynomial one. By their concept of a *polynomial-like map*, DOUADY and HUBBARD have given an exact mathematical description of this phenomenon (see [44] and compare [108, 12, 25]). For our purposes, we need the special case of quadratic-like maps.

Definition 1.9. (Quadratic-like maps)
Let f be a proper holomorphic map from a simply connected domain $U \subsetneq \mathbb{C}$ onto another one $V \subsetneq \mathbb{C}$. (Proper means that the preimage of each compact set is compact and implies the existence of a finite degree.)

Then f is said to be quadratic-like *if f has degree 2 and the closure of U forms a compact subset of V. If f is a quadratic-like map, then the set $K(f) = \bigcap_{j=1}^{\infty} f^{-j}(V)$ is called the* filled-in Julia set *and its boundary $J(f)$ the* Julia set *of f.*

Of course, a quadratic polynomial is quadratic-like in a neighborhood of its filled-in Julia set, and the two definitions of a (filled-in) Julia set are consistent. A relation of polynomial-like maps to polynomial maps was established by DOUADY and HUBBARD's Straightening Theorem, which we want to state only for the quadratic case.

Straightening Theorem ([44]) *Each quadratic-like map f is* hybrid-equivalent *to a quadratic polynomial p_c for some $c \in \mathbb{C}$: there exists a quasiconformal conjugacy ϕ from a neighborhood of $K(f)$ onto a neighborhood of K_c with $\overline{\partial}\phi = 0$ on $K(f)$. If $K(f)$ is connected, then c is unique.*

(We will only need that ϕ is a topological conjugacy preserving the orientation. Obviously, such conjugacy transforms $K(f)$ into K_c.)

One reason for the occurrence of self-similarity in quadratic iteration theory is that often a high iterate of a given quadratic polynomial has quadratic-like behavior on a part of the complex plane. If J_c is connected, equivalently, if 0 is contained in K_c, and if the map p_c^n for $n \in \mathbb{N}$ is quadratic-like anywhere, then it is quadratic-like in a neighborhood of 0. (The argumentation is as follows: p_c^n cannot be invertible at the point z mapped to 0 by the corresponding quasi-conformal conjugacy ϕ. So $(p_c^n)'(z) = 0$ and by the chain rule 0 belongs to the orbit of z.)

This leads to the concept of a renormalizable quadratic polynomial (see [108, 107, 115, 66]). Here we want to follow McMULLEN's book [108].

Definition 1.10. (Renormalizable quadratic polynomials)
For $c \in \mathbb{C}$ let p_c be a quadratic polynomial and let $n \in \mathbb{N} \setminus \{1\}$. Then p_c is said to be n-renormalizable *if there exists a simply connected domain U in $\mathbb{C}$ such that p_c^n is quadratic-like between U and $p_c^n(U)$, and it is true that $p_c^{jn}(0) \in U$ for all $j \in \mathbb{N}_0$. The pair (U, p_c^n) is called a* renormalization *of p_c. Moreover, the quadratic polynomial p_c is said to be* infinitely renormalizable *if for infinitely many $n \in \mathbb{N}$ the map p_c is n-renormalizable.*

If p_c for $c \in \mathbb{C}$ is n-renormalizable, then the corresponding (filled-in) Julia set does not depend on the domain U supporting the renormalization (U, p_c^n). This justifies to use the notation $J(p_c^n)$ $(K(p_c^n))$ for the (filled-in) Julia set corresponding to a renormalization of p_c. Moreover, by the second property in the definition, $J(p_c^n)$ is connected, hence there exists a unique $d \in \mathbb{C}$ such that p_d and p_c^n are hybrid-equivalent.

p_c^n maps the set $K^{(0)} := K(p_c^n)$ onto itself, and its orbit with respect to p_c consists of n copies of $K^{(0)}$, the so-called *small filled-in Julia sets* belonging to the renormalization. Two different small filled-in Julia sets have at most one common point. If there are two which cross one another at this point, the renormalization is called *crossed* and otherwise *simple* (see [108]).

In Chapter 4.1.3 we will say more about renormalization for quadratic polynomials, but the few concepts given are necessary to motivate the investigations in Chapter 3.3. Therein and partially in Chapter 3.2 we will discuss renormalization from our abstract point of view. For quadratic-like behavior in the complex plane, we will look for 'angle-doubling-like' behavior of an iterate of h on a part of $\mathbb{T}$ instead. The results are related to concrete renormalization in Chapter 4.1.3.

Besides quadratic polynomials possessing an irrationally indifferent periodic orbit, infinitely renormalizable ones are those with the most complicated structure. This is supported by the following famous result on the local connectivity of quadratic Julia sets by YOCCOZ (see [115, 66]):

Yoccoz's Theorem. *Let p_c be a quadratic polynomial which does not possess an indifferent periodic orbit. If J_c fails to be locally connected, then p_c is infinitely renormalizable.*

1.2 The Mandelbrot set

1.2.1 Julia equivalences and the Mandelbrot set (I)

The celebrated Mandelbrot set, which was independently found by BROOKS and MATELSKI [22] and by MANDELBROT [104], consists of all complex parameters c for which J_c is connected. As mentioned above, this is equivalent to the statement that c lies in the filled-in Julia set K_c or, in other words, that the postcritical orbit is bounded.

We will be especially interested in the boundary ∂M of the Mandelbrot set M. (In Figure 1.3 the boundary of the Mandelbrot set is drawn black and its interior gray.)

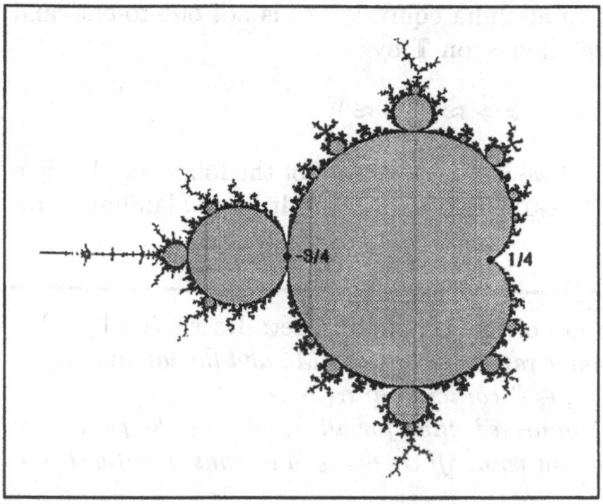

Fig. 1.3. The Mandelbrot set

A parameter c is contained in $\mathbb{C} \setminus M$ iff $c \in \mathbb{C} \setminus K_c$, and so a map Φ with

$$\Phi(c) = \Phi_c(c)$$

for $c \in \mathbb{C}\setminus M$ and $\Phi(\infty) = \infty$ is well-defined on $\overline{\mathbb{C}}\setminus M$. DOUADY and HUBBARD [43] have shown that Φ forms a conformal isomorphism onto the complement of the unit disk in $\overline{\mathbb{C}}$, implying that M is compact and connected. As the parametric counterpart to Definition 1.5, the following concept is used:

Definition 1.11. (Parameter rays)
For $\alpha \in \mathbb{T}$, the set $\mathbf{R}^\alpha = \Phi^{-1}(\mathbf{R}_0^\alpha)$ is called a parameter ray. *In a similar manner as in the dynamic plane, one says that $\mathbf{R}^\alpha$* lands *at a point $c \in \partial M$ and that c is the* landing point *of $\mathbf{R}^\alpha$ and α is an* external angle *of c (see Definition 1.5).*

It is conjectured that Φ^{-1} extends continuously to the unit circle $\mathbb{T}$, which by Caratheodory's Theorem coincides with the famous

MLC Conjecture. *The Mandelbrot set M (and equivalently its boundary ∂M) is locally connected.*

This conjecture is supported by results of YOCCOZ (see [66]) and LYUBICH [101, 99, 100]. If the MLC Conjecture is valid, then all parameter rays land at ∂M in a continuous way, and therefore ∂M is a factor of the circle $\mathbb{T}$ with respect to of a planar equivalence relation being closed in $\mathbb{T} \times \mathbb{T}$. Closely related to this fact, there are some different models of the Mandelbrot set (see LAVAURS [93], THURSTON [167], DOUADY [41], MILNOR [117], LAU and SCHLEICHER [92]) We discuss the matter in Chapter 4.1).

In order to get an abstract model of the Mandelbrot set near to the description by THURSTON and containing some dynamical information, we will develop our Julia equivalence approach into a further direction: the map $\alpha \mapsto \approx^\alpha$ from $\mathbb{T}$ onto the set of all Julia equivalences is not one-to-one, and so it defines an equivalence relation $\sim$ on $\mathbb{T}$ by

$$\alpha_1 \sim \alpha_2 :\Longleftrightarrow \approx^{\alpha_1} = \approx^{\alpha_2}$$

for $\alpha_1, \alpha_2 \in \mathbb{T}$. In Chapter 4.1.2 we will give a proof of the following theorem containing DOUADY and HUBBARD's well-known result on the landing of rational parameter rays.

Theorem 1.12. (The structure of the Mandelbrot set: description by $\sim$)
For rational $\alpha \in \mathbb{T}$ the parameter ray $\mathbf{R}^\alpha$ lands at ∂M, and the landing points of $\mathbf{R}^{\alpha_1}, \mathbf{R}^{\alpha_2}$ for rational $\alpha_1, \alpha_2 \in \mathbb{T}$ coincide iff $\alpha_1 \sim \alpha_2$.

Moreover, if M is locally connected, then for all $\alpha_1, \alpha_2 \in \mathbb{T}$ the parameter rays $\mathbf{R}^{\alpha_1}, \mathbf{R}^{\alpha_2}$ land at a common point iff $\alpha_1 \sim \alpha_2$, and consequently ∂M is homeomorphic to $\mathbb{T}/\sim$.

The identification of $\mathbb{T}$ with respect to $\sim$ is indicated by Figure 1.4. We will refer to the factor space $\mathbb{T}/\sim$ as the ***Abstract Mandelbrot set*** - although it actually models the boundary of the Mandelbrot set -, and also to the 'space' of all Julia equivalences.

A point in $\mathbb{T}$ is rational iff it is periodic or preperiodic. This relatively simple fact will be discussed in Chapter 2.1.2. However, on the periodic rationals the identification given by Theorem 1.12 is a kind of involution:

Theorem 1.13 ([43]). (The structure of the Mandelbrot set: Pairwise landing of 'periodic' parameter rays)
For periodic $\alpha \in \mathbb{T} \setminus \{0\}$ there exists a unique periodic point $\overline{\alpha} \in \mathbb{T}$ different from α such that $\mathbf{R}^\alpha, \mathbf{R}^{\overline{\alpha}}$ land at the same point. $\overline{\alpha}$ has the same period as α.

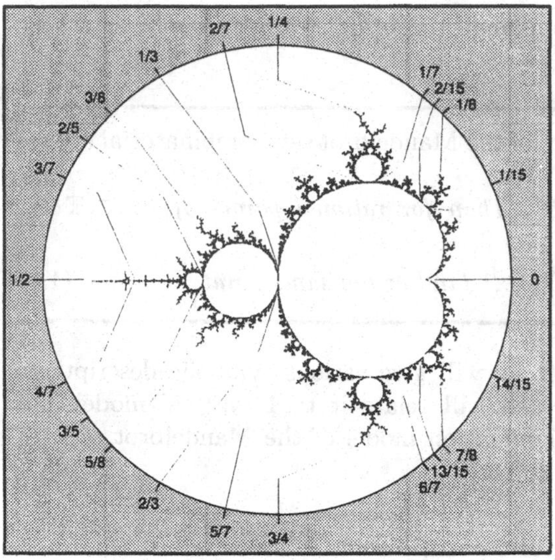

Fig. 1.4. ∂M and $\mathbb{T}/\sim$

Theorem 1.13 is covered by Corollaries 4.14 and 4.15 in Chapter 4.1.2. One of the reasons for studying the Mandelbrot set is to understand the change of combinatorics in the family $p_c; c \in \mathbb{C}$. Since for $c \notin M$ the Julia set (J_c, p_c) is always conjugate to the shift space $(\{0,1\}^{\mathbb{N}}, \sigma)$, we can indeed restrict our discussion to the parameters $c \in M$.

The pairwise landing of 'periodic' parameter rays provides splittings of the complex plane and the Mandelbrot set into parts which are equivalent from the combinatorial viewpoint in a certain sense.

Definition 1.14. (Wakes and combinatorial classes)
For periodic $\alpha \in \mathbb{T} \setminus \{0\}$ let $\overline{\alpha}$ be given as by Theorem 1.13, and for $\alpha = 0$ let formally $\overline{\alpha} = 1$, having the identification of 0 and 1 in mind. The curve consisting of the parameter rays $\mathbf{R}^{\alpha}, \mathbf{R}^{\overline{\alpha}}; \alpha \in \mathbb{T}$ and their common landing point divides the complex plane into two open parts.

The union of that one not containing $\mathbf{R}^0$ with the landing point obtained is called a wake *and denoted by $\mathbf{Wake}^{\alpha\overline{\alpha}}$. Further, the set*

$$Comb^{\gamma} = M \cap \Big(\bigcap_{\substack{\alpha \text{ periodic}, \\ \alpha < \overline{\alpha}, \\ \gamma \in [\alpha, \overline{\alpha}]}} \mathbf{Wake}^{\alpha\overline{\alpha}} \setminus \bigcup_{\substack{\alpha \text{ periodic}, \\ \alpha < \overline{\alpha}, \\ \gamma \notin [\alpha, \overline{\alpha}]}} \mathbf{Wake}^{\alpha\overline{\alpha}} \Big)$$

for $\gamma \in \mathbb{T}$ is called a combinatorial class *of M.*

We will see that M is locally connected iff all combinatorial classes for $\gamma \in \mathbb{T}$ with non-periodic kneading sequence consist of only one point.

It is important to note the simple fact that two given wakes are disjoint or one of them is contained in the other one, and that consequently combinatorial classes are equal or disjoint. In Chapter 4.1.2 we will prove the following statement, which reflects the combinatorics of quadratic dynamics in a very clear way:

Theorem 1.15. (The structure of the Mandelbrot set: combinatorial classes and dynamic landing patterns)
Let $\alpha \in \mathbb{T}$ and let $c \in Comb^\alpha$. Then for rational points $\beta_1, \beta_2 \in \mathbb{T}$ the following holds:

$$\beta_1 \approx^\alpha \beta_2 \iff \mathbf{R}_c^{\beta_1}, \mathbf{R}_c^{\beta_2} \text{ land at the same point.} \tag{1.6}$$

Furthermore, in Chapter 3.1 we will give various symbolic descriptions of the equivalence relation $\sim$ and will relate it to LAVAURS' model and THURSTON's quadratic minor lamination model of the Mandelbrot set. (In particular, see Theorems 3.1 and 3.6.)

1.2.2 The system of hyperbolic components

A periodic point z for given p_c can uniquely be continued as a locally invertible holomorphic function depending on the parameter c, as long as the multiplier is different from 1. If the map g defined by $g(c,z) = p_c^m(z) - z$ satisfies $g(c_0, z_0) = 0$ and $\frac{\partial g}{\partial z}(c_0, z_0) \neq 0$, then the implicit function theorem provides a solution $z = z(c) = z_{(c_0,z_0)}(c)$ of $g(c,z) = 0$ in some neighborhood of c_0 satisfying $z(c_0) = z_0$ and being a periodic point for p_c of the same period as that of z_0 for p_{c_0}. Moreover, one easily sees that in a neighborhood of c_0 the map which assigns to each c the multiplier of $z(c)$ is holomorphic.

In particular, this shows that the set of all *hyperbolic* parameters $c \in \mathbb{C}$, which are characterized by the fact that p_c possesses an attractive periodic orbit, is open. Moreover, since an attractive periodic orbit attracts the post-critical orbit, this set is contained in M. The following is conjectured:

Conjecture of hyperbolicity. *The interior of the Mandelbrot set is equal to the set of all hyperbolic parameters.*

We will say more about this conjecture and its relation to the MLC Conjecture in Chapter 4.1.2, but at the moment we are only interested in the connectedness components of the set of all hyperbolic parameters, which are called the *hyperbolic components.* By the above considerations the period of the attractive orbit belonging to a hyperbolic parameter is locally constant and so it does not change on a hyperbolic component. This justifies speaking of the *period* of the hyperbolic component.

Example 1.16. (The main hyperbolic component)
A fixed point of some p_c for a given $c \in \mathbb{C}$ is equal to $z = \frac{1+w}{2}$ where w is a root of $1 - 4c$. Its multiplier is equal to $2z$. If $c = \frac{1}{4}$, the fixed point is unique and its multiplier equal to 1. Otherwise, there are two roots of

$1-4c$ and so two fixed points. The set $\{(1-(re^{2\pi\nu i}-1)^2)/4 \mid \nu, r \in [0,1[\}$ forms a hyperbolic component, the only one of period 1. It is called the *main* hyperbolic component, and its boundary is parameterized by $\nu \in [0,1[\mapsto (1-(e^{2\pi\nu i}-1)^2)/4$.

(Note that the unique hyperbolic component of period 2 is the disk of radius $\frac{1}{4}$ around -1, and that GIARRUSSO and FISHER [55] gave a parameterization of the three hyperbolic components of period 3.)

The example shows that from the conformal viewpoint the main hyperbolic component is a disk. This generalizes to all hyperbolic components as it was shown by DOUADY and HUBBARD:

Theorem 1.17 ([43]). (The structure of the Mandelbrot set: hyperbolic components)
The multiplier map associated with a hyperbolic component W, which assigns to each parameter $c \in W$ the multiplier of the unique attractive orbit for p_c, forms a conformal isomorphism from W onto the open unit disk and has a unique continuous extension $Mult_W$ to a homeomorphic map from the closure of W onto the closed unit disk.

For completeness, we will provide a proof of Theorem 1.17 in Chapter 4.1.2. Note that the unique point c in a hyperbolic component W with $Mult_W(c) = 1$ is called the *root* of W.

Theorem 1.18 ([43]). (The structure of the Mandelbrot set: roots of hyperbolic components and 'periodic' parameter rays)
The parameter rays $\mathbf{R}^\alpha$ and $\mathbf{R}^{\alpha}$ for a periodic $\alpha \in \mathbb{T}$ land at the root of a unique hyperbolic component $W^{\alpha\overline{\alpha}}$ satisfying the following properties:

(i) The periods of $W^{\alpha\overline{\alpha}}$ and α coincide,
(ii) $W^{\alpha\overline{\alpha}}$ lies in $\mathbf{Wake}^{\alpha\overline{\alpha}}$.

In particular, $\mathbf{R}^0$ lands at the root of the main hyperbolic component, which is $\frac{1}{4}$. Conversely, each root of a hyperbolic component is the landing point of $\mathbf{R}^\alpha$ for some periodic α.

Much of the Mandelbrot set can be described by the mutual position of the hyperbolic components labelled by their periods. We want to use the last part of the present chapter to illustrate this and to motivate some discussion in Chapter 3.

Bifurcation and order. Clearly, the hyperbolic components are disjoint. Nevertheless, the closures of two hyperbolic components can have a common point, and the following definition is useful for describing this fact:

Definition 1.19. (Internal angles)
A point on the boundary of a hyperbolic component W is said to have the internal angle $\nu \in [0,1[$ *if it is mapped to* $e^{2\pi\nu i}$ *by the map* $Mult_W$ *defined in Theorem 1.17. (So the root of a hyperbolic component has internal angle 0.)*

The points on the boundary of hyperbolic components with rational internal angles are substantial for understanding the structure of M. Hyperbolic components meet only there. Interpreting results of Chapter 3, we will derive the following theorem of DOUADY and HUBBARD:

Theorem 1.20 ([43]). (The structure of the Mandelbrot set: bifurcation)
Two different hyperbolic components have at most one common boundary point z. Such z is the root of one of the components and has a non-zero rational internal angle with respect to the other one.

Conversely, each point z having internal angle $\frac{p}{q} \neq 0$ with respect to a hyperbolic component W of period m is the root of a hyperbolic component U of period qm. (Here $\frac{p}{q}$ is assumed to be reduced.) If z is the landing point of $\mathbf{R}^\alpha$ *for some periodic $\alpha \in \mathbb{T}$, then* $W \cap \mathbf{Wake}^{\alpha\overline{\alpha}} = \emptyset$.

If W, U and z are given as in Theorem 1.20, one says that U *bifurcates directly from* W, and that component of $M \setminus \{z\}$ containing U is called the $\frac{p}{q}$-sublimb of W. If only the denominator q is of interest, we speak of a *sublimb of denominator* q.

Theorem 1.20 shows that the set of hyperbolic components is ordered in a natural way: a hyperbolic component U is said to be *behind* a hyperbolic component W if U lies in some $\frac{p}{q}$-sublimb of W and *separates* two hyperbolic components if one of them is behind U, and U is behind the other one.

Hyperbolic components, tuning and renormalization. DOUADY and HUBBARD's *tuning* says that for each hyperbolic component W of period m there exists a homeomorphism $d \mapsto W * d$ from M into M mapping the main hyperbolic component W_0 onto W and satisfying the following properties (for a complete proof, see HAÏSSINSKY [61]):

- For $d \in W_0$, the multipliers of the attractive orbits for p_d and p_c with $c = W * d$ coincide,
- p_c with $c = W * d; d \in M$ is m-renormalizable, and p_c^m (restricted to the corresponding renormalization) is hybrid-equivalent to p_d. Here in some cases it must be excluded that d is equal to the root $\frac{1}{4}$ of W_0.

$d \mapsto W * d$ must transform each d with an attractive orbit of period k into a parameter with an attractive orbit of period km, and so the tuning procedure multiplies periods of hyperbolic components by m (see MILNOR [113]), which is illustrated by Figure 1.2.2 (consider the hyperbolic component W of period 3 on the left). The statement we will prove is

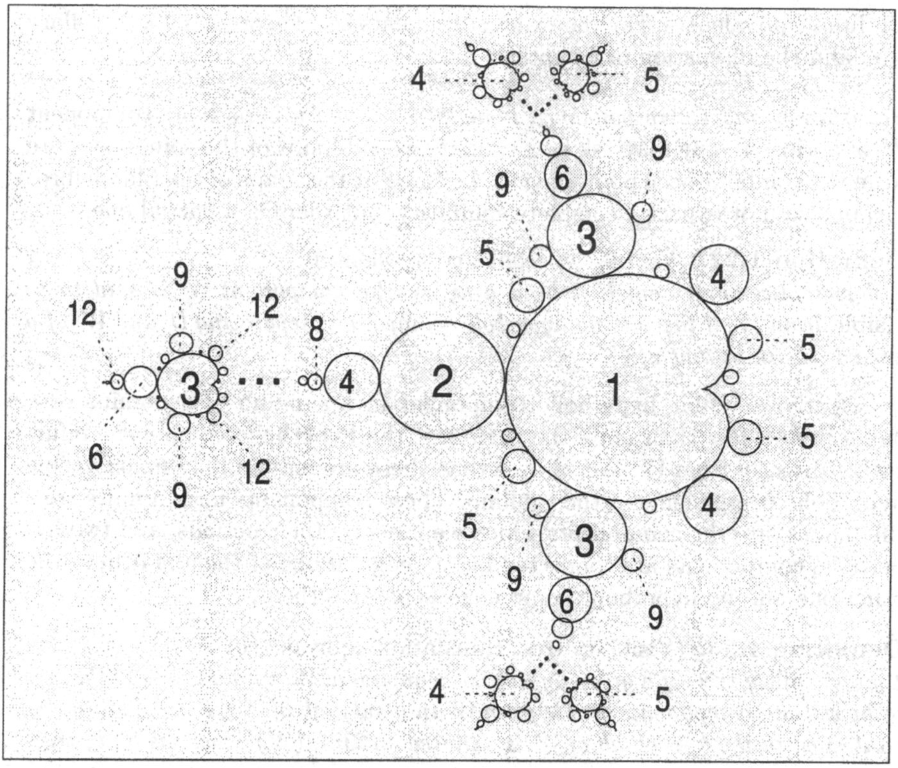

Fig. 1.5. Hyperbolic components and tuning

Theorem 1.21. (The structure of the Mandelbrot set: hyperbolic components and tuning)
For each hyperbolic component W of some period m, there exists a injection from the set of all hyperbolic components into itself preserving order and embedding into the plane, mapping the main hyperbolic component to W and hyperbolic components of period $k \in \mathbb{N}$ to ones of period km, and satisfying the following property: its image contains with each two hyperbolic components all hyperbolic components separating them.

The proof will be provided in Chapter 3.2.2 (see proof of Theorem 3.32) in a context briefly discussed at the end of this chapter.

For a long time it was assumed that each renormalization is due to a tuning $d \mapsto W * d$, but McMullen found a new type - the crossed renormalization mentioned above. It is a particular task of our approach to discuss the relations between simple and crossed renormalization, and tuning.

At this place note that simple renormalization and tuning are near to the inclusion order on the Abstract Mandelbrot set - here considered as the set of all Julia equivalences. For the exact statements see Theorem 3.33, Corollary 3.52 and the discussion in Chapter 4.1.3.

Translation and Correspondence Principles. Given a hyperbolic component W of period m, one may ask how similar two sublimbs of the component are from the combinatorial point of view. Looking at the shape of the Mandelbrot set, one cannot expect a complete similarity, but there is a partial one.

Definition 1.22. (Visible hyperbolic components)
A hyperbolic component U behind a hyperbolic component W is said to be visible *from W if there is no hyperbolic component separating U and W of a period less than that of U.*

Clearly, a visible hyperbolic component in a sublimb of denominator q has a period less than qm - the period of the hyperbolic component bifurcating directly from W. We will see that there are only finitely many visible hyperbolic components in the sublimb. These components form a finite tree: the 'nodes' are the components which we label by their periods, and two hyperbolic components are connected by an 'edge' if one of them is behind the other one and no hyperbolic component separates them.

Definition 1.23. (Visibility tree, translation-equivalence)
By a visibility tree *of a given hyperbolic component W we understand the tree of hyperbolic components visible from W in a sublimb of W. The visibility tree contained in the $\frac{p}{q}$-sublimb of W is denoted by $\mathfrak{Vis}_{\frac{p}{q}}(W)$. Two visibility trees $\mathfrak{Vis}_{\frac{p_1}{q_1}}(W)$ and $\mathfrak{Vis}_{\frac{p_2}{q_2}}(W)$ are called* translation-equivalent *if they coincide, including the embedding into the plane, when all periods in $\mathfrak{Vis}_{\frac{p_1}{q_1}}(W)$ are increased by $(q_2 - q_1)m$.*

LAU and SCHLEICHER (see [148, 92]) observed that for certain hyperbolic components W all visibility trees are translation-equivalent. This phenomenon - what they called *Translation Principle* - is illustrated by Figure 1.6 for the hyperbolic component $W = W_{\frac{5}{31}\frac{6}{31}}$ of period 5 (compare Example 3.63).

The phenomenon appears for a wide class of hyperbolic components. However, there are counter-examples to the general validity of the Translation Principle. In chapter 3.4.2 we will prove that the following weaker statement is valid in the whole (see Theorem 3.78):

Theorem 1.24. (The structure of the Mandelbrot set: Partial Translation Principle)
Each visibility tree other than $\mathfrak{Vis}_{\frac{1}{2}}(W)$ for a hyperbolic component W is translation-equivalent to $\mathfrak{Vis}_{\frac{1}{3}}(W)$ or to $\mathfrak{Vis}_{\frac{2}{3}}(W)$.

The discussion of the Translation Principle will mainly be based on 'Correspondence Principles' relating dynamical structure and structure in the pa-

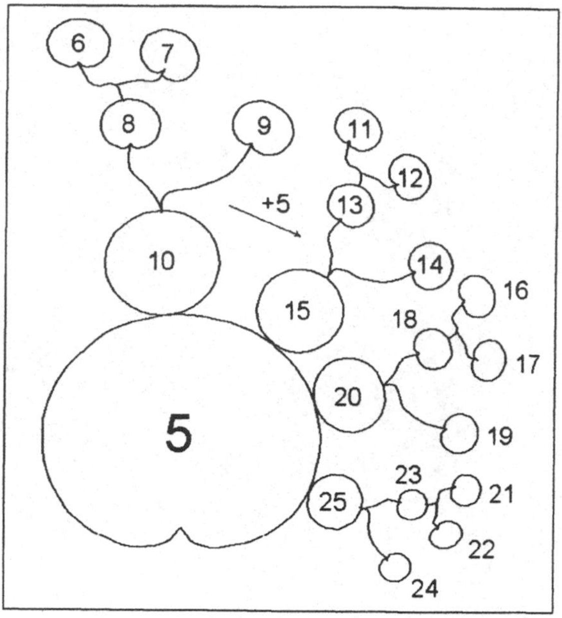

Fig. 1.6. A Translation Principle

rameter space. It follows DOUADY's philosophy 'to plough in the dynamical plane and then to harvest in the parameter space'.

Hyperbolic components and 'periodic parameter chords'. Theorem 1.18 provides a one-to-one correspondence

$$W^{\alpha\overline{\alpha}} \longleftrightarrow \alpha\overline{\alpha} \tag{1.7}$$

between the system of all hyperbolic components and the system $\mathcal{B}_* := \{\alpha\overline{\alpha} \mid \alpha \in \mathbb{T},\ \alpha \text{ periodic}\}$ of mutually non-intersecting chords $\alpha\overline{\alpha}$ lying in the closed unit disk and possessing the endpoints $\alpha, \overline{\alpha}$ in its boundary $\mathbb{T}$. The embedding structure of the component system in the plane and of the chord system B_* in the unit disk are combinatorially equivalent.

Therefore, some statements concerning the combinatorial structure of the system of hyperbolic components will only be given in the language of the 'periodic parameter chords'. On the one hand, the statements will be deduced within $\mathcal{B}_*$, and on the other hand, we mainly want to avoid giving them twice in the hyperbolic component language and in that of $\mathcal{B}_*$.

We emphasize that via the correspondence (1.7) Theorems 1.21 and 1.24 are covered by Theorems 3.32 and 3.78, respectively.

2. Abstract Julia sets

In this chapter we discuss and classify Julia equivalences. The base of our investigations is the combination of THURSTON's quadratic invariant laminations with the two concepts 'itinerary' and 'kneading sequence' given by Definition 1.3.

2.1 Symbolic dynamics for the angle-doubling map

2.1.1 Generalized binary expansion

In the following we deal with sequences of the symbols 0,1 and $*$. Nevertheless, let us give the necessary notations for the case of a general finite symbol set. (This is useful in view of Appendix A.2.)

Definition 2.1. (Sequences and words)
1. For a finite set $\mathbb{S}$ of symbols, let $\mathbb{S}^{\mathbb{N}}$ be the set of all sequences with members in $\mathbb{S}$. We want to consider $\mathbb{S}^{\mathbb{N}}$ as being equipped with the product topology, in other words, as a Cantor discontinuum (see [142]).

2. Denote the set of all words (i.e. finite sequences) over $\mathbb{S}$ of length $n \in \mathbb{N}$ by $\mathbb{S}^n$, and by $\mathbb{S}^*$ the set of all words, including the empty one λ.

3. Moreover, for $\mathbf{w} \in \mathbb{S}^*$ and $\mathbf{s} \in \mathbb{S}^{\mathbb{N}}$, we use the following notations: $\mathbf{w}^n$ for the n-fold concatenation of $\mathbf{w}$ and $\overline{\mathbf{w}}$ instead of $\mathbf{www}\ldots$; $|\mathbf{w}|$ for the length of $\mathbf{w}$; $\mathbf{s}(n)$ and $\mathbf{w}(n)$ for the n-th symbol of $\mathbf{s}$ and $\mathbf{w}$; $\mathbf{s}|n$ for the initial subword of $\mathbf{s}$ of length n.

4. The map on $\mathbb{S}^{\mathbb{N}}$ we are interested in is the shift map σ defined by $\sigma(s_1 s_2 s_3 \ldots) = s_2 s_3 s_4 \ldots$ for $s_1 s_2 s_3 \ldots \in \mathbb{S}^{\mathbb{N}}$.

For $\alpha = 0$ the itinerary $I^\alpha(\beta)$ of a point β is equal to its binary expansion, with the restriction that now the points 0 and $\frac{1}{2}$ are not distinguished from the symbolic point of view. We have assigned to them the 'wild card'-symbol $*$.

By use of the fact that the i-th members of $I^\alpha(\beta)$ and $I^0(\beta)$ differ iff $h^{i-1}(\beta)$ lies in one of the intervals $[0, \frac{\alpha}{2}]$ or $[\frac{1}{2}, \frac{\alpha+1}{2}]$, one obtains the following relation between the itinerary and the binary expansion ('>' means greater in the lexicographic ordering):

$$I^\alpha(\beta) = s_1 s_2 s_3 \ldots \quad \text{with} \quad s_i = \begin{cases} b_i & \text{for} \quad \sigma^i(\mathbf{b}(\beta)) > \mathbf{b}(\alpha) \\ 1 - b_i & \text{for} \quad \sigma^i(\mathbf{b}(\beta)) < \mathbf{b}(\alpha) \\ * & \text{for} \quad \sigma^i(\mathbf{b}(\beta)) = \mathbf{b}(\alpha) \end{cases} . \qquad (2.1)$$

There is a substantial difference between the usual and the generalized binary expansion: in the generalized case, there exist itineraries which (do not contain $*$ and) represent more than one point. For example, $I^{1/2}(\frac{1}{3}) = I^{1/2}(\frac{2}{3}) = 000\ldots = \overline{0}$.

As already mentioned, this disadvantage will turn into a very good tool for describing Julia equivalences. In accordance to the use of the symbol $*$ as a 'wild card'-symbol, let us introduce the following concept:

Definition 2.2. (Substantially equal sequences)
We call two sequences $\mathbf{s} = s_1 s_2 s_3 \ldots, \mathbf{t} = t_1 t_2 t_3 \ldots \in \{0, 1, *\}^{\mathbb{N}}$ substantially equal *and write* $\mathbf{s} \cong \mathbf{t}$ *if* $s_i = t_i$ *or* $s_i = *$ *or* $t_i = *$ *for all* $i \in \mathbb{N}$. *Analogously we use the symbol* $\cong$ *for 0-1-words.*

By $\alpha \mapsto \hat{\alpha}$ for variable $\alpha \in \mathbb{T}$ and by $\beta \mapsto I^\alpha(\beta)$ for a fixed $\alpha \in \mathbb{T}$ and variable $\beta \in \mathbb{T}$, there are defined maps from $\mathbb{T}$ into the space $\{0, 1, *\}^{\mathbb{N}}$, which we think to be equipped with the product topology. Obviously, these maps are continuous exactly at those points whose image sequences do not contain the symbol $*$. Nevertheless, the clockwise limits $\lim_{\epsilon \to 0, \epsilon > 0} I^\alpha(\beta + \epsilon)$, $\lim_{\epsilon \to 0, \epsilon > 0} \widehat{\alpha + \epsilon}$ and also the counter-clockwise limits $\lim_{\epsilon \to 0, \epsilon > 0} I^\alpha(\beta - \epsilon)$, $\lim_{\epsilon \to 0, \epsilon > 0} \widehat{\alpha - \epsilon}$ exist. (In each formula, $\beta \pm \epsilon$ is taken modulo 1.) For the four limit sequences, which are contained in $\{0, 1\}^{\mathbb{N}}$, we will use the notations $\overset{\curvearrowright}{I^\alpha}(\beta), \hat{\alpha}^{\curvearrowright}, \overset{\curvearrowleft}{I^\alpha}(\beta), \hat{\alpha}^{\curvearrowleft}$ successively.

2.1.2 Points with periodic kneading sequence

The itinerary of a point in $\mathbb{T}$ with finite orbit must be a periodic or preperiodic sequence. If $\alpha \in \mathbb{T}$ is periodic with period $PER(\alpha) = m$, then let $\mathbf{v}^\alpha := \hat{\alpha}|m-1$, such that $\hat{\alpha} = \overline{\mathbf{v}^\alpha *}$.

Moreover, a point $\alpha \in \mathbb{T}$ is periodic or preperiodic iff it is rational: $2^n \alpha \equiv \alpha \bmod 1$, hence $(2^n - 1)\alpha \bmod 1 = 0$, implies $\alpha = \frac{p}{2^n - 1}$ for some $p \in \mathbb{N}$. From this one easily sees that a point is periodic and preperiodic, iff it can be represented by a reduced fraction with odd and even denominator, respectively. In the periodic case one gets a representation $\alpha = \frac{p}{2^{PER(\alpha)} - 1}$.

We denote the set of all periodic points in $\mathbb{T}$ by $\mathbb{P}_*$, and by $\mathbb{P}$ the set of all points with periodic kneading sequence. Clearly, $\mathbb{P}_*$ is a subset of $\mathbb{P}$. At first glance, it seems to be possible that the two sets coincide. We will show that this is false, but first let us verify that no preperiodic point is contained in the set $\mathbb{P}$. (For a complete discussion of the relation between the rational points in $\mathbb{T}$ and the points with a periodic or preperiodic itinerary we refer to Corollary 2.23.)

Proposition 2.3. *If a point $\alpha \in \mathbb{T}$ is preperiodic with respect to h, then also its kneading sequence $\hat{\alpha}$ is preperiodic.*

Proof: Let α be preperiodic with binary expansion $\mathbf{b}(\alpha) = b_1 b_2 b_3 \ldots$, and let m be the period of the periodic part of $\hat{\alpha}$. Then one finds a multiple n of m such that $\mathbf{b}(\alpha) = b_1 b_2 \ldots b_n \overline{b_{n+1} b_{n+2} \ldots b_{2n}}$. If i is the greatest index less than $n+1$ with $b_i \neq b_{i+n}$, then formula (2.1) yields that the i-th and the $i+n$-th members of $\hat{\alpha}$ are different. Therefore, $\hat{\alpha}$ cannot be periodic. ■

We come to the question whether there are non-periodic points with periodic kneading sequence. The set of all those points is denoted by $\mathbb{P}_\infty$, and we will present a complete characterization of $\mathbb{P}_\infty$ in Section 3.2 (see Theorem 3.34). As a first step, we are interested in all points of $\mathbb{T}$ whose kneading sequence is substantially equal to $\overline{0}$. Let us start our discussion by introducting some notations. In particular, let us define a special type of 0-1-sequences. (An element of $\{0,1\}^{\mathbb{N}}$ and $\{0,1\}^*$ is said to be a 0-1-sequence and a 0-1-word, respectively.)

By $\mathbf{a}^-$ we denote the 0-1-sequence (0-1-word) obtained from a given 0-1-sequence (0-1-word) $\mathbf{a}$ substituting all symbols 1 by 0 and all symbols 0 by 1 in it. Obviously, the sum of the two numbers having binary expansions $\mathbf{b}$ and $\mathbf{b}^-$ is equal to one, and the corresponding kneading sequences coincide.

In the following, a sequence $(\mathbf{q}_1, \mathbf{q}_2, \mathbf{q}_3, \ldots)$ of 0-1-words is said to be a *partition* of $\mathbf{s} \in \{0,1\}^{\mathbb{N}}$ with i-th part $\mathbf{q}_i$ if $\mathbf{s} = \mathbf{q}_1 \mathbf{q}_2 \mathbf{q}_3 \ldots$.

To each word $\mathbf{k} = k_1 k_2 \ldots k_n \in \mathbb{N}^*$, consisting of positive integers, we assign a 0-1-word $\mathbf{a}(\mathbf{k})$. For this, let $\mathbf{a}_{-1} = 1$ and $\mathbf{a}(\lambda) = \mathbf{a}_0 = 0$. (Recall that λ denotes the empty word.) The definition of $\mathbf{a}(\mathbf{k})$ is given recursively.

If $\mathbf{a}(\mathbf{k})$ has been already defined for all words $\mathbf{k}$ of length less than n, let $\mathbf{a}(k_1 k_2 \ldots k_n) \in \{0,1\}^*$ be given by

$$\mathbf{a}(k_1 k_2 \ldots k_n) = \mathbf{a}_n = \mathbf{a}_{n-1}^{k_n} \mathbf{a}_{n-2} \tag{2.2}$$

with $\mathbf{a}_{n-1} = \mathbf{a}(k_1 k_2 \ldots k_{n-1})$ and $\mathbf{a}_{n-2} = \mathbf{a}(k_1 k_2 \ldots k_{n-2})$.

Finally, for an infinite sequence $\mathbf{k} = k_1 k_2 k_3 \ldots$ of positive integers $\mathbf{a}_i$ forms an initial subword of $\mathbf{a}_j$ if $i < j$, and so the following definition makes sense:

$$\mathbf{a}(\mathbf{k}) = lim_{n \to \infty} \mathbf{a}(k_1 k_2 \ldots k_n).$$

Now we come to the characterization of the points in $\mathbb{T}$ with kneading sequence substantially equal to $\overline{0}$.

Theorem 2.4. (α with $\hat{\alpha} \cong \overline{0}$)
For $\alpha \in \mathbb{T}$ the following statements are valid:

(i) $\hat{\alpha} = \overline{0} \iff$ *There exists a sequence* $k_1 k_2 k_3 \ldots \in \mathbb{N}^{\mathbb{N}}$ *with* $\mathbf{b}(\alpha) = \mathbf{a}(k_1 k_2 k_3 \ldots)$ *or* $\mathbf{b}(\alpha) = \mathbf{a}(k_1 k_2 k_3 \ldots)^-$.

(ii) $\hat{\alpha} = \overline{00 \ldots 0*} \iff$ *There exists a word* $k_1 k_2 \ldots k_n \in \mathbb{N}^*$ *for which* $\mathbf{b}(\alpha) = \overline{\mathbf{a}(k_1 k_2 \ldots k_n)}$ *or* $\mathbf{b}(\alpha) = \overline{\mathbf{a}(k_1 k_2 \ldots k_n)}^-$.

Proof: Since the kneading sequences of $\alpha \in \mathbb{T}$ and $1-\alpha$ coincide, we can assume that $\alpha \leq \frac{1}{2}$. First of all, note that $\mathbf{a}(k_1k_2\dots k_n)$ ends with 1 iff n is odd, which can easily be shown by induction.

Let $k_1k_2k_3\dots \in \mathbb{N}^{\mathbb{N}}$ and let $\mathbf{a}_n = \mathbf{a}(k_1k_2\dots k_n)$ for $n = 1,2,3,\dots$. We call a partition $(\mathbf{w}_1, \mathbf{w}_2, \mathbf{w}_3, \dots)$ of a 0-1-sequence $(k_1k_2\dots k_n)$-*regular* or simple n-*regular* if $\mathbf{w}_j \in \{\mathbf{a}_n, \mathbf{a}_{n-1}\}$ for all $j = 1,2,3,\dots$, and if there is no j with $\mathbf{w}_j = \mathbf{w}_{j+1} = \mathbf{a}_{n-1}$.

If $Z = (\mathbf{w}_1, \mathbf{w}_2, \mathbf{w}_3, \dots)$ is an n-regular partition with $n > 0$, then by decomposing all $\mathbf{w}_j = \mathbf{a}_n = \mathbf{a}_{n-1}^{k_n}\mathbf{a}_{n-2}$ into the words $\mathbf{a}_{n-1}, \dots, \mathbf{a}_{n-1}, \mathbf{a}_{n-1}$, $\mathbf{a}_{n-2}$, one obtains an $n-1$-regular partition. Repeated application of this procedure leads to an m-regular partition for each m with $0 \leq m < n$, which we want to call the m-regular partition *derived from* Z.

Let $\mathbf{d}$ and $\mathbf{e}$ be 0-1-sequences having n-regular partitions $(\mathbf{v}_1, \mathbf{v}_2, \mathbf{v}_3, \dots)$ and $(\mathbf{w}_1, \mathbf{w}_2, \mathbf{w}_3, \dots)$ with $\mathbf{v}_1 = \mathbf{a}_n$ and $\mathbf{w}_1 = \mathbf{a}_{n-1}$. By induction on n, we verify the statement that $\mathbf{d} > \mathbf{e}$ for odd n and $\mathbf{d} < \mathbf{e}$ for even n.

For $n = 0$ the statement is obvious, and so let us assume that it is valid for $n = l$. To show its validity for $n = l+1$, let $\mathbf{d} = \mathbf{v}_1\mathbf{v}_2\mathbf{v}_3\dots, \mathbf{e} = \mathbf{w}_1\mathbf{w}_2\mathbf{w}_3\dots$ as above. Then $\mathbf{d} = \mathbf{a}_l^{k_{l+1}}\mathbf{a}_{l-1}\mathbf{v}_2\mathbf{v}_3\mathbf{v}_4\dots$ and $\mathbf{e} = \mathbf{a}_l^{k_{l+1}+1}\mathbf{a}_{l-1}\mathbf{w}_3\mathbf{w}_4\mathbf{w}_5\dots$. The sequences $\mathbf{d}' = \mathbf{a}_l\mathbf{a}_{l-1}\mathbf{w}_3\mathbf{w}_4\mathbf{w}_5\dots$ and $\mathbf{e}' = \mathbf{a}_{l-1}\mathbf{v}_2\mathbf{v}_3\mathbf{v}_4\dots$ satisfy the induction hypothesis, thus the assertion follows immediately.

Let us verify '$\Longleftarrow$' in (ii). For this, assume that $\mathbf{b}(\alpha) = \overline{\mathbf{a}(k_1k_2\dots k_n)} = \overline{\mathbf{a}_n}$ for a word $k_1k_2\dots k_n \in \mathbb{N}^*$. Then by the periodicity of $\mathbf{b}(\alpha)$ and by formula (2.1) it suffices to show that $\hat{\alpha}(i) = 0$ for each i less than the length of $\mathbf{a}_n$. Given such an i, consider the n-regular partition $(\mathbf{a}_n, \mathbf{a}_n, \mathbf{a}_n, \dots)$ of $\mathbf{b}(\alpha)$, and determine the maximal m for which the m-regular partition $(\mathbf{q}_1, \mathbf{q}_2, \mathbf{q}_3, \dots)$ derived from $(\mathbf{a}_n, \mathbf{a}_n, \mathbf{a}_n, \dots)$ satisfies the following condition: There exists a j such that the word $\mathbf{q}_1\mathbf{q}_2\dots\mathbf{q}_j$ has length i.

Then it follows that $\mathbf{q}_j = \mathbf{a}_m$ and that one of the words $\mathbf{q}_{j+1}, \mathbf{q}_{j+2}, \dots,$ $\mathbf{q}_{j+k_n}$ coincides with $\mathbf{a}_{m-1}$. Otherwise, m could not be maximal with respect to the condition above.

We only deal with the case that m is odd. (The even case has to be dealt with as the odd one.) In this case, $\mathbf{q}_j$ ends with $b_i = 1$. Since $\mathbf{b}(\alpha)$ has the initial subword $\mathbf{a}_m^{k_m+1}$, by the previous discussion $\mathbf{q}_{j+1}\mathbf{q}_{j+2}\mathbf{q}_{j+3}\dots < \mathbf{b}(\alpha)$ and by formula (2.1) one concludes that $\hat{\alpha}(i) = 0$.

In order to show '$\Longleftarrow$' in (i), take into consideration that a point with binary expansion $\mathbf{a}(k_1k_2k_3\dots)$ forms an accumulation point of the points with binary expansions $\overline{\mathbf{a}(k_1k_2\dots k_n)}; n \in \mathbb{N}$ and that the map $\alpha \mapsto \hat{\alpha}$ is continuous at non-periodic α.

Now let $\hat{\alpha} = \overline{0}$. By induction, we define a $k_1k_2k_3\dots \in \mathbb{N}^{\mathbb{N}}$ and a sequence $(Z_n)_{n=0}^{\infty}$ consisting of partitions of $\mathbf{b}(\alpha) = b_1b_2b_3\dots$, such that the following conditions are satisfied:

1. Each partition Z_n is $(k_1k_2\dots k_n)$-regular.

2. For each n, the first word in Z_n is equal to $\mathbf{a}_n = \mathbf{a}(k_1k_2\ldots k_n)$.

Obviously, the existence of the $k_1k_2k_3\ldots$ and the $(Z_n)_{n=0}^{\infty}$ with the conditions required yield the validity of the implication '$\Longrightarrow$' in (i).

Let Z_0 be the trivial partition whose members are $\mathbf{a}_{-1} = 1$ and $\mathbf{a}_0 = 0$. There is no i such that both the i-th and the $i+1$-th member of Z_0 are equal to 1. Otherwise, by formula (2.1) one would have $\hat{\alpha}(i) = 1$ for such i, contradicting the above assumptions. Therefore, 1. and 2. are valid for $n = 0$.

Now our induction hypothesis is that $k_1, k_2, \ldots, k_l$ and the partition $Z_l = \{\mathbf{q}_1, \mathbf{q}_2, \mathbf{q}_3, \ldots\}$ have been already defined and that 1. and 2. are valid for $n = l$. Further, let k_{l+1} be the largest number k with $\mathbf{q}_1 = \mathbf{q}_2 = \ldots = \mathbf{q}_k$.

For an s with $\mathbf{q}_s = \mathbf{a}_{l-1}$ we define m to be the maximum of all r with $\mathbf{q}_{s+1} = \mathbf{q}_{s+2} = \ldots = \mathbf{q}_{s+r}$ and show that $m = k_{l+1}$ or $m = k_{l+1} + 1$.

For this, let $i = |\mathbf{q}_1\mathbf{q}_2\ldots\mathbf{q}_s|$ and $j = |\mathbf{q}_1\mathbf{q}_2\ldots\mathbf{q}_{s+1}|$. If $l+1$ is even, then $\mathbf{q}_s$ ends with $b_i = 0$ and $\mathbf{q}_{s+1}$ with $b_j = 1$. On the one hand, $m < k_{l+1}$ would imply $\mathbf{q}_{s+1}\mathbf{q}_{s+2}\mathbf{q}_{s+3}\ldots < \mathbf{b}$, and on the other hand, from $m > k_{l+1} + 1$ it would follow $\mathbf{q}_{s+2}\mathbf{q}_{s+3}\mathbf{q}_{s+4}\ldots > \mathbf{b}$.

Let us conclude. By formula (2.1), one would obtain $\hat{\alpha}(i) = 1$ in the first case, and $\hat{\alpha}(j) = 1$ in the second case, but both would be contrary to the assumptions made. Analogous arguments show that $m = k_{l+1}$ or $m = k_{l+1}+1$ if $l+1$ is odd.

Connecting all finite word sequences $\mathbf{q}_j, \mathbf{q}_{j+1}, \ldots, \mathbf{q}_{j+k_{l+1}}$ with $\mathbf{q}_j = \mathbf{q}_{j+1} = \ldots = \mathbf{q}_{j+k_{l+1}-1} = \mathbf{a}_l$ and $\mathbf{q}_{j+k_{l+1}} = \mathbf{a}_{l-1}$, one obtains a new partition Z_{l+1}. Obviously, 1. and 2. for $n = l+1$ are satisfied.

If α is periodic, i.e. if $\hat{\alpha} = \overline{0\ldots0*}$, one gets a proof of '$\Longrightarrow$' in (ii) by the same method as above, but then the construction ends in a finite number of steps. ■

Using induction, one easily shows that in $\mathbf{a}(k_1k_2\ldots k_n)$ for $k_1k_2\ldots k_n \in \mathbb{N}^* \setminus \{\lambda\}$, hence in $\mathbf{a}(k_1k_2k_3\ldots)$ for $k_1k_2k_3\ldots \in \mathbb{N}^{\mathbb{N}}$, the first symbol must be different from 1, a symbol 1 cannot be followed by another symbol 1, and each maximal subword of symbols 0 must have length k_1 or $k_1 + 1$ (compare also the proof of Theorem 2.4). Substituting all subwords of the form $0^{k_1}1$ by 0 and the remaining symbols 0 by the symbol 1, one obtains no more than $\mathbf{a}(k_2k_3\ldots k_n)$ or $\mathbf{a}(k_2k_3k_4\ldots)$.

Supplementary to Theorem 2.4, by this fact we want to point out the following property having sequences of type $\overline{\mathbf{a}(k_1k_2\ldots k_n)}$ or $\overline{\mathbf{a}(k_1k_2\ldots k_n)^-}$ and $\mathbf{a}(k_1k_2k_3\ldots)$ or $\mathbf{a}(k_1k_2k_3\ldots)^-$: The numbers of symbols 1 in two subwords of the same length differ by at most one.

Let us mention that 0-1-sequences possessing this property are known as *Sturmian sequences* (see [119]). A comprehensive discussion of Sturmian sequences and their role in the dynamics of angle-doubling is contained in Section 3.2.

Call the number of symbols 1 in a 0-1-word its 1-*length*, and suppose that in a sequence as considered above, there exist two subwords $\mathbf{w}_0, \mathbf{w}_1$ such that the 1-length of $\mathbf{w}_1$ is greater than that of $\mathbf{w}_0$ by at least two. We can assume that these words are contained in a sequence $\mathbf{a}(k_1 k_2 \ldots k_n)$ for some $k_1 k_2 \ldots k_n \in \mathbb{N}^* \setminus \{\lambda\}$ (and sufficiently large $n \in \mathbb{N}$) and have minimal length. Then, the first as well as the last symbol of $\mathbf{w}_0$ are 0, and for $\mathbf{w}_1$ these symbols are 1, and the difference of the 1-lengths of $\mathbf{w}_0$ and $\mathbf{w}_1$ is two.

We will come to a contradiction as follows: Also in $\mathbf{a}(k_2 k_3 \ldots k_n)$ there are two subwords whose lengths are equal, but whose 1-lengths differ by two, and arguing inductively, there must be two such words in $\mathbf{a}(\lambda) = \overline{0}$, which is impossible.

Indeed, in accordance with the assumptions above, we find words $\mathbf{w}_0', \mathbf{w}_1'$ of common 1-lengths, such that $\mathbf{w}_0 = 0^{k_1+1}\mathbf{w}_0' 0^{k_1+1}$ and $\mathbf{w}_1 = 10^{k_1}\mathbf{w}_1' 0^{k_1}1$. Moreover, one easily sees that both $\mathbf{w}_0$ and $\mathbf{w}_1$ begin and end with 1, and that also $0^{k_1+1}\mathbf{w}_0' 0$ and $0^{k_1}10^{k_1}\mathbf{w}_1' 0^{k_1}1$ are contained in $\mathbf{a}(k_1 k_2 \ldots k_n)$. After substituting by the rules $0^{k_1}1 \longrightarrow 0, 0 \longrightarrow 1$ as described above, one obtains subwords of the same lengths whose 1-lengths differ by two.

2.1.3 Symbolic classification of the points in $\mathbb{T}$

For a fixed $\alpha \in \mathbb{T}$, the map h is invertible on the open semi-circles $]\frac{\alpha}{2}, \frac{\alpha+1}{2}[$ and $]\frac{\alpha+1}{2}, \frac{\alpha}{2}[$. We denote the inverse maps obtained and defined on $\mathbb{T} \setminus \{\alpha\}$ by l_0^α and l_1^α, respectively.

More generally, for a 0-1-word $\mathbf{w} = w_1 w_2 \ldots w_n \in \{0,1\}^*$, by $l_{\mathbf{w}}^\alpha = l_{w_1}^\alpha \circ l_{w_2}^\alpha \circ \ldots \circ l_{w_n}^\alpha$ there is given a map $l_{\mathbf{w}}^\alpha$ whose domain of definition includes the set $\mathbb{T} \setminus \{\alpha, h(\alpha), h^2(\alpha) \ldots h^{n-1}(\alpha)\}$. (For the empty word $\mathbf{w} = \lambda$, let $l_{\mathbf{w}}^\alpha$ be the identity.) Further, let $\mathbb{T}_{\mathbf{w}}^\alpha = \mathrm{cl}\, l_{\mathbf{w}}^\alpha(\mathbb{T} \setminus \{\alpha, h(\alpha), h^2(\alpha) \ldots h^{n-1}(\alpha)\})$.

In the case that α is periodic with respect to h, one of the points $\frac{\alpha}{2}, \frac{\alpha+1}{2}$ coincides with $h^{PER(\alpha)-1}(\alpha)$. This (periodic) point is denoted by $\dot{\alpha}$, the other (preperiodic) one by $\ddot{\alpha}$.

For periodic α and given $s \in \{0,1\}$, we are interested in two injective extensions of l_s^t to the whole circle. These are the maps $l_s^{\alpha,t}$ depending on $t \in \{0,1\}$ and defined by

$$l_s^{\alpha,t}(\alpha) = \begin{cases} \dot{\alpha} \text{ for } s = t \\ \ddot{\alpha} \text{ for } s \neq t \end{cases} . \tag{2.3}$$

Furthermore, let $l_{\mathbf{w}}^{\alpha,t} = l_{w_1}^{\alpha,t} \circ l_{w_2}^{\alpha,t} \circ \ldots \circ l_{w_n}^{\alpha,t}$ for $\mathbf{w} = w_1 w_2 \ldots w_n \in \{0,1\}^*$. (Again, $l_\lambda^{\alpha,t}$ means the identical map.)

For $\alpha \in \mathbb{T}$ and $\mathbf{w} = w_1 w_2 \ldots w_n \in \{0,1\}^*$ it holds

$$\mathbb{T}_{\mathbf{w}}^\alpha = \mathrm{cl}\{\beta \in \mathbb{T} | I^\alpha(\beta)(i) = w_i \text{ for } i = 1, 2, \ldots, n\}.$$

For $\mathbf{s} \in \{0,1\}^{\mathbb{N}}$, let $\mathbb{T}_{\mathbf{s}}^\alpha := \bigcap_{n=1}^\infty \mathbb{T}_{\mathbf{s}|n}^\alpha$ and $\widetilde{\mathbb{T}_{\mathbf{s}}^\alpha} = \{\beta \in \mathbb{T} \mid I^\alpha(\beta) \cong \mathbf{s}\}$. Obviously, the interior of $\widetilde{\mathbb{T}_{\mathbf{s}}^\alpha}$ is empty, and it holds $\mathbb{T}_{\mathbf{s}}^\alpha \subseteq \widetilde{\mathbb{T}_{\mathbf{s}}^\alpha}$. Moreover, one easily sees that $\mathbb{T}_{\mathbf{s}}^{\alpha} = \{\beta \in \mathbb{T} \,|\, \overset{\frown}{I^{\alpha}}(\beta) = \mathbf{s} \text{ or } \overset{\frown}{I^{\alpha}}(\beta) = \mathbf{s}\}$. It will be important to know when the sets $\widetilde{\mathbb{T}_{\mathbf{s}}^{\alpha}}$ and $\mathbb{T}_{\mathbf{s}}^{\alpha}$ are equal.

Often, for $\alpha \in \mathbb{P}_*$, we will have to consider the special case that a word $\mathbf{w} \in \{0,1\}^*$ ends with $0\mathbf{v}^{\alpha}$ or $1\mathbf{v}^{\alpha}$, or that a shift of a sequence $\mathbf{s} \in \{0,1\}^{\mathbb{N}}$ is substantially equal to $\hat{\alpha}$. Thus, to shorten some statements, let us introduce the following concept:

Definition 2.5. *Let $\alpha \in \mathbb{T}$, let $\mathbf{w} \in \{0,1\}^*$ and $\mathbf{s} \in \{0,1\}^{\mathbb{N}}$. Then the word $\mathbf{w}$ is said to be* α-regular *if either α is non-periodic, or α is periodic and $\mathbf{w}$ does not end with $0\mathbf{v}^{\alpha}$ or $1\mathbf{v}^{\alpha}$. The sequence $\mathbf{s}$ is called* α-regular *if either α is non-periodic, or α is periodic and $\sigma^n(\mathbf{s}) \not\cong \hat{\alpha}$ for all $n \in \mathbb{N}_0$.*

Now let $\alpha \in \mathbb{T}$ and $\mathbf{s} \in \{0,1\}^{\mathbb{N}}$. Then for all $\beta \in \mathbb{T}$ and $k \in \mathbb{N}$ one has $I^{\alpha}(\beta)(k) = *$ iff $\overset{\frown}{I^{\alpha}}(\beta)(k) \neq \overset{\frown}{I^{\alpha}}(\beta)(k)$. Consequently, $\mathbb{T}_{\mathbf{s}}^{\alpha} \neq \widetilde{\mathbb{T}_{\mathbf{s}}^{\alpha}}$ is only possible if there exists a point $\beta \in \widetilde{\mathbb{T}_{\mathbf{s}}^{\alpha}}$ whose itinerary contains the symbol $*$ more than one time, in other words, if $\mathbf{s}$ fails to be α-regular.

So we can assume that α is periodic of $PER(\alpha) = m$, that $\sigma^n(\mathbf{s}) \cong \hat{\alpha}$ for some $n \in \mathbb{N}_0$, and that $\beta \in \widetilde{\mathbb{T}_{\mathbf{s}}^{\alpha}}$. If $I^{\alpha}(\beta) = \mathbf{s}$, then of course $\beta \in \mathbb{T}_{\mathbf{s}}^{\alpha}$. Otherwise, let $n \in \mathbb{N}_0$ be minimal with the property that $h^n(\beta) = \dot{\alpha}$ or $h^n(\beta) = \ddot{\alpha}$. Then there exists a sequence $t_1t_2t_3\ldots \in \{0,1\}^{\mathbb{N}}$ such that $\sigma^n(\mathbf{s}) = t_1\mathbf{v}^{\alpha}t_2\mathbf{v}^{\alpha}t_3\mathbf{v}^{\alpha}\ldots$.

If $h^n(\beta) = \dot{\alpha}$, then $\ldots = \overset{\frown}{I^{\alpha}}(\beta)(n+2m) = \overset{\frown}{I^{\alpha}}(\beta)(n+m) = \overset{\frown}{I^{\alpha}}(\beta)(n) \neq \overset{\frown}{I^{\alpha}}(\beta)(n) = \overset{\frown}{I^{\alpha}}(\beta)(n+m) = \overset{\frown}{I^{\alpha}}(\beta)(n+2m)\ldots$, and if $h^n(\beta) = \ddot{\alpha}$, then $\ldots = \overset{\frown}{I^{\alpha}}(\beta)(n+2m) = \overset{\frown}{I^{\alpha}}(\beta)(n+m) \neq \overset{\frown}{I^{\alpha}}(\beta)(n) \neq \overset{\frown}{I^{\alpha}}(\beta)(n) \neq \overset{\frown}{I^{\alpha}}(n+m) = \overset{\frown}{I^{\alpha}}(\beta)(n+2m)\ldots$. Therefore, in the first case $\beta \in \mathbb{T}_{\mathbf{s}}^{\alpha}$ iff $t_1 = t_2 = t_3\ldots$, and in the second case $\beta \in \mathbb{T}_{\mathbf{s}}^{\alpha}$ iff $t_1 \neq t_2 = t_3\ldots$. Let us summarize:

Proposition 2.6. *For $\alpha \in \mathbb{T}$ and $\mathbf{s} \in \{0,1\}^{\mathbb{N}}$, the sets $\widetilde{\mathbb{T}_{\mathbf{s}}^{\alpha}}$ and $\mathbb{T}_{\mathbf{s}}^{\alpha}$ coincide iff $\mathbf{s}$ is α-regular. In particular, $\beta \in \widetilde{\mathbb{T}_{\mathbf{s}}^{\alpha}} \setminus \mathbb{T}_{\mathbf{s}}^{\alpha}$ for $\mathbf{s}$ not α-regular iff there exist a word $\mathbf{w} \in \{0,1\}^*$ and a sequence $t_1t_2t_3\ldots \in \{0,1\}^{\mathbb{N}}$ satisfying $\mathbf{s} = \mathbf{w}t_1\mathbf{v}^{\alpha}t_2\mathbf{v}^{\alpha}t_3\mathbf{v}^{\alpha}\ldots$ and*

(i) $\beta = l_{\mathbf{w}}^{\alpha}(\ddot{\alpha})$ and not $t_1 \neq t_2 = t_3 = t_4\ldots$ or
(ii) $\mathbf{w}$ is α-regular, $\beta = l_{\mathbf{w}}^{\alpha}(\dot{\alpha})$, but not $t_1 = t_2 = t_3\ldots$.

$\widetilde{\mathbb{T}_{\mathbf{s}}^{\alpha}}$ for $\mathbf{s}$ not α-regular is infinite. ■

Since the compact sets $\mathbb{T}_{\mathbf{s}|n}^{\alpha}$ form a monotonically decreasing sequence, it holds $\widetilde{\mathbb{T}_{\mathbf{s}}^{\alpha}} \supseteq \mathbb{T}_{\mathbf{s}}^{\alpha} \neq \emptyset$. By this, one easily sees that at least 0-1-sequences whose shifts are not substantially equal to $\hat{\alpha}$, can be realized by the itinerary of a point in $\mathbb{T}$. Note that the admissibility of sequences as itineraries is discussed in Section 2.3 completely (see Proposition 2.56).

2.2 Invariant laminations

2.2.1 Some basic geometric statements

The following discussion is devoted to THURSTON's concept of an invariant lamination. First of all, we want to give some basic notions and elementary statements.

By $\mathbb{D}$ we denote the closed unit disk in the complex plane, whose boundary points $e^{2\pi\beta i}$ are regarded as the points β of $\mathbb{T}$, according to Chapter 1.1.2. Further, let $Comp(\mathbb{D})$ be the set of all closed (and so compact) subsets of $\mathbb{D}$, which we equip with the topology induced by the Hausdorff metric. (In this metric the distance of two sets $A, B \in Comp(\mathbb{D})$ is defined to be the maximum of $\sup_{a\in A}(\inf_{b\in B}|a-b|)$ and $\sup_{b\in B}(\inf_{a\in A}|b-a|)$.)

We extend the action of a map f from the circle $\mathbb{T}$ into itself to the set of all subsets of $\mathbb{D}$ forming the convex hull of subsets of $\mathbb{T}$: if $A \subseteq \mathbb{T}$, then $f(A) := \mathrm{conv}(f(A \cap \mathbb{T}))$, where $\mathrm{conv}(B)$ denotes the convex hull of a set $B \subseteq \mathbb{T}$. (Obviously, $\mathrm{conv}(B)$ is closed iff B is.)

The convex hull of two points $\beta_1, \beta_2 \in \mathbb{T}$ is said to be a *chord* and denoted by $\beta_1\beta_2$. Let us emphasize that we allow the case $\beta_1 = \beta_2$. In this case, the chord $\beta_1\beta_2$ is said to be *degenerate* and otherwise *non-degenerate*.

We say that two different chords *cross* and *touch* each other if they have a common interior point and a common endpoint, respectively.

A non-degenerate chord divides $\mathbb{D}$ into two disjoint open parts. The chord is defined to *separate* two given subsets of $\mathbb{D}$ if one of them has common points with only one part, and the other one with only the other part. In the case that one of the parts is smaller than the other one, a subset of $\mathbb{D}$ is called *behind* the given chord if it has common points only with the small part.

Analogously, a subset of $\mathbb{D}$ is said to be *between* two disjoint chords if it has at least one point in the open component between the chords, but none in the other two open parts of the disk.

Let d be the inner metric given by the arc length on $\mathbb{T}$. In accordance to the arrangements above, we normalize it in such a way that the circumference is equal to one. The length of a chord $S = \beta_1\beta_2$ is measured by $d(S) := d(\beta_1, \beta_2)$. Moreover, we say that a point $\beta_1 \in \mathbb{T}$ lies *between* two points $\beta_2, \beta_3 \in \mathbb{T}$ if $d(\beta_2, \beta_3) < \frac{1}{2}$ and β_1 is contained in the smaller open arc connecting β_2 and β_3.

For each chord it holds

$$d(h(S)) = \begin{cases} 2d(S) & \text{for} \quad d(S) \leq \frac{1}{4} \\ 1 - 2d(S) & \text{for} \quad d(S) > \frac{1}{4} \end{cases}, \tag{2.4}$$

i.e. h acts on the (double) length of the chords as the Tent map (compare Example A.7). In particular, $d(h(S)) < d(S)$ iff $d(S) > \frac{1}{3}$, and one easily obtains the following conclusions:

Lemma 2.7. *If the length of a chord S and its iterates $h^i(S); i = 1, 2, \dots, n-1$ does not exceed $\frac{1}{4}$, then $d(S) = 2^{-n}d(h^n(S))$.* ■

Lemma 2.8. *Let S be a chord satisfying $d(h(S)) < (=)\frac{1}{2} - a$ for some a with $0 < a < \frac{1}{2}$. Then either $d(S) = \frac{d(h(S))}{2}$ or S is longer than a (of length equal to a).* ■

For given points $\beta_1, \beta_2, \dots, \beta_n \in \mathbb{T}$ we write $\beta_1 \curvearrowright \beta_2 \curvearrowright \dots \curvearrowright \beta_n$ ($\beta_1 \curvearrowleft \beta_2 \curvearrowleft \dots \curvearrowleft \beta_n$) if the points lie in clockwise (counter-clockwise) orientation. Obviously, the following is valid:

Lemma 2.9. (Orientation-invariance)
If three points $\beta_1, \beta_2, \beta_3$ are contained in an open semi-circle $]\alpha, \alpha'[$ with $\alpha \in \mathbb{T}$, and if $\beta_1 \curvearrowright \beta_2 \curvearrowright \beta_3$ ($\beta_1 \curvearrowleft \beta_2 \curvearrowleft \beta_3$), then $h(\beta_1) \curvearrowright h(\beta_2) \curvearrowright h(\beta_3)$ ($h(\beta_1) \curvearrowleft h(\beta_2) \curvearrowleft h(\beta_3)$). ■

We start now to list some elementary statements, which can be found in THURSTON's paper [167]. Because of their importance for our further considerations, we will give complete proofs of the statements.

Lemma 2.10 ([167]). *Let S be a chord, and let $n \in \mathbb{N}$ be given such that none of the chords $h(S), h^2(S), \dots, h^{n-1}(S)$ is longer than S. Then $d(h^n(S)) > d(S)$ if $h^n(S)$ lies between S and S'.*

Proof: For $d(S) < \frac{1}{3}$ the statement is obvious, so assume that $d(S) \geq \frac{1}{3}$. Then either $d(h^n(S)) > d(S)$ or $d(h^n(S)) \leq \frac{1}{2} - d(S)$. In the second case there would exist a least index $i \leq n$ with $d(h^i(S)) \leq 1 - 2d(S))$. By formula (2.4), this would imply $d(h^i(S)) = 1 - 2d(h^{i-1}(S))$. Hence, in contradiction to our assumptions, $h^{i-1}(S)$ would be longer than S. ■

The following interesting fact plays a central role in THURSTON's paper. However, it does not occur explicitly in the form presented now.

Theorem 2.11 ([167]). (No wandering triangles)
There does not exist a triangle Δ with the property that the iterated triangles $h^n(\Delta); n = 1, 2, 3, \dots$ are non-degenerate and have mutually disjoint interiors.

Proof: Suppose the existence of a triangle with the properties given. Further, for each $n \in \mathbb{N}_0$, let δ_n and η_n be the minimum and the maximum of the side lengths of $\Delta_n = h^n(\Delta)$, respectively.

The sequence $(\delta_n)_{n=0}^{\infty}$ must converge to 0. Otherwise, the sum of the areas of all iterated triangles would be infinite, contrary to the assumptions made. The 'length map' $d(S) \mapsto d(h(S))$ is increasing on $[0, \frac{1}{4}]$ and decreasing on $[\frac{1}{4}, \frac{1}{2}]$, and throughout the interval $[\frac{1}{6}, \frac{1}{3}]$ it only takes on values not less than $\frac{1}{3}$. This implies the following:

If $\frac{1}{3} > \delta_{k-1} > \delta_k$ for some $k \in \mathbb{N}$, than at least one side of Δ_{k-1} is longer than $\frac{1}{3}$ and h maps the longest side of Δ_{k-1} to the shortest side of Δ_k.

Consequently, by formula (2.4) we have $\delta_k = 1 - 2\eta_{k-1}$ and so $\eta_{k-1} = \frac{1}{2} - \frac{\delta_k}{2}$. In particular, we get $\sup_{n=0}^{\infty} \eta_n = \frac{1}{2}$.

By the already shown, we can assume that $\frac{1}{3} > \delta_0 > \delta_1$ and that Δ has a longest chord S with $d(S) = \eta_0 = \frac{1}{2} - \frac{\delta_1}{2}$. Besides S, the triangle Δ possesses no chord of length η_0. The contrary is impossible by the assumptions on Δ and by $\sup_{n=0}^{\infty} \eta_n = \frac{1}{2}$. By assumption, no two sides of different triangles $\Delta_n; n \in \mathbb{N}$ cross each other. We show indirectly that S' does not cross a side of the triangles.

Suppose that U is a side of an iterated triangle which crosses S' (but not S). If U does not touch S, then there is a semi-circle containing all endpoints of S' and U. This and the orientation-invariance of h (Lemma 2.9) lead to the contradiction that also the chords $h(U)$ and $h(S') = h(S)$ cross each other.

In case that U touches S, by $\sup_{n=0}^{\infty} \eta_n = \frac{1}{2}$ infinitely many sides of iterated triangles must end in the corresponding touching point. Hence there exist an $n \in \mathbb{N}$ and a $k \in \mathbb{N}$ with $h^k(\beta) = \beta$ for an angle β of Δ_n. That side V of Δ_n which does not end in β has infinitely many iterates of length not less than $\frac{1}{3}$. Thus, for at least one $l \in \mathbb{N}$ the triangles $\Delta_{n+l+ik}; i \in \mathbb{N}_0$ could not have mutually disjoint interior.

If $\delta_k \geq \delta_1$ for some $k \in \mathbb{N}$, then by $\delta_1 + \eta_0 > \frac{1}{2}$ the triangle Δ_k cannot lie between S and S', which implies that $\eta_k \leq \eta_0$ and so that $\delta_{k+1} \geq \delta_1$. Therefore, by induction one can conclude that $\delta_n \geq \delta_1$ for all $n \in \mathbb{N}$, which is impossible. ∎

Definition 2.12. (Laminations)
By a lamination *we understand a non-empty set $\mathcal{L}$ of chords which is closed in Comp($\mathbb{D}$), and whose elements do not cross each other. $\mathcal{L}$ is said to be* invariant *if for all $S \in \mathcal{L}$, it holds $h(S), S' \in \mathcal{L}$ and there exists a chord $S^- \in \mathcal{L}$ with $h(S^-) = S$.*

If $\mathcal{L}$ is a lamination, then the closure of a component of $\mathbb{D} \setminus \bigcup \mathcal{L}$ is called a gap *of $\mathcal{L}$. A non-degenerate chord in $\mathcal{L}$ is said to* bound *a gap or called a* boundary chord *of a gap if it is contained in (the boundary of) the gap. Sometimes it is convenient to consider some chord as a boundary chord of itself, thinking of the chord as a degenerate gap.*

A gap possessing infinitely many non-degenerate boundary chords is called an infinite *gap, one with only finitely many boundary chords a* polygonal *gap. In case that $\mathcal{L}$ is invariant, a gap containing the center of the disc in its interior is said to be* critical, *and a gap forming the image of a critical one with respect to h is called* critical value gap.

The concept 'critical gap' is justified by the fact that h acts two-to-one on (the intersection of $\mathbb{T}$ with) the gap, and obviously, if a critical gap exists, it is unique.

Remarks. 1. THURSTON [167] has introduced invariant laminations in the more general context of the maps $\beta \mapsto d\beta \bmod 1; d = 2, 3, 4, \ldots$. For $d = 2$

the invariant lamination was called quadratic, and the simple formulation of the invariance in this case is due to REES [139] and equivalent to THURSTON's original formulation.

Instead of the closeness of $\mathcal{L}$ in the sense above, THURSTON required that the union of all chords in $\mathcal{L}$ is closed. This difference is not substantial because a lamination in the sense of THURSTON can be 'completed' to a lamination in our sense by adding some degenerate chords.

2. All degenerate chords $\beta\beta$ for $\beta \in \mathbb{T}$ belong to an invariant lamination (according to our definition). This can easily be seen since the backward orbit of every point in $\mathbb{T}$ with respect to h is dense in $\mathbb{T}$.

Given a gap G of an invariant lamination, let us have a look at its image with respect to h. If G is not critical, then immediately from the definition of an invariant lamination and from the orientation-invariance of h (Lemma 2.9) it follows that $h(G)$ forms a gap or a chord of the invariant lamination.

Otherwise, there exists an $\alpha \in \mathbb{T}$ with $\alpha, \alpha' \in G \cap \mathbb{T}$, such that G is divided by $\alpha\alpha'$ into two closed parts. Since the map $'$ transforms each of these parts into the other one, their image with respect to h is equal to $h(G)$. Hence, as in the non-critical case, $h(G)$ is a gap or a chord of the lamination.

If G is neither periodic nor preperiodic, then decompose G into triangles with mutually disjoint interiors. By Theorem 2.11 one iterate of each triangle obtained is bounded by some chord $\beta\beta'; \beta \in \mathbb{T}$. Thus G must be the union of not more than two triangles. Let us summarize:

Theorem 2.13 ([167]). *If G is a gap of an invariant lamination, then $h(G)$ forms a gap or a chord of the lamination. Moreover, there exists an $n \in \mathbb{N}_0$ such that*

(i) $h^n(G)$ is periodic with respect to h, or

(ii) $h^n(G)$ forms a triangle whose one side is equal to $\beta\beta'$ for some $\beta \in \mathbb{T}$, or a rectangle containing $\beta\beta'$ as its diagonal for some $\beta \in \mathbb{T}$. ■

Furthermore, we will use the following result of THURSTON:

Proposition 2.14 ([167])). (Polygonal gaps and periodic points)
Let G be a periodic polygonal gap of an invariant lamination. Then all points in $G \cap \mathbb{T}$ as well as all sides of G lie on a common periodic orbit.

Proof: Let p be the period of G with respect to h. Then obviously h^p permutes the points of $G \cap \mathbb{T}$. Thus for no $n \in \mathbb{N}_0$ does the gap $h^n(G)$ contain a chord $\beta\beta'$ with $\beta \in \mathbb{T}$, and so $h^n(G) \cap \mathbb{T}$ is a subset of an open semi-circle.

Also on the set $\mathcal{S}$ of all chords bounding any gap of the orbit of G, the map h^p acts as a permutation. Let $\overline{S}$ be the longest chord in $\mathcal{S}$. We can assume that $\overline{S}$ bounds G. Then besides $\overline{S}$ there is at most one boundary chord $\overline{\overline{S}}$ of G which is not shorter than $\frac{1}{3}$.

For an arbitrary side S of the orbit gaps there exists an $n \in \mathbb{N}_0$ with $d(h^n(S)) \geq \frac{1}{3}$. If $h^n(S)$ itself does not coincide with $\overline{S}$ or $\overline{\overline{S}}$, then G lies between $h^n(S)$ and $h^n(S)'$, and by Lemma 2.10 $h^m(S) = \overline{S}$ or $h^m(S) = \overline{\overline{S}}$ for the least $m > n$ with the property that $h^m(S)$ bounds G. Thus the chords in $\mathcal{S}$ belong to at most two different orbits.

If $\overline{\overline{S}}$ exists, than again by Lemma 2.10 $h^p(\overline{\overline{S}})$ is equal to $\overline{\overline{S}}$ or to $\overline{S}$. The first case is impossible since then both orbits in $\mathcal{S}$ would have different lengths. Hence $\mathcal{S}$ does not split into two orbits, and so also all points of $G \cap \mathbb{T}$ lie on one orbit. ■

2.2.2 Invariant laminations constructed from a 'long' chord

Our aim now is to construct an invariant lamination $\mathcal{B}^\alpha$ for each $\alpha \in \mathbb{T}$ The idea for the natural construction goes back to THURSTON [167], but our presentation is directed to a further development of the symbolic concept.

From now on, we want to use 'dynamical' notions like 'iterate', 'backward iterate', 'orbit', 'periodic' etc. only in connection with the map h, and so we do not want to indicate this further on.

For every $\alpha \in \mathbb{T}$, all proper iterates of the chord $S_*^\alpha := \frac{\alpha}{2}\frac{\alpha+1}{2}$ are degenerate, i.e. one-point chords. Therefore, one can try to construct an invariant lamination from S_*^α by successively choosing backward iterates of S_*^α in an appropriate way.

In fact, we inductively define sets $\mathcal{R}_0^\alpha, \mathcal{R}_1^\alpha, \mathcal{R}_2^\alpha, \ldots$ of backward iterates of S_*^α, where $\mathcal{R}_0^\alpha := \{S_*^\alpha\}$. Assuming that $\mathcal{R}_0^\alpha, \mathcal{R}_1^\alpha, \ldots, \mathcal{R}_n^\alpha$ have been already defined, let $\mathcal{R}_{n+1}^\alpha$ be the set of all those preimages taken from elements in $\mathcal{R}_n^\alpha$ which do not cross a chord in $\bigcup_{i=0}^n \mathcal{R}_i^\alpha$. Then, if for each $i \in \mathbb{N}$ every two chords of $\mathcal{R}_i^\alpha$ do not cross one another, the set $\bigcup_{i=0}^\infty \mathcal{R}_i^\alpha$ consists of mutually non-crossing chords.

We have to distinguish between the cases $\alpha \in \mathbb{T} \setminus \mathbb{P}_*$ (see Figures 2.1 and 2.3) and $\alpha \in \mathbb{P}_*$ (see Figures 2.2 and 2.4).

The non-periodic case. For $\alpha \notin \mathbb{P}_*$ let $\mathcal{R}^\alpha$ be the set of all backward iterates of S_*^α which do not cross S_*^α. In this case, it will turn out that $\mathcal{R}^\alpha$ consists of mutually disjoint chords (see text below Theorem 2.15), and so that $\bigcup_{i=0}^\infty \mathcal{R}_i^\alpha$ and $\mathcal{R}^\alpha$ coincide.

Since α is non-periodic, the endpoints of each chord $S = \beta_1\beta_2 \in \mathcal{R}^\alpha$ are different to α, hence it has exactly two preimages in $\mathcal{R}^\alpha$: the chords $l_0^\alpha(S) = l_0^\alpha(\beta_1)l_0^\alpha(\beta_2)$ and $l_1^\alpha(S) = l_1^\alpha(\beta_1)l_1^\alpha(\beta_2)$. By use of the notation $S_{\mathbf{w}*}^\alpha := l_{\mathbf{w}}^\alpha(S_*^\alpha)$ one obtains

$$\mathcal{R}^\alpha = \{S_{\mathbf{w}*}^\alpha \mid \mathbf{w} \in \{0,1\}^*\} \text{ for } \alpha \notin \mathbb{P}_*.$$

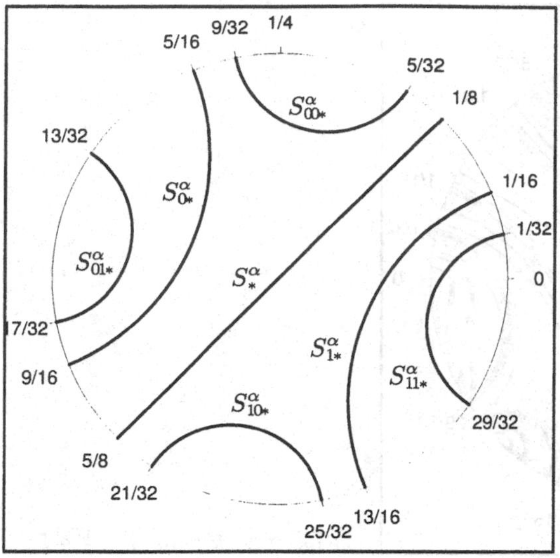

Fig. 2.1. Construction of an invariant lamination with 'long' chord: $\mathcal{R}_2^{1/4}$

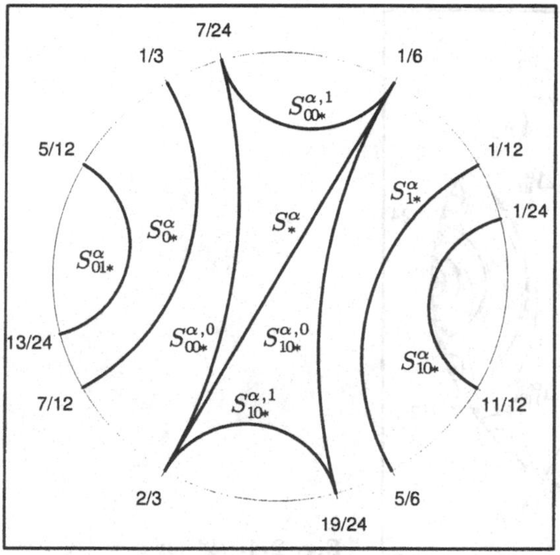

Fig. 2.2. Construction of an invariant lamination with 'long' chord: $\mathcal{R}_2^{1/3}$

The periodic case. For $\alpha \in \mathbb{P}_*$ the situation is a little more complicated. In this case, there exist backward iterates S of S_*^α with α as an endpoint. For such S the chords $l_0^\alpha(S)$ and $l_1^\alpha(S)$ are not defined. Nevertheless, the chord S has exactly four preimages, which do not cross S_*^α, but touch it at $\dot{\alpha}$ or $\ddot{\alpha}$: on the one side of S_*^α, there lie the chords $l_0^{\alpha,0}(S)$, $l_0^{\alpha,1}(S)$, and on the other side the chords $l_1^{\alpha,0}(S)$ and $l_1^{\alpha,1}(S)$.

Since $l_{\mathbf{v}}^\alpha(\dot{\alpha}) = \alpha$ for $\mathbf{v} = \mathbf{v}^\alpha$, the map $l_{\mathbf{w}}^\alpha$ is defined in S_*^α iff $\mathbf{w}$ is α-regular, i.e. does not end with $0\mathbf{v}$ or $1\mathbf{v}$.

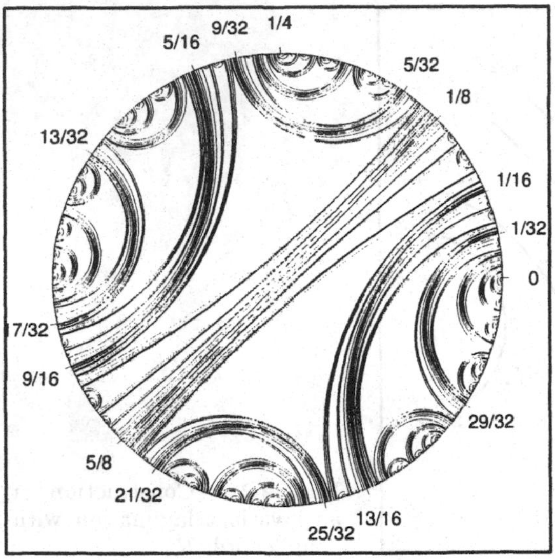

Fig. 2.3. $\mathcal{R}^\alpha$ for $\alpha = \frac{1}{4} \notin \mathbb{P}_*$

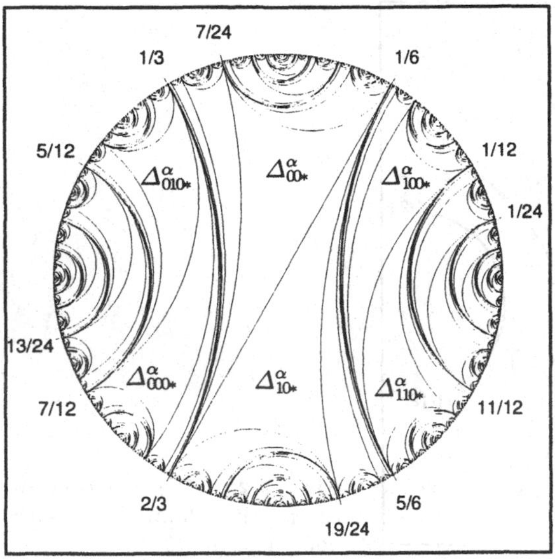

Fig. 2.4. $\mathcal{R}^\alpha$ for $\alpha = \frac{1}{3} \in \mathbb{P}_*$

In this case, let $S^\alpha_{\mathbf{w}*} = l^\alpha_{\mathbf{w}}(S^\alpha_*)$ as above. Further, for all $\mathbf{w} \in \{0,1\}^*$ and $t \in \{0,1\}$ let $S^{\alpha,t}_{\mathbf{w}*} = l^{\alpha,t}_{\mathbf{w}}(S^\alpha_*)$ (see formula (2.3)). Then one easily sees that $S^{\alpha,0}_{\mathbf{w}*}$ and $S^{\alpha,1}_{\mathbf{w}*}$ are equal (and coincide with $S^\alpha_{\mathbf{w}*}$) iff $\mathbf{w}$ is α-regular.

$S^\alpha_{\mathbf{v}*}$ is the first backward iterate of S^α_* for which l^α_0 and l^α_1 fail to be defined and which has four preimages not crossing $S^\alpha_{\mathbf{v}*}$. The two preimages $S^{\alpha,0}_{0\mathbf{v}*}, S^{\alpha,1}_{0\mathbf{v}*}$ as well as the two preimages $S^{\alpha,0}_{1\mathbf{v}*}, S^{\alpha,1}_{1\mathbf{v}*}$ together with S^α_* form a triangle (see Figures 2.2 and 2.4 for the example $\alpha = \frac{1}{3}$ with $\mathbf{v} = \mathbf{v}^\alpha = 0$).

Now the following problem arises: For further backward iterates of S_*^α with one endpoint equal to α, again there are two preimages on each side of S_*^α, but in each case one of them crosses one of the four chords above. Therefore, the set of all backward iterates of S_*^α which do not cross S_*^α is bigger than $\bigcup_{i=0}^\infty \mathcal{R}_i^\alpha$. In particular, if the four chords above were elements of $\mathcal{R}_m^\alpha$ with $m = PER(\alpha)$, then each chord in $\bigcup_{i=0}^\infty \mathcal{R}_i^\alpha$ different from $l_{\mathbf{v}}^\alpha(S_*^\alpha)$ would have at most two preimages in $\bigcup_{i=0}^\infty \mathcal{R}_i^\alpha$.

A set of chords containing $S_{\mathbf{0v}*}^{\alpha,0}, S_{\mathbf{0v}*}^{\alpha,1}, S_{\mathbf{1v}*}^{\alpha,0}, S_{\mathbf{1v}*}^{\alpha,1}$, with each chord its image, and with each chord different from $S_{\mathbf{v}*}^\alpha$ exactly two preimages, is given by $\mathcal{R}^\alpha$ being defined as follows:

$$\mathcal{R}^\alpha := \mathcal{R}^{\alpha,0} \cup \mathcal{R}^{\alpha,1} \text{ with } \mathcal{R}^{\alpha,t} := \{S_{\mathbf{w}*}^{\alpha,t} \mid \mathbf{w} \in \{0,1\}^{\mathbb{N}}\}; t = 0,1 \quad \text{for } \alpha \in \mathbb{P}_*. \tag{2.5}$$

It will turn out that the elements of $\mathcal{R}^\alpha$ do not cross each other (see below Theorem 2.15), which implies $\mathcal{R}^\alpha = \bigcup_{i=0}^\infty \mathcal{R}_i^\alpha$ also in the present case.

The notation in (2.5) is a little misleading. Namely, the sets $\mathcal{R}^{\alpha,0}$ and $\mathcal{R}^{\alpha,1}$ are not disjoint, and by the above considerations, $\mathcal{R}^\alpha$ divides into the set of all chords $S_{\mathbf{w}*}^\alpha$ for α-regular $\mathbf{w} \in \{0,1\}^*$ and into pairs of chords $S_{\mathbf{utv}*}^{\alpha,0}$ and $S_{\mathbf{utv}*}^{\alpha,1}$ for all $\mathbf{u} \in \{0,1\}^*, t \in \{0,1\}$.

Let us have a closer look at such a pair of chords for given $\mathbf{u} \in \{0,1\}^*$ and $t \in \{0,1\}$. Both chords considered end in $l_{\mathbf{utv}}^\alpha(\ddot{\alpha})$, and the other endpoints $l_{\mathbf{utv}}^{\alpha,t}(\dot{\alpha}) = l_{\mathbf{u}}^{\alpha,t}(\dot{\alpha})$ and $l_{\mathbf{utv}}^{\alpha,1-t}(\dot{\alpha}) = l_{\mathbf{u}}^\alpha(\ddot{\alpha})$ are connected by the chord $S_{\mathbf{u}*}^{\alpha,t} \in \mathcal{R}^\alpha$. Finally, we denote the triangle with sides $S_{\mathbf{utv}*}^{\alpha,0}, S_{\mathbf{utv}*}^{\alpha,1}$ and $S_{\mathbf{u}*}^{\alpha,t}$ by $\Delta_{\mathbf{utv}*}^\alpha$. It can be considered as a 'blown up' chord, which takes the place of $S_{\mathbf{utv}*}^\alpha$ in the periodic case.

We are able now to describe the geometry of the set $\mathcal{R}^\alpha$ for an arbitrarily given $\alpha \in \mathbb{T}$. In the following, let $\mathbb{D}_{\mathbf{w}}^\alpha := \mathrm{conv}(\mathbb{T}_{\mathbf{w}})$ for $\mathbf{w} \in \{0,1\}^*$.

Theorem 2.15. *For $\alpha \in \mathbb{T}$ and $\mathbf{w} \in \{0,1\}^*$ the following statements are valid:*

(i) If $\mathbf{w}$ is α-regular, then
   1. $\mathbb{D}_{\mathbf{w}}^\alpha = \mathbb{D}_{\mathbf{w}0}^\alpha \cup \mathbb{D}_{\mathbf{w}1}^\alpha$ *and*
   2. $\mathbb{D}_{\mathbf{w}0}^\alpha \cap \mathbb{D}_{\mathbf{w}1}^\alpha = S_{\mathbf{w}*}^\alpha$.

(ii) If $\mathbf{w}$ is not α-regular, then
   1. $\mathbb{D}_{\mathbf{w}}^\alpha = \mathbb{D}_{\mathbf{w}0}^\alpha \cup \mathbb{D}_{\mathbf{w}1}^\alpha \cup \Delta_{\mathbf{w}*}^\alpha$ *and*
   2. $\mathbb{D}_{\mathbf{w}0}^\alpha \cap \Delta_{\mathbf{w}*}^\alpha = S_{\mathbf{w}*}^{\alpha,0}$ *and* $\mathbb{D}_{\mathbf{w}1}^\alpha \cap \Delta_{\mathbf{w}*}^\alpha = S_{\mathbf{w}*}^{\alpha,1}$.

Proof: If $\mathbf{w} \in \{0,1\}^*$ is α-regular, then the points $l_{\mathbf{w}}^\alpha(\frac{\alpha}{2})$ and $l_{\mathbf{w}}^\alpha(\frac{\alpha+1}{2})$ belong to the interior of $\mathbb{T}_{\mathbf{w}}^\alpha$, hence the chord $S_{\mathbf{w}*}^\alpha$ cuts the interior of $\mathbb{D}_{\mathbf{w}}^\alpha$. Now let n be the length of $\mathbf{w}$. By application of Lemma 2.9 to the set $\mathbb{T}_{\mathbf{w}}^\alpha$ and its first n iterates, one easily obtains that $S_{\mathbf{w}*}^\alpha$ separates the points $\beta \in \mathbb{T}_{\mathbf{w}}^\alpha$ with $I^\alpha(\beta)(n+1) = 0$ and with $I^\alpha(\beta)(n+1) = 1$. This shows (i).

We now verify (ii) for a given $\alpha \in \mathbb{P}_*$, by induction on the length of $\mathbf{w}$ and by using (i). If $\mathbf{w} = \lambda$, then (ii) is trivially satisfied since the empty word is α-regular. So we can assume that (ii) is valid for all words shorter than a given $n \in \mathbb{N}$.

Let $\mathbf{v} = \mathbf{v}^\alpha$, and let $t \in \{0,1\}$ and $\mathbf{u} \in \{0,1\}^*$ be given such that $\mathbf{w} = \mathbf{u}t\mathbf{v}$ has length n. Then, instead of the chord $S^\alpha_{\mathbf{w}*}$, consider the triangle $\Delta^\alpha_{\mathbf{w}*}$ spanned by $\beta_1 = l^\alpha_{\mathbf{w}}(\ddot{\alpha}), \beta_2 = l^{\alpha,1-t}_{\mathbf{w}}(\dot{\alpha}) = l^\alpha_{\mathbf{u}}(\ddot{\alpha})$ and $\beta_3 = l^{\alpha,t}_{\mathbf{w}}(\dot{\alpha})$.

Obviously, the point β_1 lies in the interior of $\mathbb{T}^\alpha_{\mathbf{w}}$, and one has the equality $\{\overset{\frown}{I^\alpha}(\beta_2), \overset{\frown}{I^\alpha}(\beta_2)\} = \{\overline{\mathbf{u}t\mathbf{v}(1-t)}, \mathbf{u}(1-t)\overline{\mathbf{v}t}\}$. This implies $\beta_2 \in \mathbb{T}^\alpha_{\mathbf{w}(1-t)} \subset \mathbb{T}^\alpha_{\mathbf{w}}$. Moreover, there exist an $\mathbf{r} \in \{0,1\}^*$ and a $k \in \mathbb{N}$ with the property that either $\mathbf{w} = \mathbf{r}(t\mathbf{v})^k$ and $\mathbf{r}$ does not end with $0\mathbf{v}$ or $1\mathbf{v}$, or $\mathbf{w} = \mathbf{r}(1-t)\mathbf{v}(t\mathbf{v})^k$. In the first case it follows $\beta_3 = l^\alpha_{\mathbf{r}}(\dot{\alpha})$, and in the second case $\beta_3 = l^\alpha_{\mathbf{r}}(\ddot{\alpha})$. So $\overset{\frown}{I^\alpha}(\beta_3)$ or $\overset{\frown}{I^\alpha}(\beta_3)$ coincides with $\mathbf{w}\overline{t\mathbf{v}}$ and we obtain $\beta_3 \in \mathbb{T}^\alpha_{\mathbf{w}t} \subset \mathbb{T}^\alpha_{\mathbf{w}}$.

The chord $S^{\alpha,t}_{\mathbf{u}*}$, which connects β_2 and β_3, is a boundary chord of $\mathbb{D}^\alpha_{\mathbf{u}t}$ by the induction hypothesis, and by $\mathbb{D}^\alpha_{\mathbf{u}t} \supset \mathbb{D}^\alpha_{\mathbf{w}}$ it must also bound $\mathbb{D}^\alpha_{\mathbf{w}}$. The other boundary chords $S^{\alpha,0}_{\mathbf{w}*}$ and $S^{\alpha,1}_{\mathbf{w}*}$ of $\Delta^\alpha_{\mathbf{w}*}$ cross the interior of $\mathbb{D}^\alpha_{\mathbf{w}}$. Hence $\Delta^\alpha_{\mathbf{w}*}$ is completely contained in $\mathbb{D}^\alpha_{\mathbf{w}}$. As above, by use of Lemma 2.9 one concludes that $\mathbb{D}^\alpha_{\mathbf{w}} \setminus \Delta^\alpha_{\mathbf{w}*}$ splits into the sets $\mathbb{D}^\alpha_{\mathbf{w}t}$ and $\mathbb{D}^\alpha_{\mathbf{w}(1-t)}$ with the only common point β_1. Clearly, $\mathbb{D}^\alpha_{\mathbf{w}t} \cap \Delta^\alpha_{\mathbf{w}*} = S^{\alpha,t}_{\mathbf{w}*}$ and $\mathbb{D}^\alpha_{\mathbf{w}(1-t)} \cap \Delta^\alpha_{\mathbf{w}*} = S^{\alpha,1-t}_{\mathbf{w}*}$. ■

Immediately from Theorem 2.15 it follows that the chords in $\mathcal{R}^\alpha$ do not cross each other for both $\alpha \in \mathbb{T} \setminus \mathbb{P}_*$ and $\alpha \in \mathbb{P}_*$, and as announced above, this implies $\mathcal{R}^\alpha = \bigcup_{i=0}^\infty \mathcal{R}^\alpha_i$.

In the case $\alpha \in \mathbb{T} \setminus \mathbb{P}_*$ the elements of $\mathcal{R}^\alpha$ are even mutually disjoint. For each endpoint β of a chord in $\mathcal{R}^\alpha$ there exists a $\mathbf{w}_\beta \in \{0,1\}^*$ with $I^\alpha(\beta) = \mathbf{w}_\beta * \hat{\alpha}$. Moreover, for two such endpoints β_1, β_2 it holds $\mathbf{w}_{\beta_1} = \mathbf{w}_{\beta_2}$ iff they belong to the same chord. Hence different chords in $\mathcal{R}^\alpha$ do not touch each other.

We denote the closure of the set $\mathcal{R}^\alpha$ in $Comp(\mathbb{D})$ by $\mathcal{B}^\alpha$. Moreover, for $\alpha \in \mathbb{T}$ and $\mathbf{s} \in \{0,1\}^{\mathbb{N}}$ let $\mathbb{D}^\alpha_{\mathbf{s}} := \mathrm{conv}(\mathbb{T}^\alpha_{\mathbf{s}})$. Then $\mathbb{D}^\alpha_{\mathbf{s}} = \bigcap_{n=1}^\infty \mathbb{D}^\alpha_{\mathbf{s}|n}$, and the statements above provide the following theorem:

Theorem 2.16. *For all $\alpha \in \mathbb{T}$, the set $\mathcal{B}^\alpha$ forms an invariant lamination, and each gap of $\mathcal{B}^\alpha$ is equal to a set $\mathbb{D}^\alpha_{\mathbf{s}}$ for some $\mathbf{s} \in \{0,1\}^{\mathbb{N}}$ or (in case that $\alpha \in \mathbb{P}_*$) to a set $\Delta^\alpha_{\mathbf{u}t\mathbf{v}^\alpha}$ for some $\mathbf{u} \in \{0,1\}^{\mathbb{N}}$ and $t \in \{0,1\}$.* ■

2.2.3 The structure graph

In the following we want to determine the cardinality of the sets $\widetilde{\mathbb{T}^\alpha_{\mathbf{s}}}$ and $\mathbb{T}^\alpha_{\mathbf{s}}$ in dependence on $\alpha \in \mathbb{T}$ and on a given $\mathbf{s} \in \{0,1\}^{\mathbb{N}}$. For this we will develop a technique which will tell us by how many chords a set $\mathbb{D}^\alpha_{\mathbf{s}}$ is bounded.

Let us start with some notations. For $\alpha \in \mathbb{T}$ and $\mathbf{w} \in \{0,1\}^*$ let

$$A^\alpha_{\mathbf{w}*} = \begin{cases} S^\alpha_{\mathbf{w}*} & \text{if } \mathbf{w} \text{ is } \alpha\text{-regular} \\ \Delta^\alpha_{\mathbf{w}*} & \text{else} \end{cases}.$$

Further, we say that $A^\alpha_{\mathbf{w}*}$ bounds a subset B of $\mathbb{D}$ if $A^\alpha_{\mathbf{w}*}$ has a non-degenerate chord in common with B. (This generalizes the concept of a boundary chord of some gap.)

Now let $\mathbf{s} = s_1s_2s_3\ldots \in \{0,1\}^{\mathbb{N}}$ be given, and let $\hat{\alpha} = t_1t_2t_3\ldots$. Recall that the set $\mathbb{D}^\alpha_{\mathbf{s}}$ forms the intersection of the decreasing sequence $\mathbb{D}^\alpha_{s_1}, \mathbb{D}^\alpha_{s_1 s_2}, \ldots, \mathbb{D}^\alpha_{\mathbf{s}|n}, \ldots$. By Theorem 2.15, $A^\alpha_{(\mathbf{s}|n-1)*}$ for a given $n \in \mathbb{N}$ bounds the set $\mathbb{D}^\alpha_{\mathbf{s}|n}$, but none of the sets $\mathbb{D}^\alpha_{\mathbf{s}|k}$ with $k < n$.

If $\mathbf{w} = \mathbf{s}|n-1$ is α-regular, then for all $k \in \mathbb{N}$ the points $l^\alpha_{\mathbf{w}}(\frac{\alpha}{2})$ and $l^\alpha_{\mathbf{w}}(\frac{\alpha+1}{2})$ are contained in $\mathbb{T}^\alpha_{\mathbf{w}s_n t_1 t_2\ldots t_k}$. Therefore, $A^\alpha_{\mathbf{w}*} = S^\alpha_{\mathbf{w}*}$ also bounds the sets $\mathbb{D}^\alpha_{\mathbf{s}|n+1}, \mathbb{D}^\alpha_{\mathbf{s}|n+2}, \ldots \mathbb{D}^\alpha_{\mathbf{s}|n+k}$, until $s_{n+1}s_{n+2}\ldots s_{n+k+1}$ fails to be an initial subword of $\hat{\alpha}$.

Indeed, if $s_{n+k+1} \neq t_{k+1} \in \{0,1\}$, then $l^\alpha_{\mathbf{w}}(\frac{\alpha}{2}), l^\alpha_{\mathbf{w}}(\frac{\alpha+1}{2}) \notin \mathbb{T}^\alpha_{\mathbf{s}|n+k+1}$, hence $A^\alpha_{\mathbf{w}*}$ does not bound $\mathbb{D}^\alpha_{\mathbf{s}|n+k+1}$. On the other hand, if $t_{k+1} = *$, then α is periodic of period $k+1$, and $A^\alpha_{\mathbf{w}*} = S^\alpha_{\mathbf{w}*}$ turns out to be that side of $\Delta^\alpha_{(\mathbf{s}|n+k)*} \subset \mathbb{D}^\alpha_{\mathbf{s}|n+k}$, which bounds neither $\mathbb{D}^\alpha_{(\mathbf{s}|n+k)0}$ nor $\mathbb{D}^\alpha_{(\mathbf{s}|n+k)1}$ (see Theorem 2.15(ii)).

Let us come to the case that $\mathbf{w}$ fails to be α-regular. In other words, let α be periodic, let $\mathbf{v} = \mathbf{v}^\alpha, t = s_n$, and suppose that $\mathbf{w}$ ends with $t\mathbf{v}$ or $(1-t)\mathbf{v}$. By Theorem 2.15(ii), $\Delta^\alpha_{\mathbf{w}*}$ bounds the set $\mathbb{D}^\alpha_{\mathbf{w}t}$, and it holds $\Delta^\alpha_{\mathbf{w}*} \cap \mathbb{D}^\alpha_{\mathbf{w}t} = S^{\alpha,t}_{\mathbf{w}*}$.

If $\mathbf{v}'$ forms an initial subword of $\mathbf{v}$, then $\mathbb{D}^\alpha_{\mathbf{w}t\mathbf{v}'}$ is bounded by $A^\alpha_{\mathbf{w}*} = \Delta^\alpha_{\mathbf{w}*}$. On the other hand, $\mathbb{D}^\alpha_{\mathbf{w}t\mathbf{v}t}$ and $\mathbb{D}^\alpha_{\mathbf{w}t\mathbf{v}(1-t)}$ are not bounded by $A^\alpha_{\mathbf{w}*} = \Delta^\alpha_{\mathbf{w}*}$ but by $\Delta^\alpha_{\mathbf{w}t\mathbf{v}*}$. The latter triangle has the side $S^{\alpha,t}_{\mathbf{w}*}$. Let us summarize:

Lemma 2.17. *For $\alpha \in \mathbb{T}, \mathbf{s} = s_1s_2s_3\ldots \in \{0,1\}^{\mathbb{N}}$ and $n, l \in \mathbb{N}$, the set $\mathbb{D}^\alpha_{\mathbf{s}|l}$ is bounded by $A^\alpha_{(\mathbf{s}|n-1)*}$ if $l \geq n$ and $s_{n+1}s_{n+2}\ldots s_l = \hat{\alpha}|l-n$.* ■

So the question of up to which l a set of the form $\mathbb{D}^\alpha_{\mathbf{s}|l}$ is bounded by a 'cutting set' of the form $A^\alpha_{(\mathbf{s}|n-1)*}$, only depends on the structure of the sequence $\mathbf{s}$. Let us describe the structure of a sequence $\mathbf{s}$ in the following more general context.

Definition 2.18. (Structure graph of a symbol sequence)
Let $\mathbb{S}$ be a set of symbols, and let $\mathbf{s} = s_1s_2s_3\ldots$ and $\mathbf{t}$ be sequences in $\mathbb{S}^{\mathbb{N}}$. Then define a map $\chi^{\mathbf{t}}_{\mathbf{s}} : \mathbb{N} \mapsto \mathbb{N}$ in the following way:

$$\chi^{\mathbf{t}}_{\mathbf{s}}(n) = \begin{cases} n & \text{if } s_{n+1}s_{n+2}\ldots = \mathbf{t} \\ 1 + \max\{l \in \mathbb{N}_0 \mid s_{n+1}s_{n+2}\ldots s_l = \mathbf{t}_{|l-n}\} & \text{else} \end{cases}.$$

For $\alpha \in \mathbb{T}$ and $\mathbf{s} \in \{0,1,*\}^{\mathbb{N}}$, the directed graph $\mathfrak{G}^\alpha(\mathbf{s}) = (\mathbb{N}, E)$ with edge set $E = \{(n, \chi^{\hat{\alpha}}_{\mathbf{s}}(n)) \mid n \in \mathbb{N}\}$ is said to be the α structure graph *of $\mathbf{s}$. Finally, by $CARD^\alpha(\mathbf{s})$ we denote the cardinality of components of $\mathfrak{G}^\alpha(\mathbf{s})$.*

Remark. What we call α structure graph was introduced by C. Penrose [123, 124] and independently by Bandt and the author (see [6]).

There is a slight difference in PENROSE's papers. He includes a point ∞ into the graph and defines $\chi_{\mathbf{s}}^{\mathbf{t}}(n) = \infty$ if $s_{n+1}s_{n+2}\ldots = \mathbf{t}$. PENROSE calls the map corresponding to our $\chi_{\mathbf{s}}^{\mathbf{t}}$ *non-periodicity function* and the members of the sequence $\chi_{\mathbf{t}}^{\mathbf{t}}(1), \chi_{\mathbf{t}}^{\mathbf{t}}(\chi_{\mathbf{t}}^{\mathbf{t}}(1)), \chi_{\mathbf{t}}^{\mathbf{t}}(\chi_{\mathbf{t}}^{\mathbf{t}}(\chi_{\mathbf{t}}^{\mathbf{t}}(1))), \ldots$ the *principal non-periodicities.*

If $\mathbf{t}$ is the kneading sequence of a point in $\mathbb{T}$, it will be of special interest that $\mathbf{t}$ can be got back from the latter sequence (see Chapter 3.1.3).

The contents of the following Theorem 2.19 and of some of its consequences can also be found in PENROSE' papers, but in the (rather different locking) context of his 'glueing space' approach (compare text at the end of Chapter 4.2.1), which is more general than the lamination approach.

To do not lose track of things, for a moment we write $\mathbb{D}_n$ and A_n instead of $\mathbb{D}_{\mathbf{s}|n}^{\alpha}$ and $A_{(\mathbf{s}|n-1)*}^{\alpha}$. Further, let $\chi = \chi_{\mathbf{s}}^{\hat{\alpha}}$. If $n \in \mathbb{N}$ is given, then Lemma 2.17 shows the following: A_n bounds $\mathbb{D}_n, \mathbb{D}_{n+1}, \ldots, \mathbb{D}_{\chi(n)-1}$ but not $\mathbb{D}_{\chi(n)}$, $A_{\chi(n)}$ bounds $\mathbb{D}_{\chi(n)}, \mathbb{D}_{\chi(n)+1}, \ldots, \mathbb{D}_{\chi(\chi(n))-1}$ but not $\mathbb{D}_{\chi(\chi(n))}$, $A_{\chi(\chi(n))}$ bounds $\mathbb{D}_{\chi(\chi(n))}, \mathbb{D}_{\chi(\chi(n))+1}, \ldots, \mathbb{D}_{\chi(\chi(\chi(n)))-1}$ etc.

So, as simple geometric arguments show, the sequence $(A_{\chi^j(n)})_{j\in\mathbb{N}}$ converges to a boundary chord of $\mathbb{D}_{\mathbf{s}}^{\alpha}$. If this chord is degenerate, then $\mathbb{D}_{\mathbf{s}}^{\alpha}$ forms a one-point set. Moreover, if two sequences $(A_{\chi^j(n_1)})_{j\in\mathbb{N}}$ and $(A_{\chi^j(n_2)})_{j\in\mathbb{N}}$ determine the same limit chord, then either the sequences coincide except for finitely many members or they converge to a non-degenerate chord from different sides. In the latter case, $\mathbb{D}_{\mathbf{s}}^{\alpha}$ is equal to the limit chord. So we have the following statement:

Theorem 2.19. (Chords bounding a gap and components of the α structure graph)
Let $\alpha \in \mathbb{T}$ and $\mathbf{s} \in \{0,1\}^{\mathbb{N}}$. Then for each component K of $\mathfrak{G} = \mathfrak{G}^{\alpha}(\mathbf{s})$ a chord $S_{\mathbf{s}}^{\alpha}(K) = \lim_{n\to\infty, n\in K} A_{(\mathbf{s}|n-1)*}^{\alpha}$ is well-defined and bounds $\mathbb{D}_{\mathbf{s}}^{\alpha}$. In particular, $S_{\mathbf{s}}^{\alpha}(K)$ is degenerate if $K = \mathbb{N}$, and for two different components K_1, K_2 of $\mathfrak{G}$, the chords $S_{\mathbf{s}}^{\alpha}(K_1)$ and $S_{\mathbf{s}}^{\alpha}(K_2)$ coincide iff $\mathfrak{G}$ has no further component.

It holds $CARD(\mathbb{T}_{\mathbf{s}}^{\alpha}) = CARD^{\alpha}(\mathbf{s})$. In particular, $\mathbb{D}_{\mathbf{s}}^{\alpha}$ forms a single set if $CARD^{\alpha}(\mathbf{s}) = 1$, a chord if $CARD^{\alpha}(\mathbf{s}) = 2$, and an n-side polygon if $n = CARD^{\alpha}(\mathbf{s}) > 2$. ■

From Theorem 2.19 and the discussion concerning general invariant laminations we draw some conclusions now, which will be very important for the subsequent considerations.

Corollary 2.20. *For $\alpha \in \mathbb{T}$ and $\mathbf{s} \in \{0,1\}^{\mathbb{N}}$ the following statements are valid:*

(i) If $CARD(\mathbb{T}_{\mathbf{s}}^{\alpha}) > 2$, then
  (a) $\mathbf{s}$ is periodic or preperiodic, or
  (b) there exists some $n \in \mathbb{N}_0$ with $\sigma^n(\mathbf{s}) \cong \hat{\alpha}$

(ii) $\mathbb{T}_{\mathbf{s}}^{\alpha}$ is infinite iff $\alpha \in \mathbb{P}_{\infty}$ and there exists some $n \in \mathbb{N}_0$ with $\sigma^n(\mathbf{s}) = \hat{\alpha}$.

Proof: (i): If $CARD(\mathbb{T}^\alpha_{\mathbf{s}}) > 2$, then we apply Theorem 2.13 to the gap $G = \mathbb{D}^\alpha_{\mathbf{s}}$. There exists a $k \in \mathbb{N}_0$, such that $h^n(G)$ is periodic or bounded by $\frac{\alpha}{2}\frac{\alpha+1}{2}$. In the first case $\mathbf{s}$ is periodic oder preperiodic; in the second case, in which $\mathbb{T}^\alpha_{\mathbf{s}}$ must be finite, $\alpha \in h^{k+1}(G) \cap \mathbb{T}$, hence $\sigma^{k+1}(\mathbf{s}) \cong \hat{\alpha}$.

(ii): Clearly, $\mathbb{T}^\alpha_{\mathbf{s}}$ is infinite if $\alpha \in \mathbb{P}_\infty$ and there exists an $n \in \mathbb{N}$ with $\sigma^n(\mathbf{s}) = \hat{\alpha}$. By the considerations above, we only need to show that $CARD^\alpha(\mathbf{s})$ is finite for a periodic $\mathbf{s}$ not ending with $\hat{\alpha}$.

Obviously, for such an $\mathbf{s}$ there exists a $k \in \mathbb{N}$ with $\chi^{\hat{\alpha}}_{\mathbf{s}}(n) - n < k$ for all $n \in \mathbb{N}$. Assuming that $\mathfrak{G}^\alpha(\mathbf{s})$ possesses infinitely many components, then fix k of them and an $n_0 \in \mathbb{N}$ with the property that each of these components contains an element less than n_0. By $\chi^{\hat{\alpha}}_{\mathbf{s}}(n) - n < k$ for all $n \in \mathbb{N}$, in each case we find also an element lying in $\{n_0+1, n_0+2, \ldots, n_0+k-1\}$. Therefore, the components cannot be different, such that it follows $CARD^\alpha(\mathbf{s}) < k$. ■

Example 2.21. The point $\alpha = \frac{9}{56}$ has kneading sequence $001\overline{0}$, and as illustrated in Figure 2.5 the graph $\mathfrak{G}^\alpha_{001\overline{0}}$ possesses exactly three components. In fact, $\mathbb{T}^\alpha_{001\overline{0}} = \{\frac{9}{56}, \frac{11}{56}, \frac{15}{56}\}$, which can be deduced from different statements following subsequently.

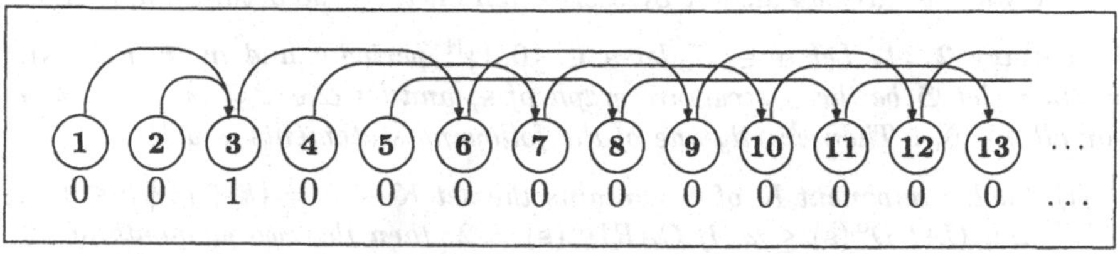

Fig. 2.5. $\mathfrak{G}^\alpha_{001\overline{0}}$ for $\alpha = \frac{9}{56}$

If $\mathbf{s} \in \{0,1\}^\mathbb{N}$ is α-regular for a given $\alpha \in \mathbb{T}$, then $\widetilde{\mathbb{T}^\alpha_{\mathbf{s}}}$ and $\mathbb{T}^\alpha_{\mathbf{s}}$ coincide (see Proposition 2.6). Taking into consideration that $\widetilde{\mathbb{T}^\alpha_{\mathbf{s}}}$ is infinite if $\mathbf{s}$ fails to be α-regular, one obtains the following result immediately from Corollary 2.20(ii).

Corollary 2.22. *For $\alpha \in \mathbb{T}$ and $\mathbf{s} \in \{0,1\}^\mathbb{N}$, the set $\widetilde{\mathbb{T}^\alpha_{\mathbf{s}}}$ is infinite iff $\alpha \in \mathbb{P}$ and there exists some $n \in \mathbb{N}_0$ with $\sigma^n(\mathbf{s}) \cong \hat{\alpha}$.* ■

As announced prior to Proposition 2.3, we want to give a final statement concerning the relation of periodicity and preperiodicity between points and their itineraries now.

Corollary 2.23. *Let $\alpha \in \mathbb{T}$ and $\mathbf{s} \in \{0,1\}^\mathbb{N}$, and let $\alpha \notin \mathbb{P}_\infty$ or $\sigma^n(\mathbf{s}) \neq \hat{\alpha}$ for all $n \in \mathbb{N}_0$. Then the following statements are valid:*

(i) If $\mathbf{s}$ is periodic, then all points in $\mathbb{T}^\alpha_{\mathbf{s}}$ are periodic with the same period. If, moreover, $CARD(\mathbb{T}^\alpha_{\mathbf{s}}) > 2$, then all $\beta \in \mathbb{T}^\alpha_{\mathbf{s}}$ lie on a common orbit.

Finally, if $\mathbb{T}_{\mathbf{s}}^{\alpha}$ *is contained in the orbit of a periodic point* $\beta \in \mathbb{T}$, *then* $PER(\beta) = CARD(\mathbb{T}_{\mathbf{s}}^{\alpha})PER(\mathbf{s})$.

(ii) *If* $\mathbf{s}$ *is preperiodic, then all points in* $\mathbb{T}_{\mathbf{s}}^{\alpha}$ *are preperiodic.*

Proof: Let $\mathbf{w} \in \{0,1\}^*$ with $\mathbf{s} = \overline{\mathbf{w}}$ and $m = PER(\mathbf{s}) = |\mathbf{w}|$ be given, and let $\beta \in \mathbb{T}_{\mathbf{s}}^{\alpha}$. Then $h^{km}(\beta) \in \mathbb{T}_{\mathbf{s}}^{\alpha}$ for all $k \in \mathbb{N}_0$. If $\mathbb{T}_{\mathbf{s}}^{\alpha}$ consists of only one point, then this point must be periodic. Further, if there are exactly two points in $\mathbb{T}_{\mathbf{s}}^{\alpha}$ which lie on different orbits, they also must be periodic. Clearly, then the lengths of the orbits must coincide with $PER(\mathbf{s})$.

If $\mathbb{T}_{\mathbf{s}}^{\alpha}$ is a set of two elements β_1, β_2 on a common orbit, we can assume that $\beta_2 = h^m(\beta_1)$ and that β_2 is periodic of period m. Supposing that the point β_1 is only preperiodic, it is impossible that simultaneously it lies on the orbit of α and α is periodic. Hence $l_{\mathbf{w}}^{\alpha}(\beta_1)$ is defined and a third point of $\mathbb{T}_{\mathbf{s}}^{\alpha}$, in contradiction to our assumption.

Finally, if $CARD(\mathbb{T}_{\mathbf{s}}^{\alpha}) > 2$, then $\mathbb{D}_{\mathbf{s}}^{\alpha}$ forms a polygonal gap by Corollary 2.20(ii), and by Proposition 2.14 all points of $\mathbb{T}_{\mathbf{s}}^{\alpha}$ lie in a common orbit. The last statement of (i) is obvious now, and (ii) follows from (i). ■

Corollary 2.23 leaves one question open: When is a two-element set $\mathbb{T}_{\mathbf{s}}^{\alpha}$ for periodic $\mathbf{s}$ contained in the orbit of a point in $\mathbb{T}$?

We want to give an answer by a criterion on the α structure graph of $\mathbf{s}$.

Corollary 2.24. *Let* $\alpha \in \mathbb{T}$, *let* $\mathbf{s} \in \{0,1\}^{\mathbb{N}}$ *periodic and* $m = PER(\mathbf{s})$. *Further, let* $\mathfrak{G}$ *be the* α *structure graph of* $\mathbf{s}$, *and let* $\alpha \notin \mathbb{P}_\infty$ *or* $\sigma^n(\mathbf{s}) \neq \hat{\alpha}$ *for all* $n \in \mathbb{N}_0$. *Then exactly one of the following statements is valid:*

(i) *Each component* K *of* $\mathfrak{G}$ *contains the set* $K + m = \{k + m \mid k \in K\}$, *and* $CARD^{\alpha}(\mathbf{s}) \leq 2$. *If* $CARD^{\alpha}(\mathbf{s}) = 2$, *then the two elements of* $\mathbb{T}_{\mathbf{s}}^{\alpha}$ *yield different periodic orbits of period* m.

(ii) *Each component* K *of* $\mathfrak{G}$ *is disjoint to* $K + m$, *and all points of* $\mathbb{T}_{\mathbf{s}}^{\alpha}$ *lie on a common periodic orbit. If* K_0 *is a fixed component of* $\mathfrak{G}$, *then for every component* K *of* $\mathfrak{G}$ *there exists a* $j \in N_0$ *such that* $K_0 + jm \subseteq K$.

Proof: The case with the only component $\mathbb{N}$ is obvious. Therefore, we first assume that there are exactly two components K_0 and K_1 of $\mathfrak{G}$ and that α, γ are the points in $\mathbb{T}_{\mathbf{s}}^{\alpha}$. Then $S_{\mathbf{s}}^{\alpha}(K_0) = S_{\mathbf{s}}^{\alpha}(K_1) = \alpha\gamma = h^m(\alpha\gamma)$ according to Theorem 2.19.

Obviously, $h^m(A_{(\mathbf{s}|n+m+1)*}^{\alpha}) = A_{(\mathbf{s}|n+1)*}^{\alpha}$ for all $n \in \mathbb{N}$, and together with the orientation-invariance of h (Lemma 2.9) this implies the following: $h^m(\alpha) = \alpha$ and $h^m(\gamma) = \gamma$ iff the sets $A_{(\mathbf{s}|n+1)*}^{\alpha}$ for $n \in K_0$ lie on the same side of $\alpha\gamma$ as those for $n \in K_0 + m$, and $h^m(\alpha) = \gamma, h^m(\gamma) = \alpha$ iff they lie on different sides. So in the special case of two components we are finished by the discussion preceding Theorem 2.19.

If $l = CARD^{\alpha}(\mathbf{s}) > 2$, then by Corollary 2.23 the set $\mathbb{T}_{\mathbf{s}}^{\alpha}$ is contained in a periodic orbit, hence $\mathbb{T}_{\mathbf{s}}^{\alpha} = \{\beta, h^m(\beta), h^{2m}(\beta), \ldots, h^{(l-1)m}(\beta)\}$ for all $\beta \in \mathbb{T}_{\mathbf{s}}^{\alpha}$.

Let K_0 be a fixed component of $\mathfrak{G}$, and for each $j = 1, 2, \ldots, l-1$ let K_j be the component containing $K_0 + jm$. Then obviously $S^\alpha_{\mathbf{s}}(K_j) = h^{jm}(S^\alpha_{\mathbf{s}}(K_0))$.

By Proposition 2.14, in the case $l > 2$ we can conclude that the boundary chords of $\mathbb{D}^\alpha_{\mathbf{s}}$ are given by $S^\alpha_{\mathbf{s}}(K_j); j = 0, 1, 2, \ldots, l-1$, that the components $K_j; j = 0, 1, 2, \ldots, l-1$ are different and that their union is $\mathbb{N}$. ■

2.2.4 Construction of further invariant laminations

Later we will construct Julia equivalences from a given invariant lamination by identifying points on $\mathbb{T}$ which are connected by a finite sequence of chords in the lamination. Not each invariant lamination is suitable for this construction, as the example $\mathcal{B}^\alpha$ for $\alpha \in \mathbb{P}_*$ shows: here $\dot{\alpha}$ and $\ddot{\alpha}$ would be identified and the obtained equivalence relation would be degenerate.

One procedure for getting a new invariant lamination from a given one $\mathcal{L}$, is to take out the isolated chords. Indeed, one easily sees that the set of all accumulation chords of $\mathcal{L}$ forms an invariant lamination again. In general, we denote the set of accumulation chords of a given set $\mathcal{L}$ of chords by $\mathfrak{d}\mathcal{L}$.

Some invariant laminations constructed as described are shown by Figures 2.6, 2.7, 2.12 and 2.13.

In dependence on $\alpha \in \mathbb{T}$, we want to investigate the structure of the invariant laminations $\mathfrak{d}\mathcal{B}^\alpha$. Let us start with a simple but rather useful statement.

Proposition 2.25. *For $\alpha \in \mathbb{T}$ and $\mathbf{s} \in \{0,1\}^{\mathbb{N}}$ the following statements are valid:*

(i) If $\sigma^n(\mathbf{s}) \neq \hat{\alpha}$ for all $n \in \mathbb{N}$, then $\mathbb{D}^\alpha_{\mathbf{s}}$ is not bounded by a chord in $\mathcal{R}^\alpha$.

(ii) If $\alpha \notin \mathbb{P}$, and $\mathbf{s} = \mathbf{w}t\hat{\alpha}$ for some $\mathbf{w} \in \{0,1\}^*$ and some $t \in \{0,1\}$, then $S^\alpha_{\mathbf{w}*}$ is the only chord in $\mathcal{R}^\alpha$ which bounds $\mathbb{D}^\alpha_{\mathbf{s}}$.

Proof: (i): Let us consider the case that $\alpha \in \mathbb{P}_*$, and set $\mathbf{v} = \mathbf{v}^\alpha$. Then a chord $S^\alpha_{\mathbf{w}*}$ for α-regular $\mathbf{w} \in \{0,1\}^*$ is no more than the intersection of $\Delta^\alpha_{\mathbf{w}0\mathbf{v}*}$ and $\Delta^\alpha_{\mathbf{w}1\mathbf{v}*}$. On the other hand, a chord $S^{\alpha,t}_{\mathbf{u}s\mathbf{v}*}$ with $\mathbf{u} \in \{0,1\}^*$ and $s, t \in \{0,1\}$ bounds both $\Delta^\alpha_{\mathbf{u}s\mathbf{v}*}$ and $\Delta^\alpha_{\mathbf{u}s\mathbf{v}t\mathbf{v}*}$ (see text below Theorem 2.15). By Theorem 2.15, the chords of $\mathcal{R}^\alpha$ are isolated in $\mathcal{B}^\alpha$ and do not bound a set $\mathbb{D}^\alpha_{\mathbf{s}}, \mathbf{s} \in \{0,1\}^{\mathbb{N}}$.

If $\alpha \notin \mathbb{P}_*$ and $\sigma^n(\mathbf{s}) \neq \hat{\alpha}$ for all $n \in \mathbb{N}$, then from the equality $\mathbb{D}^\alpha_{\mathbf{s}} = \operatorname{conv}(\widetilde{\mathbb{T}^\alpha_{\mathbf{s}}})$ it follows that $\mathbb{D}^\alpha_{\mathbf{s}}$ is not bounded by a chord of $\mathcal{R}^\alpha$.

(ii): Let $\alpha \notin \mathbb{P}$ and $\mathbf{s} = \mathbf{w}t\hat{\alpha}$ for some $\mathbf{w} \in \{0,1\}^*$ and some $t \in \{0,1\}$. Since $\mathbb{D}^\alpha_{\mathbf{s}} = \operatorname{conv}(\widetilde{\mathbb{T}^\alpha_{\mathbf{s}}})$, Theorem 2.15(ii) applies that $S^\alpha_{\mathbf{w}*}$ forms a boundary chord of $\mathbb{D}^\alpha_{\mathbf{s}}$.

If a further boundary chord lying in $\mathcal{R}^\alpha$ existed, then $\widetilde{\mathbb{T}^\alpha_{\mathbf{s}}}$ would contain two points each having $*$ in its itinerary, but for different coordinates. Since a symbol $*$ is followed by $\hat{\alpha}$, the sequence $\hat{\alpha}$ must be periodic, which contradicts our assumption. ■

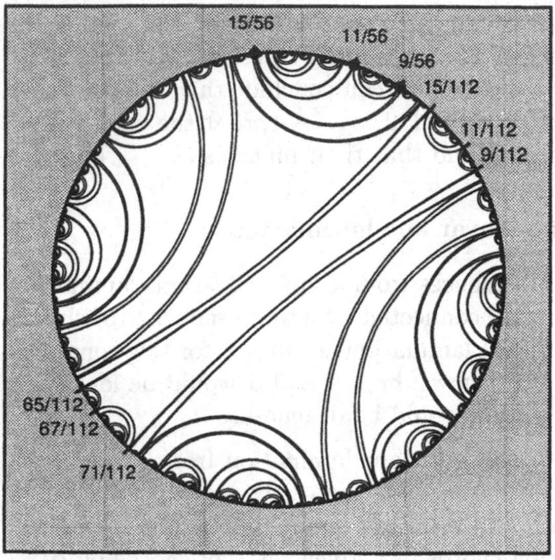

Fig. 2.6. Invariant lamination defined by accumulation: $\eth\mathcal{B}^\alpha$ for $\alpha = \frac{9}{56}, \frac{11}{56}, \frac{15}{56}$

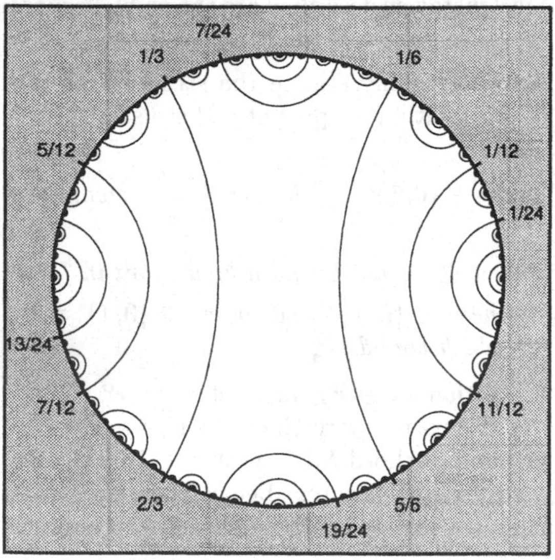

Fig. 2.7. Invariant lamination defined by accumulation: $\eth\mathcal{B}^\alpha$ for $\alpha = \frac{1}{3}, \frac{2}{3}$

We want to clear up now when $\mathcal{B}^\alpha = \mathrm{cl}\,\mathcal{R}^\alpha$ and $\eth\mathcal{B}^\alpha = \eth\mathcal{R}^\alpha$ coincide and when not. For $\alpha \in \mathbb{P}_*$ all chords in $\mathcal{R}^\alpha$ are isolated in $\mathcal{B}^\alpha$, as noted in the proof of Proposition 2.25.

In the case $\alpha \notin \mathbb{P}_*$ one easily sees that all chords of $\mathcal{R}^\alpha$ are isolated if one among them is; in particular, if the two sets $\mathbb{D}^\alpha_{0\hat{\alpha}}$ and $\mathbb{D}^\alpha_{1\hat{\alpha}}$ with intersection S^α_* form gaps. The latter is valid iff $\mathbb{T}^\alpha_{\hat{\alpha}}$ consists of more than one point.

$\mathbb{T}^\alpha_{\hat{\alpha}}$ contains infinitely many elements if $\alpha \in \mathbb{P}_\infty$ (see Corollary 2.20), and from Theorem 2.19 one obtains the following statement:

Proposition 2.26. *$\mathcal{B}^\alpha = \mathfrak{d}\mathcal{B}^\alpha \Longleftrightarrow \alpha \in \mathbb{T} \setminus \mathbb{P}$ and $CARD^\alpha(\hat{\alpha}) = 1$.* ■

We want to give two examples, one for $\mathcal{B}^\alpha = \mathfrak{d}\mathcal{B}^\alpha$ and one for $\mathcal{B}^\alpha \neq \mathfrak{d}\mathcal{B}^\alpha$.

Example 2.27. ($\mathcal{B}^\alpha = \mathfrak{d}\mathcal{B}^\alpha$ for $\alpha = \frac{1}{2}$)
In the case $\alpha = \frac{1}{2}$, the structure of $\mathcal{B}^\alpha$ is very simple: one has $S_*^\alpha = \frac{1}{4}\frac{3}{4}$, and by induction one easily shows that $\mathcal{R}^\alpha = \{\frac{k}{2^n}\frac{2^n-k}{2^n} \mid n = 2,3,4,\ldots; k = 1,2,3,\ldots,2^n-1\}$. Besides the degenerate chords, $\mathcal{B}^\alpha$ contains exactly all chords $\beta(1-\beta)$ for $\beta \in]0,\frac{1}{2}[$. Clearly, no chord in $\mathcal{B}^\alpha$ is isolated, in other words, $\mathcal{B}^\alpha = \mathfrak{d}\mathcal{B}^\alpha$.

Example 2.28. ($\mathcal{B}^\alpha \neq \mathfrak{d}\mathcal{B}^\alpha$ for $\alpha = \frac{9}{56}, \frac{11}{56}, \frac{15}{56}$)
Figure 2.6 shows $\mathfrak{d}\mathcal{B}^\alpha$ for the three preperiodic points $\alpha = \frac{9}{56}, \frac{11}{56}, \frac{15}{56}$. In each case, $\mathbb{T}_{\hat{\alpha}}^\alpha = \mathbb{T}_{001\overline{0}}^\alpha = \{\frac{9}{56}, \frac{11}{56}, \frac{15}{56}\}$ (compare Example 2.21) and by Proposition 2.26 the laminations $\mathcal{B}^\alpha$ and $\mathfrak{d}\mathcal{B}^\alpha$ are different. Namely, $\mathfrak{d}\mathcal{B}^\alpha$ does not contain a 'long' chord.

Already in this context, we want to refer to the fact that the three invariant laminations $\mathcal{B}^{9/56}, \mathcal{B}^{11/56}, \mathcal{B}^{15/56}$ accumulate in the same set of chords (compare the discussion after Theorem 2.47).

Also if $\alpha \in \mathbb{T} \setminus \mathbb{P}$ but $CARD^\alpha(\hat{\alpha})$ is greater than 1, the laminations $\mathcal{B}^\alpha$ and $\mathfrak{d}\mathcal{B}^\alpha$ are near in some sense, as the following proposition shows:

Proposition 2.29. *Let $\alpha \in \mathbb{T} \setminus \mathbb{P}$, and let $S = \gamma\delta$ be a chord in $\mathcal{B}^\alpha$. Then there exist points $\beta_1, \beta_2, \ldots, \beta_n \in \mathbb{T}$ such that the chords $\gamma\beta_1, \beta_1\beta_2, \ldots, \beta_{n-1}\beta_n, \beta_n\delta$ lie in $\mathfrak{d}\mathcal{B}^\alpha$.*

Proof: We can assume that $CARD^\alpha(\hat{\alpha}) > 1$ and that $S = \gamma\delta = S_{\mathbf{w}*}^\alpha$ for some $\mathbf{w} \in \{0,1\}^*$. Then $CARD^\alpha(\mathbf{w}0\hat{\alpha}) = CARD^\alpha(\mathbf{w}1\hat{\alpha}) > 2$ and so $\mathbb{D}_{\mathbf{w}0\hat{\alpha}}^\alpha$ and $\mathbb{D}_{\mathbf{w}1\hat{\alpha}}^\alpha$ are polygonal gaps both bounded by S. According to Proposition 2.25(ii), all boundary chords of $\mathbb{D}_{\mathbf{w}0\hat{\alpha}}^\alpha$ different from S do not lie in $\mathcal{R}^\alpha$. Thus they are contained in $\mathfrak{d}\mathcal{B}^\alpha$, which implies the above statement. ■

In contrast to the case $\alpha \in \mathbb{T} \setminus \mathbb{P}$, if $\alpha \in \mathbb{T}$ has a periodic kneading sequence, then $\mathfrak{d}\mathcal{B}^\alpha = \mathcal{B}^\alpha \setminus \mathcal{R}^\alpha$ and $\mathfrak{d}\mathcal{B}^\alpha$ is far away from $\mathcal{B}^\alpha$. We want to have a closer look at the cases $\alpha \in \mathbb{P}_\infty$ and $\alpha \in \mathbb{P}_*$ now.

The case $\alpha \in \mathbb{P}_\infty$. Let $\alpha \in \mathbb{P}_\infty$ with $\hat{\alpha} = \overline{\mathbf{u}}, \mathbf{u} = u_1u_2\ldots u_m \in \{0,1\}^m$ and $m = PER(\hat{\alpha})$ be given. Further, let $\mathbf{w} = w_1w_2\ldots w_n \in \{0,1\}^n; n \in \mathbb{N}_0$ and $\mathbf{s} = \mathbf{w}\hat{\alpha}$, and consider the α structure graph $\mathfrak{G} = \mathfrak{G}^\alpha(\mathbf{s})$ of $\mathbf{s}$. To each $k \in \mathbb{N}_0$, there corresponds a cycle $n + km \longrightarrow n + km$ in $\mathfrak{G}$, and of course two cycles lie in different components of $\mathfrak{G}$. Besides the components generated by the cycles, there could be at most finitely many further components of $\mathfrak{G}$.

Therefore, by Theorem 2.19 we have infinitely many boundary chords of $\mathbb{D}_{\mathbf{w}\hat{\alpha}}^\alpha$, among them the chords $S_{w_1w_2\ldots w_{n-1}*}^\alpha$ and $S_{\mathbf{w}\mathbf{u}^k u_1u_2\ldots u_{m-1}*}^\alpha$ for $k \in \mathbb{N}_0$. Actually, there are no further boundary chords; in other words, all components of $\mathfrak{G}$ are generated by cycles. In order to show this, consider the set $\mathbb{P}_\infty^\alpha := \mathbb{T}_{\hat{\alpha}}^\alpha$. Moreover, let $\beta_1 = l_{u_1u_2\ldots u_{m-1}}^\alpha(\frac{\alpha}{2})$ and $\beta_2 = l_{u_1u_2\ldots u_{m-1}}^\alpha(\frac{\alpha+1}{2})$.

Since $\mathbb{D}^\alpha_{\hat{\alpha}}$ is bounded by the mutually disjoint chords $S^\alpha_{\mathbf{u}^k u_1 u_2 \ldots u_{m-1} *} = l^\alpha_{\mathbf{u}^k}(\beta_1\beta_2); k \in \mathbb{N}_0$ and possibly by finitely many chords $\gamma_1\delta_1, \gamma_2\delta_2, \ldots, \gamma_l\delta_l$ in $\mathcal{R}^\alpha$, the set $\mathbb{P}^\alpha_\infty$ is obtained from $\mathbb{T}$ by cutting out infinitely many open intervals and it has infinitely many accumulation points. Further, $\alpha \in \mathbb{P}^\alpha_\infty \supseteq h^m(\mathbb{P}^\alpha_\infty)$, and the sets $h^n(\mathbb{P}^\alpha_\infty); n = 0, 1, \ldots, m-1$ are mutually disjoint.

Consider an exceptional chord $\gamma_i\delta_i$. Obviously, $h^m(\gamma_i\delta_i)$ can only be degenerate in the case that $h^{m-1}(\gamma_i\delta_i) = \frac{\alpha}{2}\frac{\alpha+1}{2}$ (and $h^m(\gamma_i\delta_i) = \alpha$). If $h^m(\gamma_i\delta_i)$ coincides with α or $l^\alpha_{\mathbf{u}^k}(\beta_1\beta_2)$ for some $k \in \mathbb{N}_0$, then by the definition of $\mathbb{P}^\alpha_\infty$ one easily sees that $\gamma_i\delta_i = \beta_1\beta_2$ or $\gamma_i\delta_i = l^\alpha_{\mathbf{u}^{k+1}}(\beta_1\beta_2)$. $\mathbb{P}^\alpha_\infty$ forms the union of the sets $A_1 = \mathrm{cl}(\{l^\alpha_{\mathbf{u}^k}(\beta_1) \mid k \in \mathbb{N}_0\} \cup \{l^\alpha_{\mathbf{u}^k}(\beta_2) \mid k \in \mathbb{N}_0\})$ and $A_2 = \{\gamma_1, \delta_1, \gamma_2, \delta_2, \ldots, \gamma_l, \delta_l\}$. As just seen, it holds the inclusion $h^m(\{\gamma_1\delta_1, \gamma_2\delta_2, \ldots, \gamma_l\delta_l\}) \subseteq \{\gamma_1\delta_1, \gamma_2\delta_2, \ldots, \gamma_l\delta_l\}$, and so $h^m(A_2) \subseteq A_2$.

If the set A_2 were not empty, then not all points of A_2 could be isolated in $\mathbb{P}^\alpha_\infty$ since $\mathbb{D}^\alpha_{\hat{\alpha}}$ fails to be a polygonal gap. So there would be a periodic point in $A_1 \cap A_2$. This is impossible.

Indeed, if $\beta = h^{qm}(\beta)$ for some $q \in \mathbb{N}$, then α would not lie on the orbit of β and each sufficiently small neighborhood of β would contain its image with respect to $l^\alpha_{\mathbf{w}^q}$. Every such neighborhood contains the points $l^\alpha_{\mathbf{u}^r}(\beta_1)$ and $l^\alpha_{\mathbf{u}^r}(\beta_2)$ for some at $r \in \mathbb{N}$.

Therefore, also the points $l^\alpha_{\mathbf{u}^r+km}(\beta_1), l^\alpha_{\mathbf{u}^r+km}(\beta_2); k \in \mathbb{N}$ would be in the neighborhood regarded. Hence all accumulation points of A_1 would fall into the orbit of β, but $\mathbb{P}^\alpha_\infty$ possesses infinitely many accumulation points.

By Corollary 2.20(ii), the following statement is obvious now:

Proposition 2.30. (The infinite gaps of $\mathcal{B}^\alpha$ for $\alpha \in \mathbb{P}_\infty$)
Let $\alpha \in \mathbb{P}_\infty$ and $\mathbf{u} = u_1 u_2 \ldots u_m \in \{0,1\}^m$ with $m = PER(\hat{\alpha})$ and $\hat{\alpha} = \overline{\mathbf{u}}$. Then for each 0-1-word $\mathbf{w} = w_1 w_2 \ldots w_n; n \in \mathbb{N}$, the set $\mathbb{D}^\alpha_{\mathbf{w}\hat{\alpha}}$ forms an infinite gap of $\mathcal{B}^\alpha$. This gap is bounded exactly by the chords $S^\alpha_{w_1 w_2 \ldots w_{n-1} *}$ and $S^\alpha_{\mathbf{w}\mathbf{u}^k u_1 u_2 \ldots u_{n-1} *}; k \in \mathbb{N}_0$. Each infinite gap in $\mathcal{B}^\alpha$ is equal to $\mathbb{D}^\alpha_{\mathbf{w}\hat{\alpha}}$ for some 0-1-word $\mathbf{w}$. ■

In the following, let $\equiv$ be the equivalence relation on $\mathbb{P}^\alpha_\infty$ which identifies the endpoints of the open intervals taken out from $\mathbb{T}$ pairwisely. In each case, these are the endpoints $l^\alpha_{\mathbf{u}^k}(\beta_1)$ and $l^\alpha_{\mathbf{u}^k}(\beta_2)$ of a boundary chord of $\mathbb{D}^\alpha_{\hat{\alpha}}$. Factorizing $\mathbb{P}^\alpha_\infty$ by $\equiv$, the restriction of h^m to $\mathbb{P}^\alpha_\infty$ is transferred to a map H from the 'circle' $\mathbb{P}^\alpha_\infty/\equiv$ to itself, which by the orientation-invariance of h (Lemma 2.9) forms an orientation-preserving homeomorphism.

It is easy to see that the backward orbit of $[\alpha]_\equiv$ with respect to H is dense in $\mathbb{P}^\alpha_\infty/\equiv$. Since further $\mathbb{P}^\alpha_\infty/\equiv$ does not possess elements periodic with respect to H^{-1}, by DENJOY's theorem the map H^{-1} and so also H can be regarded as an irrational rotation. (e.g., see [168], IV,(6.6)).

We know that the set $\mathbb{P}^\alpha_\infty$ is compact and nowhere dense, and since it arises from $\mathbb{T}$ by cutting out open intervals with mutually disjoint endpoints, it is a perfect subset of the interval $[0, 1[(= \mathbb{T})$. (A closed subset of a topological space is said to be perfect if it does not contain isolated points.) In the present work, such a subset of an interval is called a *Cantor set.*

Moreover, we want to use the following two concepts from topological dynamics: Let f be a map on a set X. Then a subset A of X is called f-invariant if $f(A) \subseteq A$. If f is continuous on a topological space X, then a closed f-invariant A is said to be *minimal* if every proper closed invariant subset of A is empty. Let us summarize:

Proposition 2.31. (Properties of $\mathbb{P}^{\alpha}_{\infty}$, part I)
Let $\alpha \in \mathbb{P}_{\infty}$ and let $m = PER(\hat{\alpha})$. Then the following statements are valid:

(i) $\mathbb{P}^{\alpha}_{\infty}(= \mathbb{T}^{\alpha}_{\hat{\alpha}})$ *is a Cantor set.*
(ii) $\mathbb{P}^{\alpha}_{\infty}$ *is h^m-invariant and minimal with respect to h^m. The sets $h^n(\mathbb{P}^{\alpha}_{\infty})$; $n = 0, 1, 2, \ldots, m-1$ are mutually disjoint.*
(iii) The equivalence relation $\equiv$ on $\mathbb{P}^{\alpha}_{\infty}$ (defined to identify two points if they are connected by a boundary chord of $\mathbb{D}^{\alpha}_{\hat{\alpha}}$) is completely invariant with respect to h^m, and $(\mathbb{P}^{\alpha}_{\infty}/\equiv, h^m/\equiv)$ is conjugate to an irrational rotation on the unit circle. ■

We want to refer to Theorem 3.36 in Section 3.2, in which we will describe how the rotation number of the irrational rotation in (iii) can be determined. The statement following now will play an important role for the classification of Julia equivalences in Section 2.3.

Corollary 2.32. *Let $\approx$ be an h-invariant planar equivalence relation on $\mathbb{T}$, which is non-degenerate or has only finite equivalence classes. Further, let $\alpha \in \mathbb{P}_{\infty}$, and let $\beta_1, \beta_2 \in \mathbb{T}$ be different points with $\beta_1 \approx \beta_2$. If $I^{\alpha}(\beta_1), I^{\alpha}(\beta_2) \cong \hat{\alpha}$; then the chord $\beta_1\beta_2$ lies in $\mathcal{R}^{\alpha}$.*

Proof: If $\beta_1, \beta_2 \in \mathbb{T}$ are given as above, then $\beta_1, \beta_2 \in \mathbb{P}^{\alpha}_{\infty}$. A non-trivial planar equivalence relation on the circle which is invariant with respect to an irrational rotation identifies all points of the circle.

So one easily sees the following: The identification of two different points in $\mathbb{P}^{\alpha}_{\infty}$ not connected by a chord in $\mathcal{R}^{\alpha}$ - only such a chord can be a boundary chord of $\mathbb{D}^{\alpha}_{\hat{\alpha}}$ - leads to the identification of all points of $\mathbb{P}^{\alpha}_{\infty}$. Thus $\approx$ would possess an infinite $\approx$-equivalence class whose $m-1$-th iterate contains $\frac{\alpha}{2}, \frac{\alpha+1}{2}$ and $h^{m-1}(\alpha)$, where $m = PER(\alpha)$. Since this would contradict the assumption that $\approx$ is non-degenerate, it holds $\beta_1\beta_2 \in \mathcal{R}^{\alpha}$. ■

Example 2.33. ($\eth\mathcal{B}^{\alpha}$ for $\alpha \in \mathbb{P}_{\infty}$ with $\hat{\alpha} = \overline{0}$)
For $\alpha \in \mathbb{P}_{\infty}$ with $\hat{\alpha} = \overline{0}$, the set $\mathbb{D}^{\alpha}_{\overline{0}}$ is bounded exactly by the chords $S^{\alpha}_{0^n*}, n \in \mathbb{N}_0$. It holds $h(S^{\alpha}_{0^n*}) = S^{\alpha}_{0^{n-1}*}$ for all $n \in \mathbb{N}$. Consequently, by the disjointness of the chords above and by formula (2.4), $d(S^{\alpha}_{0^n*}) = \frac{1}{2^{n+1}}$ for all $n \in \mathbb{N}_0$.

Since none of the chords $\mathcal{R}^{\alpha}$ cuts the interior of the sets $\mathbb{D}^{\alpha}_{\overline{0}}$ and $\mathbb{D}^{\alpha}_{\overline{10}}$ bounded by S^{α}_{*}, besides S^{α}_{*} there is no chord in $\mathcal{R}^{\alpha}$ longer than $\frac{1}{4}$. Moreover,

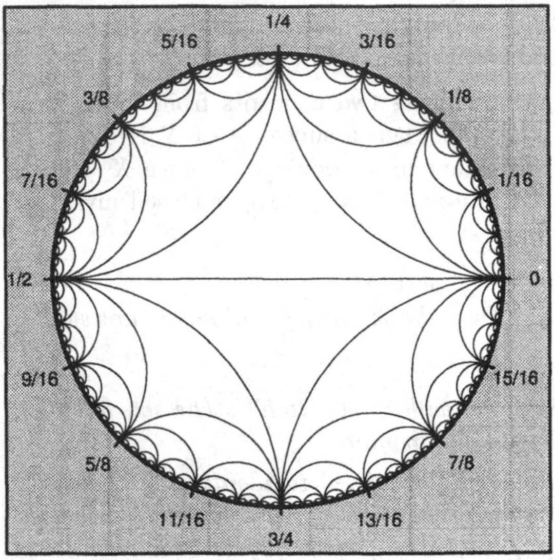

Fig. 2.8. Trivial accumulation set: $\mathcal{B}^\alpha$ for $\alpha = 0$

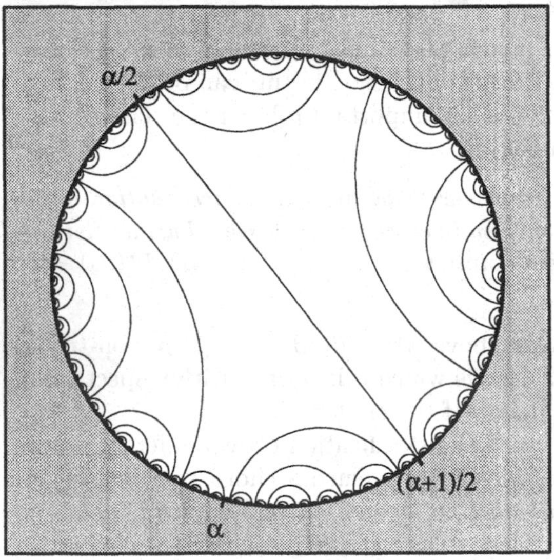

Fig. 2.9. Trivial accumulation set: $\mathcal{B}^\alpha$ for $\alpha = 0,709803\ldots$

by Lemma 2.7 there are only finitely many chords in $\mathcal{B}^\alpha$ of length greater than ϵ for each given $\epsilon > 0$. Therefore, $\eth\mathcal{B}^\alpha$ does not contain a non-degenerate chord, in other words, $\eth\mathcal{B}^\alpha$ is no more than the trivial invariant lamination $\{\{\beta\} \mid \beta \in \mathbb{T}\}$.

Figure 2.9 shows $\mathcal{B}^\alpha$ for α with binary expansion $\mathbf{a}(\overline{1}) = 0100101001001\ldots$ (see Section 2.1.2), where $\hat{\alpha} = \overline{0}$.

The case $\alpha \in \mathbb{P}_*$. This case is a little more complicated from the technical point of view, but in the following chapter it will play a key role.

For $\alpha \in \mathbb{P}_*, \mathbf{v} = \mathbf{v}^\alpha$ and a given symbol $t \in \{0,1\}$ it holds $\mathcal{R}^{\alpha,t} \setminus \mathcal{R}^{\alpha,1-t} = \{S^{\alpha,t}_{\mathbf{wv}} \mid \mathbf{w} \in \{0,1\}^* \setminus \{\lambda\}\}$. We want to verify now that the latter set for one of the symbols $t = 0$ or $t = 1$ does not make a substantial contribution to forming accumulation chords of $\mathcal{R}^\alpha$.

Look at the chords $S^{\alpha,0}_{0\mathbf{v}*}$ and $S^{\alpha,1}_{0\mathbf{v}*}$, which together with S^α_* bound the triangle $\Delta^\alpha_{0\mathbf{v}*}$. If $\alpha \neq 0$, then $S^\alpha_{\mathbf{v}*}$ is different from S^α_*, such that one of the chords $S^{\alpha,0}_{0\mathbf{v}*}, S^{\alpha,1}_{0\mathbf{v}*}$ is longer and one shorter than $\frac{1}{4}$ (see Figure 2.10). This justifies the following definition:

Definition 2.34. (The characteristic symbol)
Let $\alpha \in \mathbb{P}_* \setminus \{0\}$ and let $\mathbf{v} = \mathbf{v}^\alpha$. Then the symbol t with $d(S^{\alpha,t}_{0\mathbf{v}*}) < \frac{1}{4}$ is said to be the characteristic symbol *of α and denoted by e^α. For $\alpha = 0$ we define the characteristic symbol $e^\alpha = 0$.*

Let us discuss consequences of the property $d(S^{\alpha,e}_{0\mathbf{v}*}) < \frac{1}{4}$ with $e = e^\alpha$, and let us give a simple way to determine the characteristic symbol. Later we will provide further ways for determining the characteristic symbol.

Lemma 2.35. *Let $\alpha \in \mathbb{P}_* \setminus \{0\}$ and let $\mathbf{v} = \mathbf{v}^\alpha$. Then*

$$e = e^\alpha = \begin{cases} 0 \text{ if } l^\alpha_{\mathbf{v}*}(\ddot{\alpha}) \text{ is between } \alpha \text{ and } \ddot{\alpha} \\ 1 \text{ if } l^\alpha_{\mathbf{v}*}(\ddot{\alpha}) \text{ is between } \alpha \text{ and } \dot{\alpha} \end{cases} \tag{2.6}$$

and $d(S^{\alpha,e}_{\mathbf{uv}*}) = 2^{-|\mathbf{u}|} d(S^\alpha_{\mathbf{v}*}) \leq \frac{2^{-|\mathbf{u}|}}{3}$ for all $\mathbf{u} \in \{0,1\}^*$. Moreover, in the case $\alpha = 0$ it holds $d(S^{\alpha,t}_{\mathbf{w}*}) = 2^{-(|\mathbf{w}|+1)}$ for all $\mathbf{w} \in \{0,1\}^*$ and $t \in \{0,1\}$.

Proof: $\mathbf{v}$ begins with the symbol 0, and $S^\alpha_{\mathbf{v}*}$ does not cross $S^{\alpha,e}_{0\mathbf{v}*}$. Thus by formula (2.4) we have $\frac{1}{2} - d(S^{\alpha,e}_{0\mathbf{v}*}) \geq d(S^\alpha_{\mathbf{v}*}) = 2d(S^{\alpha,e}_{0\mathbf{v}*})$. This implies $d(S^{\alpha,e}_{0\mathbf{v}*}) \leq \frac{1}{6}$ and so $d(S^{\alpha,e}_{1\mathbf{v}*}) \leq \frac{1}{6}$, hence $d(S^\alpha_{\mathbf{v}*}) \leq \frac{1}{3}$ and $d(S^{\alpha,1-e}_{0\mathbf{v}*}), d(S^{\alpha,1-e}_{e\mathbf{v}*}) \geq \frac{1}{3}$.

Since $S^{\alpha,e}_{\mathbf{uv}*}$ for $\mathbf{u} \in \{0,1\}^* \setminus \{\lambda\}$ does not cross one of the chords $S^{\alpha,e}_{0\mathbf{v}*}, S^{\alpha,e}_{1\mathbf{v}*}, S^{\alpha,1-e}_{0\mathbf{v}*}$ or $S^{\alpha,1-e}_{1\mathbf{v}*}$, the successive application of Lemma 2.8 to the $S = S^{\alpha,e}_{\mathbf{uv}*}$ with $a = d(S^{\alpha,1-e}_{0\mathbf{v}*})$ leads to the first statement after the formula.

$S^\alpha_{\mathbf{v}*}$ must lie behind $S^{\alpha,1-e}_{0\mathbf{v}*}$, and since the whole interval behind $S^{\alpha,e}_{0\mathbf{v}*}$ is mapped onto the whole interval behind $S^\alpha_{\mathbf{v}*}$ in a homeomorphic and orientation-preserving way, it follows $\alpha \curvearrowright l^\alpha_{\mathbf{v}}(\ddot{\alpha}) \curvearrowright l^{\alpha,1-e}_0(\alpha)$ or $\alpha \curvearrowleft l^\alpha_{\mathbf{v}}(\ddot{\alpha}) \curvearrowleft l^{\alpha,1-e}_0(\alpha)$, which implies formula (2.6).

The last of the statements can be shown as above, by application of the following fact: One easily sees that the lengths of $S^{0,0}_{0*}, S^{0,1}_{0*}, S^{0,0}_{1*}$ and $S^{0,1}_{1*}$ are equal to $\frac{1}{4}$. If one chord does not cross one of the four chords above, it cannot be longer than $\frac{1}{4}$. ■

The following statement is an immediate consequence of Lemma 2.35:

Corollary 2.36. *For $\alpha \in \mathbb{P}_* \setminus \{0\}$ the invariant lamination $\eth\mathcal{B}^\alpha$ consists of all accumulation chords of the set $\mathcal{R}^{\alpha,1-e^\alpha}$. Moreover, $\eth\mathcal{B}^0$ is equal to the set of all degenerate chords $\{\beta\}; \beta \in \mathbb{T}$.* ■

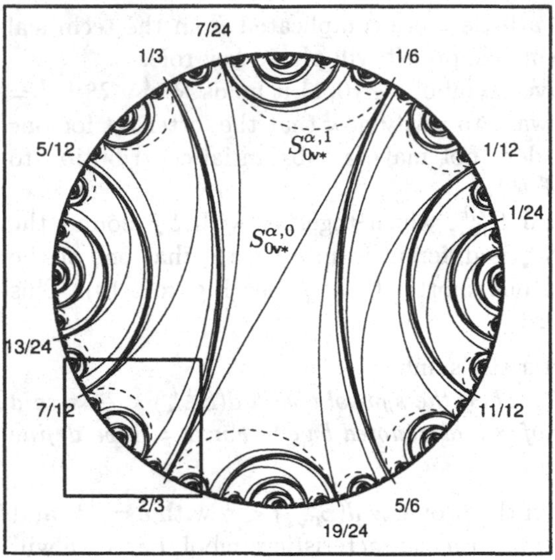

Fig. 2.10. Definition of the point associated with a periodic one: $\mathcal{R}^\alpha$ for $\alpha = \frac{1}{3}$

For $\alpha = \frac{1}{3}$, the set $\mathcal{R}^\alpha$ is shown by Figure 2.10 and $\eth\mathcal{B}^\alpha = \mathcal{R}^\alpha$ by Figure 2.7. Figure 2.11 illustrates that the chords in $\mathcal{R}^{\alpha,1} \setminus \mathcal{R}^{\alpha,0}$, which are represented by broken lines, do not accumulate in non-degenerate chords. (In the case $\alpha = \frac{1}{3}$ one has $e^\alpha = 1$.) For $\alpha = 0$ compare the corresponding statement in Corollary 2.36 with Figure 2.8.

Looking back to the discussion before Theorem 2.15, we get the following: If $\mathbf{u}$ is a α-regular 0-1-word, then $S^\alpha_{\mathbf{u}*}$ bounds both the triangle $\Delta^\alpha_{\mathbf{u}0\mathbf{v}*}$ and the triangle $\Delta^\alpha_{\mathbf{u}1\mathbf{v}*}$. Moreover, for a 0-1-word $\mathbf{r}$ which fails to be an α-regular 0-1-word, the triangle $\Delta^\alpha_{\mathbf{r}*}$ bounds $\Delta^\alpha_{\mathbf{r}0\mathbf{v}*}$ and $\Delta^\alpha_{\mathbf{r}1\mathbf{v}*}$.

By this fact, to each α-regular 0-1-word there corresponds a binary tree of triangles: each 0-1-sequence $s_1 s_2 s_3 \ldots$ is related to a path $\Delta^\alpha_{\mathbf{u}s_1\mathbf{v}*}, \Delta^\alpha_{\mathbf{u}s_1\mathbf{v}s_2\mathbf{v}*}, \Delta^\alpha_{\mathbf{u}s_1\mathbf{v}s_2\mathbf{v}s_3\mathbf{v}*}, \ldots$, where $\Delta^\alpha_{\mathbf{u}s_1\mathbf{v}s_2\ldots\mathbf{v}s_n\mathbf{v}*}, \Delta^\alpha_{\mathbf{u}s_1\mathbf{v}s_2\ldots\mathbf{v}s_{n+1}\mathbf{v}*}$ for all $n \in \mathbb{N}$ intersect in the chord $S^{\alpha,s_{n+1}}_{\mathbf{u}s_1\mathbf{v}s_2\ldots\mathbf{v}s_n\mathbf{v}*}$ (compare also Figures 2.10 and 2.11).

Both the sequence of triangles contained in the path and the sequence $S^\alpha_{\mathbf{u}*}, S^{\alpha,s_2}_{\mathbf{u}s_1\mathbf{v}*}, S^{\alpha,s_3}_{\mathbf{u}s_1\mathbf{v}s_2\mathbf{v}*}, S^{\alpha,s_4}_{\mathbf{u}s_1\mathbf{v}s_2\mathbf{v}s_3\mathbf{v}*}, \ldots$ converge to a chord, which by Lemma 2.35 is degenerate if infinitely many of the symbols $s_1, s_2, s_3, \ldots$ are equal to e. Clearly, the limit chord obtained coincides with $\mathbb{D}^\alpha_{s_1 s_2 s_3\ldots}$ and $\mathbb{T}^\alpha_{s_1 s_2 s_3\ldots}$ in the degenerate case.

Let us determine the endpoints of the limit chord if almost all symbols $s_1, s_2, s_3, \ldots$ are equal to $1-e$. We want to begin with the rather simple case that $\mathbf{u} = \mathbf{v}$ and $s_1 = s_2 = s_3 = \ldots = 1-e$, and then we want to give the general statement in Proposition 2.38.

The paths $\Delta^\alpha_{\mathbf{v}e\mathbf{v}*}, \Delta^\alpha_{\mathbf{v}e\mathbf{v}e\mathbf{v}*}, \Delta^\alpha_{\mathbf{v}e\mathbf{v}e\mathbf{v}e\mathbf{v}*}, \ldots$ and $\Delta^\alpha_{\mathbf{v}(1-e)\mathbf{v}*}, \Delta^\alpha_{\mathbf{v}(1-e)\mathbf{v}(1-e)\mathbf{v}*}, \Delta^\alpha_{\mathbf{v}(1-e)\mathbf{v}(1-e)\mathbf{v}(1-e)\mathbf{v}*}, \ldots$ 'start' at the chord $S^\alpha_{\mathbf{v}*}$. All triangles lying in these paths have α as a common angle. According to the considerations above, the

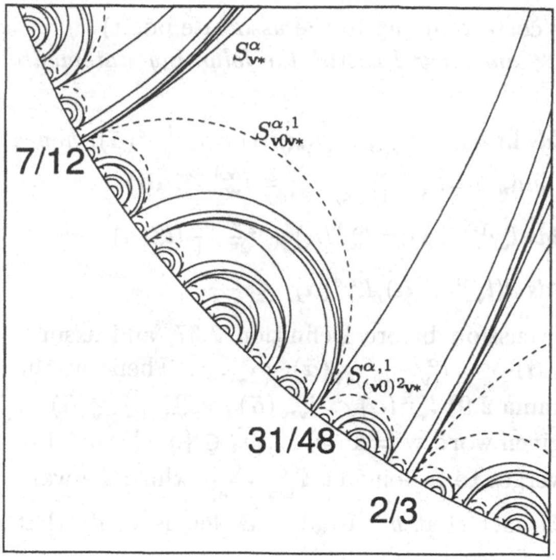

Fig. 2.11. Definition of the point associated with a periodic one: α and $\overline{\alpha}$ for $\alpha = \frac{1}{3}$

sequence $(S^{\alpha,e}_{(\mathbf{v}e)^n\mathbf{v}*})_{n\in\mathbb{N}_0}$ converges to the one-point set $\mathbb{D}^{\alpha}_{\overline{\mathbf{v}e}} = \mathbb{T}^{\alpha}_{\overline{\mathbf{v}e}} = \{\alpha\}$. Hence, in particular it holds $PER(\overline{\mathbf{v}e}) = PER(\alpha)$.

On the other hand, the length of the chords $S^{\alpha,1-e}_{(\mathbf{v}(1-e))^n\mathbf{v}*} = \alpha l^{\alpha}_{(\mathbf{v}(1-e))^n\mathbf{v}}(\ddot{\alpha})$; $n \in \mathbb{N}$, which in each case bound two triangles of the second path and do not cross $\dot{\alpha}\ddot{\alpha}$, must be monotonically increasing. Therefore, the sequence $(S^{\alpha,1-e}_{(\mathbf{v}(1-e))^n\mathbf{v}*})_{n\in\mathbb{N}}$ converges to a non-degenerate chord with endpoints α and $\lim_{n\to\infty} l^{\alpha}_{(\mathbf{v}(1-e))^n\mathbf{v}}(\ddot{\alpha})$.

It is not hard to see that both endpoints lie in $\mathbb{T}^{\alpha}_{\overline{\mathbf{v}(1-e)}}$, and by Corollary 2.23(i) the second endpoint is periodic of the same period as α.

In particular, one has $\alpha \curvearrowright l^{\alpha}_{\mathbf{v}}(\ddot{\alpha}) \curvearrowright l^{\alpha}_{\mathbf{v}(1-e)\mathbf{v}}(\ddot{\alpha}) \curvearrowright l^{\alpha}_{(\mathbf{v}(1-e))^2\mathbf{v}}(\ddot{\alpha}) \curvearrowright \ldots$ or $\alpha \curvearrowleft l^{\alpha}_{\mathbf{v}}(\ddot{\alpha}) \curvearrowleft l^{\alpha}_{\mathbf{v}(1-e)\mathbf{v}}(\ddot{\alpha}) \curvearrowleft l^{\alpha}_{(\mathbf{v}(1-e))^2\mathbf{v}}(\ddot{\alpha}) \curvearrowleft \ldots$.

Definition 2.37. (Associated periodic point)
For $\alpha \in \mathbb{P}_* \setminus \{0\}, \mathbf{v} = \mathbf{v}^{\alpha}, e = e^{\alpha}$, the point $\overline{\alpha} := \lim_{n\to\infty} l^{\alpha}_{(\mathbf{v}(1-e))^n\mathbf{v}}(\ddot{\alpha}) = \lim_{n\to\infty} l^{\alpha,e}_{(\mathbf{v}(1-e))^n}(\alpha)$ is said to be associated *with α. (After Lemma 3.3 we will define the point associated with 0.)*

Let us illustrate the construction of the associated point by the example $\alpha = \frac{1}{3}$ (see Figures 2.10 and 2.11). We have $e = e^{\alpha} = 1$ and $\mathbf{v} = \mathbf{v}^{\alpha} = 0$, and further $l^{\alpha}_{\mathbf{v}}(\ddot{\alpha}) = l^{\alpha}_{0}(\frac{1}{6}) = \frac{7}{12}, l^{\alpha}_{\mathbf{v}(1-e)\mathbf{v}}(\ddot{\alpha}) = l^{\alpha}_{000}(\frac{1}{6}) = \frac{31}{48}$. Moreover, $d(S^{\alpha}_{\mathbf{v}*}) = d(S^{\alpha,1}_{0*}) = \frac{1}{4}, d(S^{\alpha,e}_{\mathbf{v}(1-e)\mathbf{v}*}) = d(S^{\alpha,1}_{000*}) = \frac{1}{16}, \ldots, d(S^{\alpha,e}_{(\mathbf{v}(1-e))^n\mathbf{v}*}) = d(S^{\alpha,1}_{0^{2n+1}*}) = \frac{1}{2^{n+1}}$. So it follows $\overline{\frac{1}{3}} = \frac{1}{3} + \sum_{n=1}^{\infty} \frac{1}{2^{2n}} = \frac{2}{3}$.

We are able now to give formulae determining the endpoints of the limit chord $\lim_{n\to\infty} \Delta^{\alpha}_{\mathbf{u}\mathbf{v}s_1\mathbf{v}s_2\ldots\mathbf{v}s_n\mathbf{v}*}$ in the non-degenerate case and, in particular, to compute the point associated with a periodic one.

Proposition 2.38. (Formulae corresponding to the associate point)
For $\alpha \in \mathbb{P}_* \setminus \{0\}, \mathbf{v} = \mathbf{v}^\alpha, e = e^\alpha$ and $m = PER(\alpha)$ the following statements are valid:

(i) For all $\mathbf{w} \in \{0,1\}^*$ it holds $\lim_{n\to\infty} l^\alpha_{\mathbf{w}(\mathbf{v}(1-e))^n\mathbf{v}}(\ddot{\alpha}) = l^{\alpha,1-e}_{\mathbf{w}}(\overline{\alpha})$, hence $\lim_{n\to\infty} \Delta^\alpha_{\mathbf{w}(\mathbf{v}(1-e))^n\mathbf{v}*} = \lim_{n\to\infty} S^{\alpha,1-e}_{\mathbf{w}(\mathbf{v}(1-e))^n\mathbf{v}*} = l^{\alpha,1-e}_{\mathbf{w}}(\alpha\overline{\alpha})$.

(ii) For all $\mathbf{w} \in \{0,1\}^*$ it holds $l^{\alpha,1-e}_{\mathbf{w}}(\overline{\alpha}) = l^{\alpha,e}_{\mathbf{w}}(\alpha) + \frac{2^{m-|\mathbf{w}|}}{2^m-1}(l^\alpha_{\mathbf{v}}(\ddot{\alpha}) - \alpha)$.

(iii) For all $\mathbf{w} \in \{0,1\}^*$ it holds $d(l^{\alpha,1-e}_{\mathbf{w}}(\overline{\alpha}), l^{\alpha,e}_{\mathbf{w}}(\alpha)) \le \frac{2^{-|\mathbf{w}|}}{3}$.

Proof: Let us return to the discussion before Definition 2.37, and assume that $\alpha \curvearrowright l^\alpha_{\mathbf{v}}(\ddot{\alpha}) \curvearrowright l^\alpha_{\mathbf{v}(1-e)\mathbf{v}}(\ddot{\alpha}) \curvearrowright l^\alpha_{(\mathbf{v}(1-e))^2\mathbf{v}}(\ddot{\alpha}) \curvearrowright \ldots$. Then by the orientation-invariance of h (Lemma 2.9) $l^{\alpha,e}_{\mathbf{w}}(\alpha) \curvearrowright l^\alpha_{\mathbf{wv}}(\ddot{\alpha}) \curvearrowright l^\alpha_{\mathbf{wv}(1-e)\mathbf{v}}(\ddot{\alpha}) \curvearrowright l^\alpha_{\mathbf{w}(\mathbf{v}(1-e))^2\mathbf{v}}(\ddot{\alpha}) \curvearrowright \ldots$ for each given word $\mathbf{w} = w_1w_2\ldots w_n \in \{0,1\}^n; n \in \mathbb{N}_0$, and $(l^\alpha_{\mathbf{w}(\mathbf{v}(1-e))^n\mathbf{v}}(\ddot{\alpha}))_{n\in\mathbb{N}_0}$ converges to a point in $\mathbb{T}^\alpha_{\overline{\mathbf{wv}(1-e)}}$, which we want to denote by $\alpha_{\mathbf{w}}$. By use of the fact that $h^{|\mathbf{w}|}(\alpha_{\mathbf{w}}) = \overline{\alpha}$, let us verify that $\alpha_{\mathbf{w}} = l^{\alpha,1-e}_{\mathbf{w}}(\overline{\alpha})$.

If $l^{\alpha,1-e}_{w_{i+1}w_{i+2}\ldots w_n}(\overline{\alpha}) \neq \alpha$ for all $i = 1,2,\ldots,n-1$, then $l^{\alpha,1-e}_{\mathbf{w}}$ is continuous in a neighborhood of $\overline{\alpha}$, which follows immediately from $\alpha_{\mathbf{w}} = l^{\alpha,1-e}_{\mathbf{w}}(\overline{\alpha})$. Otherwise, fix a minimal $i \in \{1,2,\ldots,n-1\}$ with $l^{\alpha,1-e}_{w_{i+1}w_{i+2}\ldots w_n}(\overline{\alpha}) = \alpha$. Then one gets $l^{\alpha,1-e}_{\mathbf{w}}(\overline{\alpha}) = l^\alpha_{w_1w_2\ldots w_{i-1}}(\dot{\alpha})$ if $w_i = 1-e$, and $l^{\alpha,1-e}_{\mathbf{w}}(\overline{\alpha}) = l^\alpha_{w_1w_2\ldots w_{i-1}}(\ddot{\alpha})$ if $w_i = e$. In both cases, by the considerations before Proposition 2.6, $l^{\alpha,1-e}_{\mathbf{w}}(\overline{\alpha})$ is the only point in $\mathbb{T}^\alpha_{\overline{\mathbf{wv}(1-e)}}$, whose $|\mathbf{w}|$-th iterate coincides with $\overline{\alpha}$. Hence $\alpha_{\mathbf{w}} = l^{\alpha,1-e}_{\mathbf{w}}(\overline{\alpha})$, and (i) is shown.

Since $l^{\alpha,e}_{\mathbf{w}}(\alpha), l^\alpha_{\mathbf{wv}}(\ddot{\alpha})$ are the endpoints of $S^{\alpha,e}_{\mathbf{wv}*}$ and, for all $n \in \mathbb{N}_0$, the points $l^\alpha_{\mathbf{w}(\mathbf{v}(1-e))^n\mathbf{v}}(\ddot{\alpha}), l^\alpha_{\mathbf{w}(\mathbf{v}(1-e))^{n+1}\mathbf{v}}(\ddot{\alpha})$ are connected by $S^{\alpha,e}_{\mathbf{w}(\mathbf{v}(1-e))^{n+1}\mathbf{v}*}$, Lemma 2.35 implies that $|l^{\alpha,1-e}_{\mathbf{w}}(\overline{\alpha}) - l^{\alpha,e}_{\mathbf{w}}(\alpha)| = \sum_{n=0}^\infty d(S^{\alpha,e}_{\mathbf{w}(\mathbf{v}(1-e))^n\mathbf{v}*}) = \sum_{n=0}^\infty 2^{-|\mathbf{w}|-nm} d(S^\alpha_{\mathbf{v}*}) = \frac{2^{m-|\mathbf{w}|}}{2^m-1} d(S^\alpha_{\mathbf{v}*})(*)$. This shows (ii).

The chords $\alpha\overline{\alpha}$ and $l^{\alpha,1-e}_0(\alpha\overline{\alpha})$ do not cross each other, consequently $l^\alpha_0(\overline{\alpha})$ cannot lie between α and $\overline{\alpha}$. Immediately from this fact one concludes that $d(\alpha\overline{\alpha}) \le \frac{1}{3}$. Finally, setting $\mathbf{w} = \lambda$ in formula $(*)$ leads to $d(S^\alpha_{\mathbf{v}*}) \le \frac{2^m-1}{3\,2^m}$, and this together with formula $(*)$ implies (iii). ■

We want to set off the case $\mathbf{w} = \lambda$ in (ii), which provides a formula for the associated point.

Corollary 2.39. (Formula for the associated periodic point)
The point $\overline{\alpha}$ associated with a point α with period m is given by

$$\overline{\alpha} = \alpha + \frac{2^m}{2^m-1}(l^\alpha_{\mathbf{v}^\alpha}(\ddot{\alpha}) - \alpha). \quad ■ \tag{2.7}$$

Let us draw some further simple conclusions from Proposition 2.38, which will be useful subsequently.

Corollary 2.40. *For $\alpha \in \mathbb{P} \setminus \{0\}, \mathbf{v} = \mathbf{v}^\alpha, e = e^\alpha$ the following statements are valid:*

(i) $\overline{\alpha} = l^{\alpha,1-e}_{\mathbf{v}(1-e)}(\overline{\alpha}),\ \dot{\overline{\alpha}} = l^{\alpha,1-e}_{1-e}(\overline{\alpha}),\ \ddot{\overline{\alpha}} = l^{\alpha,1-e}_{e}(\overline{\alpha})$,

(ii) $d(\alpha\overline{\alpha}) \leq \frac{1}{3},\ d(\dot{\alpha}\dot{\overline{\alpha}}) = d(\ddot{\alpha}\ddot{\overline{\alpha}}) = \frac{1}{2} - \frac{d(\alpha\overline{\alpha})}{2} \geq \frac{1}{3}$,

(iii) $\{\dot{\alpha}\dot{\overline{\alpha}}, \ddot{\alpha}\ddot{\overline{\alpha}}\} = \{\frac{\alpha}{2}\frac{\overline{\alpha}+1}{2}, \frac{\alpha+1}{2}\frac{\overline{\alpha}}{2}\}$.

Proof: (i): Let $m = PER(\alpha)$, and set $\mathbf{w} = \mathbf{v}(1-e)$ in Proposition 2.38(ii). Then $l^{\alpha,1-e}_{\mathbf{v}(1-e)} = l^\alpha_{\mathbf{v}}(\ddot{\alpha}) + \frac{1}{2^m-1}(l^\alpha_{\mathbf{v}}(\ddot{\alpha}) - \alpha) = \frac{2^m}{2^m-1} l^\alpha_{\mathbf{v}}(\ddot{\alpha}) - \frac{1}{2^m-1}\alpha = \overline{\alpha}$ by (2.7). The other equalities follow immediately.

(ii): The statement (iii) of Proposition 2.38 for $\mathbf{w} = 1-e$ yields $d(\dot{\overline{\alpha}}\dot{\alpha}) = d(l^{\alpha,1-e}_{1-e}(\overline{\alpha}), l^{\alpha,1-e}_{1-e}(\alpha)) = \frac{1}{2} - d(l^{\alpha,1-e}_{1-e}(\overline{\alpha}), l^{\alpha,e}_{1-e}(\alpha)) \geq \frac{1}{2} - \frac{1}{6} = \frac{1}{3}$, and the rest is obvious.

(iii) follows since by (ii) $d(\frac{\alpha}{2}\frac{\overline{\alpha}}{2}) < \frac{1}{3}$ but $d(\dot{\alpha}\dot{\overline{\alpha}}) > \frac{1}{3}$. ∎

Proposition 2.41. (Determination of the characteristic symbol I)
For $\alpha \in \mathbb{P}_* \setminus \{0\}$ with $PER(\alpha) = m$ it holds $e^\alpha = 1 - I^\alpha(\overline{\alpha})(m)$, and $e^\alpha = \overset{\frown}{I^\alpha}(\beta)(m)$ if $\alpha < \overline{\alpha}$ and $e^\alpha = \overset{\frown}{I^\alpha}(\alpha)(m)$ if $\overline{\alpha} < \alpha$.

Further, it holds $e^\alpha = 0$ ($e^\alpha = 1$) iff $\ddot{\alpha}\ddot{\overline{\alpha}}$ and 0 are separated by $\dot{\alpha}\dot{\overline{\alpha}}$ ($\dot{\alpha}\dot{\overline{\alpha}}$ and 0 are separated by $\ddot{\alpha}\ddot{\overline{\alpha}}$).

Proof: $\overline{\alpha} = l^{\alpha,1-e}_{\mathbf{v}(1-e)}(\overline{\alpha})$ implies $e^\alpha = 1 - I^\alpha(\overline{\alpha})(m)$. If $\ddot{\alpha}\ddot{\overline{\alpha}}$ separates the points 0 and $\dot{\alpha}\dot{\overline{\alpha}}$, then $\dot{\overline{\alpha}} = h^{m-1}(\overline{\alpha}) \in l^\alpha_0(\mathbb{T} \setminus \{\alpha\})$. If $\dot{\alpha}\dot{\overline{\alpha}}$ separates 0 and $\ddot{\alpha}\ddot{\overline{\alpha}}$, then $\dot{\overline{\alpha}} \in l^\alpha_1(\mathbb{T} \setminus \{\alpha\})$. Therefore, in the first case $e^\alpha = 1$ and in the second case $e^\alpha = 0$.

To complete the proof of the proposition, let us use the orientation-invariance of h (Lemma 2.9). If $\alpha < \overline{\alpha}$ and $e^\alpha = 0$, then by the already shown $\dot{\alpha} = \frac{\alpha}{2}$, which implies $\overset{\frown}{I^\alpha}(\beta)(m) = 0$. On the other hand, $\alpha < \overline{\alpha}$ and $e^\alpha = 1$ means $\dot{\alpha} = \frac{\alpha+1}{2}$, hence $\overset{\frown}{I^\alpha}(\beta)(m) = 1$. Analogously, one obtains $e = \overset{\frown}{I^\alpha}(\alpha)(m)$ if $\overline{\alpha} < \alpha$. ∎

Remark. For $\alpha \in \mathbb{P}_*$ with $m = PER(\alpha)$ there are unique numbers $p_1, p_2 \in \mathbb{N}$ satisfying $\alpha = \frac{p_1}{2^m-1}$ and $\overline{\alpha} = \frac{p_2}{2^m-1}$. By use of Proposition 2.41 it is not hard to show that $e^\alpha = 1$ if the minimum of p_1, p_2 is odd and $e^\alpha = 0$ otherwise.

Now from Proposition 2.38 we obtain a representation of $\eth\mathcal{B}^\alpha$ for periodic $\alpha \in \mathbb{T}$, which corresponds to the definition of $\mathcal{B}^\alpha$ for non-periodic α.

Proposition 2.42. ('Generation' of $\eth\mathcal{B}^\alpha$ from $\alpha\overline{\alpha}$)
Let $\alpha \in \mathbb{P}_* \setminus \{0\}$ and let $e = e^\alpha$ be the characteristic symbol of α. Then $\eth\mathcal{B}^\alpha$ is equal to the closure of the set $\mathcal{B}^\alpha_* := \{l^{\alpha,1-e}_{\mathbf{w}}(\alpha\overline{\alpha}) | \mathbf{w} \in \{0,1\}^*\}$.

In particular, if a sequence $(\mathbf{u}_n)_{n\in\mathbb{N}}$ of 0-1-words is given, such that $(S^{\alpha,1-e}_{\mathbf{u}_n *})_{n\in\mathbb{N}}$ converges to a chord S, then also $(l^{\alpha,1-e}_{\mathbf{u}_n(1-e)}(\alpha\overline{\alpha}))_{n\in\mathbb{N}}$ converges to S.

Proof: By Corollary 2.36, $\eth\mathcal{B}^\alpha$ contains only accumulation chords of the set $\mathcal{R}^{\alpha,1-e}$ of all chords $S^{\alpha,1-e}_{\mathbf{w}*}$; $\mathbf{w} \in \{0,1\}^*$. Thus by Proposition 2.38(i), the set $\eth\mathcal{B}^\alpha_*$ is contained in $\mathcal{B}^\alpha$.

Further, it holds $S^{\alpha,1-e}_{\mathbf{u}*} = l^{\alpha,1-e}_{\mathbf{u}(1-e)}(\alpha) l^{\alpha,e}_{\mathbf{u}(1-e)}(\alpha)$ for all $\mathbf{u} \in \{0,1\}^*$, and for each $\epsilon > 0$ by Proposition 2.38(iii) $d(l^{\alpha,1-e}_{\mathbf{u}(1-e)}(\overline{\alpha}) l^{\alpha,e}_{\mathbf{u}(1-e)}(\alpha)) < \epsilon$ with the exception of finitely many 0-1-words $\mathbf{u}$. The statements above can easily be seen now. ■

Now we want to describe the infinite gaps of $\eth\mathcal{B}^\alpha = \mathcal{B}^\alpha \setminus \mathcal{R}^\alpha$. By Theorem 2.16 and Corollary 2.20(ii), all gaps of $\mathcal{B}^\alpha$ are polygonal. Moreover, gaps which are different from $\Delta^\alpha_{\mathbf{u}}$ for 0-1-words $\mathbf{u}$ not being α-regular remain gaps of $\eth\mathcal{B}^\alpha$.

This and the discussion after Corollary 2.36 yield that on the one hand each chord from $\mathcal{R}^\alpha$ cuts the interior of an infinite gap of $\eth\mathcal{B}^\alpha$, and on the other hand each infinite gap of $\eth\mathcal{B}^\alpha$ contains a chord $S^\alpha_{\mathbf{w}*}$ with α-regular $\mathbf{w} \in \{0,1\}^*$, providing a symbolic description of the infinite gaps of $\eth\mathcal{B}^\alpha$.

Proposition 2.43. (Infinite gaps of $\eth\mathcal{B}^\alpha$ for $\alpha \in \mathbb{P}_* \setminus \{0\}$)
Let $\alpha \in \mathbb{P}_* \setminus \{0\}$, let $\mathbf{v} = \mathbf{v}^\alpha, e = e^\alpha$ and $m = PER(\alpha)$. Then to each given α-regular 0-1-word $\mathbf{w}$ there corresponds an infinite gap $Gap^\alpha_{\mathbf{w}*}$, whose boundary chords are given by $l^{\alpha,1-e}_{\mathbf{w}s_1\mathbf{v}s_2\mathbf{v}\ldots s_n}(\alpha\overline{\alpha})$; $s_1 s_2 \ldots s_n \in \{0,1\}^* \setminus \{\lambda\}$. The gaps defined by α-regular 0-1-words in this way are mutually different and form all infinite gaps of $\eth\mathcal{B}^\alpha$. Moreover, the following holds:

- *(i) $Gap^\alpha_{\mathbf{v}*}$ is the critical value gap. Its longest boundary chord is the periodic $\alpha\overline{\alpha} = l^{\alpha,1-e}_{\mathbf{v}(1-e)}(\alpha\overline{\alpha})$. Moreover, $h^{m-1}(Gap^\alpha_{\mathbf{v}*}) = Gap^\alpha_*, h^m(Gap^\alpha_{\mathbf{v}*}) = Gap^\alpha_{\mathbf{v}*}$, and $h^i(Gap^\alpha_{\mathbf{v}*})$ lies behind $\dot{\alpha}\dot{\overline{\alpha}}$ or behind $\ddot{\alpha}\ddot{\overline{\alpha}}$ for all $i = 0,1,2,\ldots,$ $m-2$ and not behind $\alpha\overline{\alpha}$ for all $i = 1,2,3,\ldots,m-1$.*
- *(ii) If some $\mathbf{w} \neq \mathbf{v}$ of length n is α-regular, then $l^{\alpha,1-e}_{\mathbf{w}0}(\alpha\overline{\alpha})$ and $l^{\alpha,1-e}_{\mathbf{w}1}(\alpha\overline{\alpha})$ form the longest boundary chords of $Gap^\alpha_{\mathbf{w}*}$. More exactly, if $\alpha\overline{\alpha}$ has length a, these chords are not shorter than $\frac{1-a}{2^{n+1}}$ and the two shortest 'sides' of the 'rectangle' spanned by their endpoints have length $\frac{a}{2^{n+1}}$.*

Proof: Fix an α-regular word $\mathbf{w}$. Then the gap G whose interior is cut by $S^\alpha_{\mathbf{w}*}$ includes all triangles of the form $\Delta^\alpha_{\mathbf{w}s_1\mathbf{v}s_2\mathbf{v}\ldots s_k\mathbf{v}*}$ for $s_1 s_2 \ldots s_k \in \{0,1\}^*$, hence by Proposition 2.38(i) also the chords $l^{\alpha,1-e}_{\mathbf{w}s_1\mathbf{v}s_2\mathbf{v}\ldots s_n}(\alpha\overline{\alpha})$; $s_1 s_2 \ldots s_n \in \{0,1\}^*$ from $\eth\mathcal{B}^\alpha$. Of course, the latter chords bound G.

Moreover, geometric arguments show that to each boundary chord of G there leads a path in the tree of triangles described above. So we conclude that there are no further boundary chords of G. Therefore, the gaps $Gap^\alpha_{\mathbf{w}*}$ for α-regular $\mathbf{w}$ are well-defined.

(ii): Obviously, Gap^α_* is the critical gap and by Corollary 2.40 it has the longest boundary chords $l^{\alpha,1-e}_0(\alpha\overline{\alpha})$ and $l^{\alpha,1-e}_1(\alpha\overline{\alpha})$ coinciding with $\dot{\alpha}\dot{\overline{\alpha}}, \ddot{\alpha}\ddot{\overline{\alpha}}$. Since these two chords have lengths $\frac{1-a}{2}$, the inequality $d(l^{\alpha,1-e}_{\mathbf{w}0}(\alpha\overline{\alpha})) = d(l^{\alpha,1-e}_{\mathbf{w}1}(\alpha\overline{\alpha})) \geq \frac{1-a}{2^{n+1}}$ is obvious for words $\mathbf{w}$ not ending with $\mathbf{v}$ and having length n.

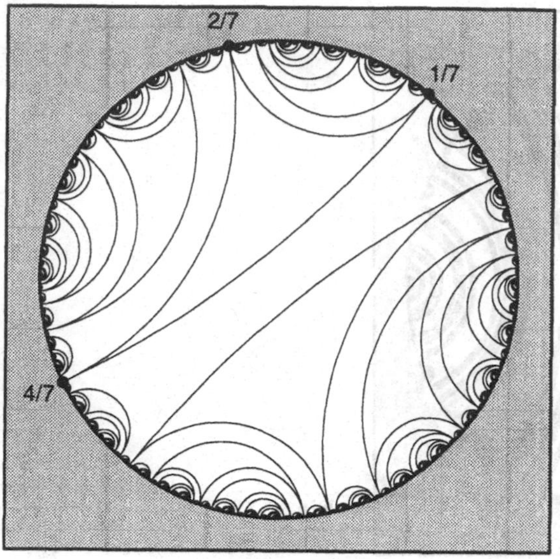

Fig. 2.12. $\mathbb{T}^{\alpha}_{\overline{\mathbf{v}^{\alpha}(1-e^{\alpha})}}$ for periodic $\alpha \in \mathbb{T}$: $\eth\mathcal{B}^{\alpha}$ for $\alpha = \frac{1}{7}, \frac{2}{7}$

Further, the chords $l_{\mathbf{w}}(\dot{\alpha}\ddot{\overline{\alpha}})$ and $l_{\mathbf{w}}(\ddot{\alpha}\dot{\overline{\alpha}})$ lie in the gap $Gap^{\alpha}_{\mathbf{w}*}$ which for non-empty $\mathbf{w}$ is different from the critical gap. Therefore, by Lemma 2.8 they have length $\frac{a}{2^n+1}$. Now (ii) is easy to see.

(i): The first iterate of $\alpha\overline{\alpha}$ which would be between $\dot{\alpha}\dot{\overline{\alpha}}$ and $\ddot{\alpha}\ddot{\overline{\alpha}}$, would be longer than $\dot{\alpha}\dot{\overline{\alpha}}$ by Lemma 2.10, but this is impossible. Since $h^n(\alpha\overline{\alpha})$ bounds the n-th iterate of the critical value gap, (i) is not hard to show now. ■

Clearly, $PER(\alpha)$ must be a multiple of $PER(\overline{\mathbf{v}(1-e)})$. The following Theorem tells us when the two periods coincide.

Theorem 2.44. *Let* $\alpha \in \mathbb{P}_* \setminus \{0\}, \mathbf{v} = \mathbf{v}^{\alpha}$, *and let* $e = e^{\alpha}$ *be the characteristic symbol of* α. *Then* $PER(\alpha) = PER(\overline{\mathbf{v}e})$, *and the following statements are equivalent:*

(i) α *and* $\overline{\alpha}$ *lie on a common orbit.*
(ii) $PER(\overline{\mathbf{v}(1-e)}) < PER(\alpha)$.
(iii) $\alpha\overline{\alpha} \notin \eth(\eth\mathcal{B}^{\alpha})$.

Proof: By the construction of $\overline{\alpha}$, each sequence of chords in $\mathcal{R}^{\alpha}$ converging to $\alpha\overline{\alpha}$ and lying behind $\alpha\overline{\alpha}$ forms a subsequence of $(S^{\alpha,1-e}_{(\mathbf{v}(1-e))^n\mathbf{v}*})_{n\in\mathbb{N}}$, with exception of finitely many members. Hence a sequence of chords in $\eth\mathcal{B}^{\alpha}$ which converges to $\alpha\overline{\alpha}$, possesses a subsequence with no member behind $\alpha\overline{\alpha}$.

If $CARD(\mathbb{T}^{\alpha}_{\overline{\mathbf{v}(1-e)}}) \geq 3$, then obviously there cannot exist a sequence in $\eth\mathcal{B}^{\alpha}$ converging to $\alpha\overline{\alpha}$, and (i), (ii) are immediate consequences of Corollary 2.23. In this case (i), (ii) and (iii) are equivalent because they are always satisfied. Therefore, we can assume that $CARD(\mathbb{T}^{\alpha}_{\overline{\mathbf{v}(1-e)}}) = 2$. On this as-

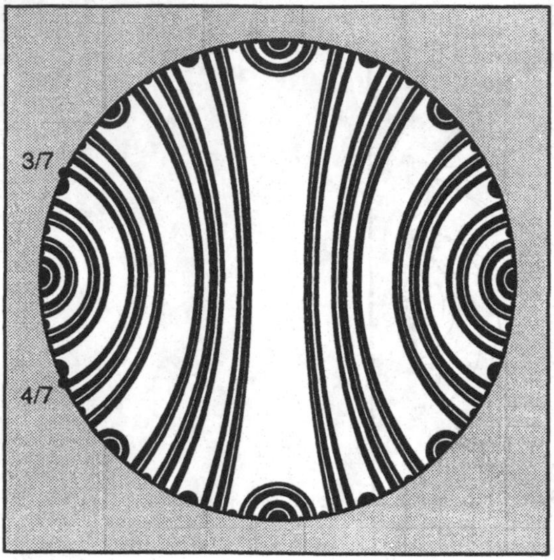

Fig. 2.13. $\mathbb{T}^{\alpha}_{\overline{\mathbf{v}^{\alpha}(1-e^{\alpha})}}$ for periodic $\alpha \in \mathbb{T}$: $\eth\mathcal{B}^{\alpha}$ for $\alpha = \frac{3}{7}, \frac{4}{7}$

sumption it holds $\mathbb{T}^{\alpha}_{\overline{\mathbf{v}(1-e)}} = \{\alpha, \overline{\alpha}\}$, which implies the equivalence of the two statements (i) and (ii).

If $h^k(\alpha) = \overline{\alpha}$ for some $k \in \mathbb{N}$, then also $h^k(\overline{\alpha}) = \alpha$, such that h^k transforms each chord sequence converging to $\alpha\overline{\alpha}$ from the one side into a sequence converging to $\alpha\overline{\alpha}$ from the other side. Hence (i) follows from (iii).

Finally, let us show that (iii) implies (ii), which completes the proof. Assume that (ii) fails to be valid, i.e. that $PER(\overline{\mathbf{v}(1-e)}) = m = PER(\alpha)$. Then consider the graph $\mathfrak{G}^{\alpha}(\mathbf{v}(1-e))$, which by Theorem 2.19 possesses two components, one of them obviously containing the set $\{m, 2m, 3m, \ldots\}$.

The elements of the other component correspond to different α-regular words $\mathbf{u}_n; n \in \mathbb{N}$, where $\alpha\overline{\alpha} = \lim_{n\to\infty} S^{\alpha,1-e}_{\mathbf{u}_n *}$. By the last statement in Proposition 2.43(ii), the sequence $(l^{\alpha,1-e}_{\mathbf{u}_n(1-e)}(\alpha\overline{\alpha}))_{n\in\mathbb{N}}$ converges to $\alpha\overline{\alpha}$, and by the α-regularity of $\mathbf{u}_n$ the chords $l^{\alpha,1-e}_{\mathbf{u}_n(1-e)}(\alpha\overline{\alpha})$ are in $\eth\mathcal{B}^{\alpha}$ and mutually different. So we obtain $\alpha\overline{\alpha} \in \eth(\eth\mathcal{B}^{\alpha})$. ∎

Example 2.45. In order to illustrate Theorem 2.44, let us have a look at $\mathbb{T}^{\alpha}_{\overline{\mathbf{v}^{\alpha}(1-e^{\alpha})}}$ for the three periodic points $\alpha = \frac{1}{3}$ (see Figure 2.7), $\alpha = \frac{1}{7}$ (see Figure 2.12) and $\alpha = \frac{3}{7}$ (see Figure 2.13).

We have $\overline{\frac{1}{3}} = \frac{2}{3}$ and $\overline{\frac{1}{7}} = \frac{2}{7}$. Moreover, $\mathbb{T}^{1/3}_{\overline{\mathbf{v}^{1/3}(1-e^{1/3})}} = \mathbb{T}^{1/3}_{\overline{0}} = \{\frac{1}{3}, \frac{2}{3}\}$ and $\mathbb{T}^{1/7}_{\overline{\mathbf{v}^{1/7}(1-e^{1/7})}} = \mathbb{T}^{1/7}_{\overline{0}} = \{\frac{1}{7}, \frac{2}{7}, \frac{4}{7}\}$.

In each of the two cases, α and $\overline{\alpha}$ lie on a common orbit and form the two endpoints of an isolated chord in $\eth\mathcal{B}^{\alpha}$. On the other hand, $\mathbb{T}^{3/7}_{\overline{\mathbf{v}^{3/7}(1-e^{3/7})}} =$

$\mathbb{T}^{3/7}_{\overline{010}} = \{\frac{3}{7}, \frac{4}{7}\}$, and $\frac{3}{7}, \overline{\frac{3}{7}} = \frac{4}{7}$ determine different orbits. Finally, $\frac{3}{7}\frac{4}{7} \in \eth(\eth\mathcal{B}^{3/7})$.

2.3 Julia equivalences

2.3.1 From invariant laminations to Julia equivalences

This section is devoted to a complete classification of Julia equivalences. First of all, we assign to each $\alpha \in \mathbb{T}$ a Julia equivalence in the following way:

Definition 2.46. (The equivalence relations $\approx^\alpha$)
For $\alpha \in \mathbb{T} \setminus \mathbb{P}_*$ and $\alpha \in \mathbb{P}_*$ let $\beta_1 \approx^\alpha \beta_2$ if there exists a finite sequence of points $\beta_1 = \gamma_0, \gamma_1, \gamma_2, \ldots, \gamma_n = \beta_2$ such that the chords $\gamma_{i-1}\gamma_i, i = 1, 2, \ldots, n$ are contained in $\mathcal{B}^\alpha$ and in $\eth\mathcal{B}^\alpha$, respectively.

The reason for defining $\approx^\alpha; \alpha \in \mathbb{T}$ in the given way will become clear in the course of the discussion following now. At the moment, we want to be content with the fact that the definition is natural in the non-periodic case, and that in the periodic case we need a slightly different definition to avoid degenerate h-invariant equivalence relations.

Of course, from each invariant lamination there can be constructed an equivalence relation in an analogous way as provided by Definition 2.46, but as the example $\mathcal{B}^\alpha; \alpha \in \mathbb{P}_*$ shows, such equivalence relations can be degenerate, and different laminations can define the same equivalence relation (compare text below Theorem 2.48).

The non-periodic case. First let $\alpha \notin \mathbb{P}$, let $\beta_1, \beta_2 \in \mathbb{T}$, and let $\beta_1 = \gamma_0, \gamma_1, \ldots, \gamma_n = \beta_2$ be a sequence of points in $\mathbb{T}$ with $\gamma_{i-1}\gamma_i \in \mathcal{B}^\alpha$ for all $i = 1, 2, \ldots, n$.

No iterates of the chords $\gamma_{i-1}\gamma_i; i = 1, 2, \ldots, n$ cross S^α_*. Therefore, in each case $I^\alpha(\gamma_{i-1}) \cong I^\alpha(\gamma_i)$. Since $\alpha \notin \mathbb{P}_*$, each itinerary contains $*$ at most one time. Furthermore, among the itineraries $I^\alpha(\gamma_i); i = 0, 1, \ldots, n$ no two contain $*$ for different coordinates. Otherwise, there would exist an $i \in \{1, 2, \ldots, n\}$ and different numbers $k_1, k_2 \in \mathbb{N}_0$ with $I^\alpha(h^{k_1}(\gamma_{i-1})) = I^\alpha(h^{k_2}(\gamma_i)) = \hat{\alpha}$, hence $\hat{\alpha}$ would be periodic.

Therefore, one obtains $I^\alpha(\beta_1) \cong I^\alpha(\beta_2)$, or $I^\alpha(\beta_1) = \mathbf{w}s\hat{\alpha}$ and $I^\alpha(\beta_2) = \mathbf{w}(1-s)\hat{\alpha}$ for some $\mathbf{w} \in \{0,1\}^*$ and some $s \in \{0,1\}$.

In the other direction, if $I^\alpha(\beta_1) \cong I^\alpha(\beta_2)$ for $\beta_1, \beta_2 \in \mathbb{T}$, then there exists an $\mathbf{s} \in \{0,1\}^{\mathbb{N}}$ with $\beta_1, \beta_2 \in \widetilde{\mathbb{T}^\alpha_{\mathbf{s}}} = \mathbb{T}^\alpha_{\mathbf{s}} = \mathbb{D}^\alpha_{\mathbf{s}} \cap \mathbb{T}$ (see Proposition 2.6). Since by Corollary 2.20(ii) the gap $\mathbb{D}^\alpha_{\mathbf{s}}$ is polygonal, it follows $\beta_1 \approx^\alpha \beta_2$.

If $I^\alpha(\beta_1) = \mathbf{w}s\hat{\alpha}$ and $I^\alpha(\beta_2) = \mathbf{w}(1-s)\hat{\alpha}$ for some $\mathbf{w} \in \{0,1\}^*$ and $s \in \{0,1\}$, then it follows $\beta_1 \approx^\alpha l_{\mathbf{w}}(\frac{\alpha}{2})$ and $\beta_2 \approx^\alpha l_{\mathbf{w}}(\frac{\alpha}{2})$, hence $\beta_1 \approx^\alpha \beta_2$. We have shown the following statement:

Theorem 2.47. (Symbolic description of $\approx^\alpha$ for $\alpha \in \mathbb{T} \setminus \mathbb{P}$)
For $\alpha \in \mathbb{T} \setminus \mathbb{P}$, it holds $\beta_1 \approx^\alpha \beta_2$ iff either $I^\alpha(\beta_1) \cong I^\alpha(\beta_2)$, or $I^\alpha(\beta_1) = \mathbf{w}s\hat{\alpha}$ and $I^\alpha(\beta_2) = \mathbf{w}(1-s)\hat{\alpha}$ for some $\mathbf{w} \in \{0,1\}^*$ and some $s \in \{0,1\}$. ■

The sets $\mathbb{T}^\alpha_\mathbf{s}; \mathbf{s} \in \{0,1\}^\mathrm{N}$ are non-empty (see the end of Section 2.1), and by looking at the proof of Theorem 2.47 one obtains another version of it.

Theorem 2.48. (Version 2 of Theorem 2.47)
$\mathbb{T}/\approx^\alpha = \{\widetilde{\mathbb{T}^\alpha_\mathbf{s}} \mid \mathbf{s} \in \{0,1\}^\mathrm{N},$ $\mathbf{s}$ does not *end with* $0\hat{\alpha}$ or $1\hat{\alpha}\} \cup \{\widetilde{\mathbb{T}^\alpha_{\mathbf{w}0\hat{\alpha}} \cup \mathbb{T}^\alpha_{\mathbf{w}1\hat{\alpha}}} \mid \mathbf{w} \in \{0,1\}^*\}$ *for* $\alpha \in \mathbb{T} \setminus \mathbb{P}$. ■

According to the above definition, for $\alpha \in \mathbb{T} \setminus \mathbb{P}_*$ two points are $\approx^\alpha$-equivalent if they can be connected by a sequence of chords in $\mathcal{B}^\alpha$. For $\alpha \in \mathbb{T} \setminus \mathbb{P}$ the chords in such a sequence can be taken even from the invariant lamination $\eth\mathcal{B}^\alpha$, which Proposition 2.29 provides.

So different invariant laminations $\mathcal{B}^\alpha$ can generate the same Julia equivalence, as the example $\alpha = \frac{9}{63}, \frac{11}{63}, \frac{15}{63}$ shows (compare Figure 2.6).

Example 2.49. (From $\mathcal{B}^0$ to $\approx^0$ and $\mathcal{B}^{\frac{1}{2}}$ to $\approx^{\frac{1}{2}}$)
According to Corollary 2.36, all non-degenerate chords in $\mathcal{B}^0$ are isolated. Therefore $\approx^0$ does not allow proper identifications, hence is trivial. Besides $\approx^0$, the equivalence relation $\approx^{\frac{1}{2}}$ has a very simple structure - opposite to all other Julia equivalences: for different $\beta_1, \beta_2 \in \mathbb{T}$, one has $\beta_1 \approx^{\frac{1}{2}} \beta_2$ iff $\beta_1 + \beta_2 = 1$ (compare Example 2.27).

Note that the latter Julia equivalence was already considered in Chapter 1.1.2 in relation to the Julia set J_{-2}.

Let us continue by discussing the case $\alpha \in \mathbb{P}_\infty$. In this case, S^α_* is the only chord in $\mathcal{B}^\alpha$ having an endpoint in $\{\frac{\alpha}{2}, \frac{\alpha+1}{2}\}$. Otherwise, α would be isolated in $\mathbb{P}^\alpha_\infty$, which would contradict Proposition 2.31. Therefore, $\beta_1 \approx^\alpha \beta_2$ implies equality of $I^\alpha(\beta_1)$ and $I^\alpha(\beta_2)$. By Proposition 2.30, for all $\mathbf{w} \in \{0,1\}^\mathrm{N}$ the boundary chords of $\mathbb{D}^\alpha_{\mathbf{w}\hat{\alpha}}$ lie in $\mathcal{R}^\alpha$. Hence $\approx^\alpha$ does not identify different points with common itinerary $\mathbf{w}\hat{\alpha}$.

Conversely, let $\beta_1 \neq \beta_2$ and $\mathbf{s} = I^\alpha(\beta_1) = I^\alpha(\beta_2) \neq \mathbf{w}\hat{\alpha}$ for all $\mathbf{w} \in \{0,1\}^*$. Then the two points β_1, β_2 are identified by the chord $S^\alpha_{\mathbf{u}*} \in \mathcal{R}^\alpha$ if both have the itinerary $\mathbf{u} * \hat{\alpha}; \mathbf{u} \in \{0,1\}^*$. Otherwise, one derives $\beta_1, \beta_2 \in \mathbb{T}^\alpha_\mathbf{s} = \mathbb{D}^\alpha_\mathbf{s} \cap \mathbb{T}$ and $\beta_1 \approx^\alpha \beta_2$ from $I^\alpha(\beta_1) = I^\alpha(\beta_2)$, as in the case $\alpha \in \mathbb{T} \setminus \mathbb{P}$. Let us summarize:

Theorem 2.50. (Symbolic description of $\approx^\alpha$ for $\alpha \in \mathbb{P}_\infty$)
For $\alpha \in \mathbb{P}_\infty$ and $\beta_1 \neq \beta_2$, it holds $\beta_1 \approx^\alpha \beta_2$ if the itineraries of β_1 and β_2 are equal but different from $\mathbf{w}\hat{\alpha}$ for each $\mathbf{w} \in \{0,1\}^*$. ■

For the given case, we have the following equivalent description of $\approx^\alpha$:

Theorem 2.51. (Version 2 of Theorem 2.50)
$\mathbb{T}/\approx^\alpha = \{\widetilde{\mathbb{T}^\alpha_{\mathbf{s}}} \mid \mathbf{s} \in \{0,1\}^{\mathbb{N}}, \mathbf{s}$ does not *end with* $\hat{\alpha}\} \cup \{\{h^n(\alpha)\} \mid n \in \mathbb{N}_0\} \cup \{l^\alpha_{\mathbf{w}}(\{\frac{\alpha}{2}, \frac{\alpha+1}{2}\}) \mid \mathbf{w} \in \{0,1\}^*\}$ *for* $\alpha \in \mathbb{P}_\infty$. ■

By use of Theorems 2.47 and 2.50, we are also able to characterize $\approx^\alpha$ for $\alpha \in \mathbb{T} \setminus \mathbb{P}_*$ as a Julia equivalence.

Proposition 2.52. (Characterization of $\approx^\alpha$ for $\alpha \in \mathbb{T} \setminus \mathbb{P}_*$)
For $\alpha \in \mathbb{T} \setminus \mathbb{P}_*$*, the equivalence relation* $\approx^\alpha$ *is a Julia equivalence, and each planar and completely invariant equivalence relation on* $(\mathbb{T}, h, ')$ *which identifies* $\frac{\alpha}{2}$ *and* $\frac{\alpha+1}{2}$ *contains* $\approx^\alpha$.

Moreover, there is no planar and invariant equivalence relation on $(\mathbb{T}, h, ')$ *which properly contains* $\approx^\alpha$ *and possesses only finite equivalence classes.*

Proof: At first, we show that $\approx^\alpha$ is a Julia equivalence. The statement that $\approx^\alpha$ is closed can be obtained as follows:

If $\alpha \in \mathbb{T} \setminus \mathbb{P}$, then by Theorem 2.48 $\beta_1 \not\approx^\alpha \beta_2; \beta_1, \beta_2 \in \mathbb{T}$ implies the existence of an $n \in \mathbb{N}$ such that $* \neq I^\alpha(\beta_1)(n) \neq I^\alpha(\beta_2)(n) \neq *$ and further that $h^n(\beta_1) \notin \mathbb{T}^\alpha_{\hat{\alpha}}$ or $h^n(\beta_2) \notin \mathbb{T}^\alpha_{\hat{\alpha}}$. Therefore, one easily sees that the set $\{(\beta_1, \beta_2) \in \mathbb{T} \times \mathbb{T} \mid \beta_1 \not\approx^\alpha \beta_2\}$ is open.

If $\alpha \in \mathbb{P}_\infty$, then according to Theorem 2.50 two different $\approx^\alpha$-equivalent points are connected by a chord which is completely contained in a $\mathbb{D}^\alpha_{\mathbf{s}}; \mathbf{s} \in \{0,1\}^{\mathbb{N}}$ with $\sigma^n(\mathbf{s}) \neq \hat{\alpha}$ for all $n \in \mathbb{N}_0$, or which is an element of $\mathcal{R}^\alpha$. Accumulation chords of sets formed by such chords do not cut the interior of a $\mathbb{D}^\alpha_{\mathbf{w}\hat{\alpha}}; \mathbf{w} \in \{0,1\}^*$. Consequently, their endpoints must be $\approx^\alpha$-equivalent too, and so $\approx^\alpha$ is closed. Now it is an immediate consequence of Theorems 2.47 and 2.50 that $\approx^\alpha$ forms a Julia equivalence.

Let $\approx$ be an arbitrary planar and completely invariant equivalence relation on $(\mathbb{T}, h, ')$ with $\frac{\alpha}{2} \approx \frac{\alpha+1}{2}$, and let us show that $\approx$ identifies the two endpoints of each chord in $\mathcal{R}^\alpha$.

Suppose that this is false. Then let $\mathbf{w} = w_1 w_2 \ldots w_n$ be a 0-1-word of minimal length with $l^\alpha_{\mathbf{w}}(\frac{\alpha}{2}) \not\approx l^\alpha_{\mathbf{w}}(\frac{\alpha+1}{2})$. Now one has $l^\alpha_{w_2 w_3 \ldots w_n}(\frac{\alpha}{2}) \approx l^\alpha_{w_2 w_3 \ldots w_n}(\frac{\alpha+1}{2})$, and since $\approx$ is completely h- and $'$-invariant, it follows $l^\alpha_{0 w_2 w_3 \ldots w_n}(\frac{\alpha}{2}) \approx l^\alpha_{1 w_2 w_3 \ldots w_n}(\frac{\alpha+1}{2})$ and $l^\alpha_{1 w_2 w_3 \ldots w_n}(\frac{\alpha}{2}) \approx l^\alpha_{0 w_2 w_3 \ldots w_{n-1}}(\frac{\alpha+1}{2})$, or $l^\alpha_{0 w_2 w_3 \ldots w_n}(\frac{\alpha}{2}) \approx l^\alpha_{1 w_2 w_3 \ldots w_n}(\frac{\alpha}{2})$ and $l^\alpha_{0 w_2 w_3 \ldots w_n}(\frac{\alpha+1}{2}) \approx l^\alpha_{1 w_2 w_3 \ldots w_n}(\frac{\alpha+1}{2})$.

Because in the first case the chords $l^\alpha_{0 w_2 w_3 \ldots w_n}(\frac{\alpha}{2}) l^\alpha_{1 w_2 w_3 \ldots w_n}(\frac{\alpha+1}{2})$ and $l^\alpha_{1 w_2 w_3 \ldots w_n}(\frac{\alpha}{2}) l^\alpha_{0 w_2 w_3 \ldots w_n}(\frac{\alpha+1}{2})$ would cross each other and because in the second case $\frac{\alpha}{2}\frac{\alpha+1}{2}$ would cross the chord $l^\alpha_{0 w_2 w_3 \ldots w_n}(\frac{\alpha}{2}) l^\alpha_{1 w_2 w_3 \ldots w_n}(\frac{\alpha}{2})$ as well as the chord $l^\alpha_{0 w_2 w_3 \ldots w_n}(\frac{\alpha+1}{2}) l^\alpha_{1 w_2 w_3 \ldots w_n}(\frac{\alpha+1}{2})$, the planarity of $\approx$ immediately implies $l^\alpha_{\mathbf{w}}(\frac{\alpha}{2}) \approx l^\alpha_{\mathbf{w}}(\frac{\alpha+1}{2})$, which contradicts the above assumption.

From the definition of $\approx^\alpha$ one obtains $\approx^\alpha \subseteq \approx$. Assuming that $\approx$ possesses only finite equivalence classes, by the planarity of $\approx$ two given $\approx$-equivalent points $\beta_1, \beta_2 \in \mathbb{T}$ can be separated by only finitely many chords of $\mathcal{R}^\alpha$. Two points separated by at most one chord in $\mathcal{R}^\alpha$ either have the same itineraries or their itineraries differ in exactly one coordinate, and after this coordinate follows $\hat{\alpha}$. In particular, by Theorem 2.47 one gets that in the case $\alpha \notin \mathbb{P}$ the two points are $\approx^\alpha$-equivalent.

If $\alpha \notin \mathbb{P}$, then from this it follows $\beta_1 \approx^\alpha \beta_2$ since the number of chords in $\mathcal{R}^\alpha$ separating β_1 and β_2 is finite. If $\alpha \in \mathbb{P}_\infty$, let us have a closer look at the position of the chord $\beta_1\beta_2$ relative to $\mathcal{R}^\alpha$.

In case that it crosses some $S^\alpha_{\mathbf{w}*}; \mathbf{w} \in \{0,1\}^*$, by the planarity of $\approx$ the set $\mathbb{T}^\alpha_{\mathbf{w}0\hat{\alpha}}$ would contain at least three points (see Proposition 2.30), which is impossible by Corollary 2.32.

If it does not cross such a chord, either $I^\alpha(\beta_1) = I^\alpha(\beta_1) \not\cong \mathbf{u}\hat{\alpha}$ for all $\mathbf{u} \in \{0,1\}^*$ or $I^\alpha(\beta_1), I^\alpha(\beta_1) \cong \mathbf{w}\hat{\alpha}$ for some $\mathbf{w} \in \{0,1\}^*$. In the first case one obtains $\beta_1 \approx^\alpha \beta_2$ (see Theorem 2.50), and in the second case by Corollary 2.32 the chord $\beta_1\beta_2$ must lie in $\mathcal{R}^\alpha$. Therefore, we have $\beta_1 \approx^\alpha \beta_2$ again. ■

The periodic case. For $\alpha \in \mathbb{P}_*$, let $\mathbf{v} = \mathbf{v}^\alpha$ and $e = e^\alpha$. In this case, it holds $\eth\mathcal{B}^\alpha = \eth\mathcal{R}^{\alpha,1-e}$ by Corollary 2.36. Thus, there is no chord in $\eth\mathcal{B}^\alpha$ which connects a point in $]\frac{\alpha}{2}, \frac{\alpha+1}{2}[$ with a point in $]\frac{\alpha+1}{2}, \frac{\alpha}{2}[$. Also $\frac{\alpha}{2}\frac{\alpha+1}{2}$ does not belong to $\eth\mathcal{B}^\alpha$.

Having a look at the discussion after Corollary 2.36, one sees that one endpoint of a chord $S \in \mathcal{R}^{\alpha,1-e}$ is $\dot{\alpha}$ iff $S = S^{\alpha,1-e}_{((1-e)\mathbf{v})^n*}$ for some $n \in \mathbb{N}_0$. The second endpoint of such a chord is always contained in $l^\alpha_{1-e}(\mathbb{T} \setminus \{\alpha\})$. Analogously, it follows that $S \in \mathcal{R}^{\alpha,1-e}$ possesses an endpoint in $l^\alpha_e(\mathbb{T} \setminus \{\alpha\})$ if the other one is equal to $\ddot{\alpha}$.

We want to modify the concept of an itinerary a little by

$$\widetilde{I^\alpha}(\beta) = s_1 s_2 s_3 \ldots \quad \text{with} \quad s_i = \begin{cases} 0 & \text{for} \quad h^{i-1}(\beta) \in \,]\frac{\alpha}{2}, \frac{\alpha+1}{2}[\\ 1 & \text{for} \quad h^{i-1}(\beta) \in \,]\frac{\alpha+1}{2}, \frac{\alpha}{2}[\\ e^\alpha & \text{for} \quad h^{i-1}(\beta) = \ddot{\alpha} \\ 1 - e^\alpha & \text{for} \quad h^{i-1}(\beta) = \dot{\alpha} \end{cases}$$

for $\alpha \in \mathbb{P}_*$ and $\beta \in \mathbb{T}$. It is called the *specified itinerary* of β with respect to α. Clearly, two $\approx^\alpha$-equivalent points in $\mathbb{T}$ have the same specified itinerary.

If β has specified itinerary $\mathbf{s} \in \{0,1\}^\mathbb{N}$, then $\beta \in \mathbb{T}^\alpha_\mathbf{s}$. In fact, if $I^\alpha(\beta)$ does not contain $*$, this is obvious. Otherwise, let $n \in \mathbb{N}_0$ be minimal with $h^n(\beta) = \dot{\alpha}$ or $h^n(\beta) = \ddot{\alpha}$. Then there exists a word $\mathbf{w} \in \{0,1\}^n$ with $\mathbf{s} = \mathbf{w}(1-e)\mathbf{v}$ or $\mathbf{s} = \mathbf{w}e\mathbf{v}(1-e)\mathbf{v}$, and according to the considerations preceding Proposition 2.6 it holds $\beta \in \mathbb{T}^\alpha_\mathbf{s} = \mathbb{D}^\alpha_\mathbf{s} \cap \mathbb{T}$. By Proposition 2.25, the boundary chords of $\mathbb{D}^\alpha_\mathbf{s}$ are not in $\mathcal{R}^\alpha$, hence they lie in $\eth\mathcal{B}^\alpha$. Since $\mathbb{D}^\alpha_\mathbf{s}$ is a polygonal gap, two points with specified itinerary $\mathbf{s}$ are $\approx^\alpha$-equivalent.

Theorem 2.53. (Symbolic description of $\approx^\alpha$ for $\alpha \in \mathbb{P}_*$)
For $\alpha \in \mathbb{P}_*$, it holds $\beta_1 \approx^\alpha \beta_2$ iff the points β_1 and β_2 have the same specified itinerary. ■

As for Theorem 2.47 and Theorem 2.50 we want to give a statement equivalent to Theorem 2.53 and describing the equivalence classes of $\approx^\alpha$ for $\alpha \in \mathbb{P}_*$. We will do this at the end of this subsection since first we need to know all admissible substantial itineraries.

Remark. For the point $\alpha = 0$ the specified itinerary is no more than its binary expansion, but (in contrast to the agreement at the beginning of Chapter 2.1.1) now sequences ending with $\overline{0}$ are excluded.

In particular, by the considerations preceding Theorem 2.53 one has $[\alpha]_{\approx^\alpha} = \mathbb{T}^\alpha_{\overline{\mathbf{v}(1-e)}}$, and each point $\beta \in \mathbb{T}^\alpha_{\overline{\mathbf{v}(1-e)}}$ is periodic with the same period as α (see Corollary 2.23).

If $\beta \neq \alpha$ satisfies this property and if $\hat{\alpha} \cong I^\alpha(\beta)$, then β must be an element of $\mathbb{T}^\alpha_{\overline{\mathbf{v}(1-e)}}$ or of $\mathbb{T}^\alpha_{\overline{\mathbf{v}e}}$. The latter is impossible since $\mathbb{T}^\alpha_{\overline{\mathbf{v}e}} = \{\alpha\}$ (see discussion before Definition 2.37), and we can summarize:

Lemma 2.54. *Let $\alpha \in \mathbb{P}_*, \mathbf{v} = \mathbf{v}^\alpha$ and let $e = e^\alpha$. Then $[\alpha]_{\approx^\alpha} = \mathbb{T}^\alpha_{\overline{\mathbf{v}(1-e)}} = \{\beta \in \mathbb{T} \mid I^\alpha(\beta) \cong \hat{\alpha}, \beta$ is periodic with $PER(\beta) = PER(\alpha)\}$.* ∎

We want to characterize $\approx^\alpha$ for $\alpha \in \mathbb{P}_*$ in a similar way as for $\alpha \notin \mathbb{P}_*$ in Proposition 2.52. For this, we assign to each equivalence relation $\approx$ on $\mathbb{T}$ the number $d(\approx) = \sup\{d(\beta_1, \beta_2) \mid \beta_1, \beta_2 \in \mathbb{T}; \beta_1 \approx \beta_2\}$ and call it the *diameter* of $\approx$.

Obviously, the diameter cannot succeed $\frac{1}{2}$, and $d(\approx^\alpha) = \frac{1}{2}$ iff α is non-periodic. If $\approx \neq \approx^0$ is h-invariant, there exist different $\approx$-equivalent points β_1, β_2, and by formula (2.4) it follows $d(\approx) \geq \frac{1}{3}$.

Further, we want to mention the simple fact that for an equivalence relation which is closed in $\mathbb{T}\times\mathbb{T}$ there exist two $\approx$-equivalent point whose distance is equal to the diameter of $\approx$.

Let $\alpha \in \mathbb{P}_* \setminus \{0\}, \mathbf{v} = \mathbf{v}^\alpha$ and $e = e^\alpha$. From Proposition 2.43 and Corollary 2.40 one easily deduces the following:

$\dot{\alpha}\dot{\overline{\alpha}}$ and $\ddot{\alpha}\ddot{\overline{\alpha}}$ are the longest boundary chords of the critical gap Gap^α_* of $\mathfrak{d}\mathcal{B}^\alpha$ and so the longest chord in $\mathfrak{d}\mathcal{B}^\alpha$. Gap^α_* is bounded by infinitely many mutually disjoint chords. Thus between $\dot{\alpha}\dot{\overline{\alpha}}$ and $\ddot{\alpha}\ddot{\overline{\alpha}}$ there is no chord in $\mathfrak{d}\mathcal{B}^\alpha$ ending with $\dot{\alpha}, \dot{\overline{\alpha}}, \ddot{\alpha}$ or $\ddot{\overline{\alpha}}$. So we have $d(\approx^\alpha) = d(\frac{\alpha}{2}, \frac{\overline{\alpha}+1}{2})$.

Obviously, the equivalence relation $\approx^\alpha$ forms a Julia equivalence, and can be characterized as follows:

Corollary 2.55. (Characterization of $\approx^\alpha$ for $\alpha \in \mathbb{P}_* \setminus \{0\}$)
If $\alpha \in \mathbb{P}_* \setminus \{0\}$, then $\approx^\alpha$ is a Julia equivalence and the only planar and completely invariant equivalence relation on $(\mathbb{T}, h, ')$ of diameter $d(\frac{\alpha}{2}, \frac{\overline{\alpha}+1}{2})$ which identifies α and $\overline{\alpha}$.

Proof: If $\approx$ is a completely invariant equivalence relation which identifies α and $\overline{\alpha}$, then it also identifies $\dot{\alpha} = h^{m-1}(\alpha)$ and $\dot{\overline{\alpha}} = h^{m-1}(\overline{\alpha})$. By Corollary 2.40, $\dot{\alpha}\dot{\overline{\alpha}}$ coincides with $\frac{\alpha}{2}\frac{\overline{\alpha}+1}{2}$ or $\frac{\overline{\alpha}}{2}\frac{\alpha+1}{2}$.

If furthermore $\approx$ is planar and if $d(\approx) = d(\frac{\alpha}{2}, \frac{\overline{\alpha}+1}{2})$, then the following holds: Two $\approx$-equivalent points are contained commonly in one of the sets $]\frac{\alpha}{2}, \frac{\overline{\alpha}}{2}[, [\frac{\overline{\alpha}}{2}, \frac{\alpha+1}{2}],]\frac{\alpha+1}{2}, \frac{\overline{\alpha}+1}{2}[$ or $[\frac{\overline{\alpha}+1}{2}, \frac{\alpha}{2}]$, if $\alpha < \overline{\alpha}$, and in one of the sets $]\frac{\overline{\alpha}}{2}, \frac{\alpha}{2}[, [\frac{\alpha}{2}, \frac{\overline{\alpha}+1}{2}],]\frac{\overline{\alpha}+1}{2}, \frac{\alpha+1}{2}[$ or $[\frac{\alpha+1}{2}, \frac{\overline{\alpha}}{2}]$, if $\overline{\alpha} < \alpha$. Otherwise, $d(\approx) = d(\frac{\alpha}{2}, \frac{\overline{\alpha}+1}{2})$ could not be valid.

This together with Theorem 2.53 implies $\approx \subseteq \approx^\alpha$. Further, as in the proof of Proposition 2.52, one shows that for each word $\mathbf{w} \in \{0,1\}^*$ the points $l_{\mathbf{w}}^{\alpha,1-e}(\alpha)$ and $l_{\mathbf{w}}^{\alpha,1-e}(\overline{\alpha})$ are identified by $\approx$. Then Proposition 2.42 yields $\approx^\alpha = \approx$. ∎

Let us finish our discussion of Julia equivalences from the symbolic viewpoint by a complete clarification of the question of which sequences can be the itinerary of a point in dependence on $\alpha \in \mathbb{T}$, and by the announced description of $\mathbb{T}/\approx^\alpha$ for $\alpha \in \mathbb{P}_*$.

Proposition 2.56. (Existence of itineraries)
Let $\alpha \in \mathbb{T}$. Then for each $\mathbf{w} \in \{0,1\}^*$ there exists a $\beta \in \mathbb{T}$ with $I^\alpha(\beta) = \mathbf{w} * \hat{\alpha}$. For a given $\mathbf{s} \in \{0,1\}^{\mathbb{N}}$, the following statements are valid:

- *(i) In case that $\alpha \notin \mathbb{P}_*$ and $CARD^\alpha(\hat{\alpha}) > 1$, there exists a $\beta \in \mathbb{T}$ with $I^\alpha(\beta) = \mathbf{s}$.*
- *(ii) In case that $\alpha \notin \mathbb{P}_*$ and $CARD^\alpha(\hat{\alpha}) = 1$, there exists a $\beta \in \mathbb{T}$ with $I^\alpha(\beta) = \mathbf{s}$ iff $\mathbf{s}$ does not end with $0\hat{\alpha}$ or $1\hat{\alpha}$.*
- *(iii) In case that $\alpha \in \mathbb{P}_*$ and $PER(\alpha) = PER(\overline{\mathbf{v}^\alpha(1-e^\alpha)})$, there exists a (unique) $\beta \in \mathbb{T}$ with $I^\alpha(\beta) = \mathbf{s}$ iff $\mathbf{s}$ does not end with $\overline{\mathbf{v}^\alpha e^\alpha}$.*
- *(iv) In case that $\alpha \in \mathbb{P}_*$ and $PER(\alpha) \neq PER(\overline{\mathbf{v}^\alpha(1-e^\alpha)})$, there exists a $\beta \in \mathbb{T}$ with $I^\alpha(\beta) = \mathbf{s}$ iff $\mathbf{s}$ does not end with $\overline{\mathbf{v}^\alpha 0}$ or $\overline{\mathbf{v}^\alpha 1}$.*
- *(v) In case that $\alpha \in \mathbb{P}_*$, there exists a $\beta \in \mathbb{T}$ with $\widetilde{I^\alpha}(\beta) = \mathbf{s}$ iff $\mathbf{s}$ does not end with $\overline{\mathbf{v}^\alpha e^\alpha}$.*

Proof: Fix a $\mathbf{w} \in \{0,1\}^*$. Then it holds $I^\alpha(\beta) = \mathbf{w} * \hat{\alpha}$ for $\beta = l_{\mathbf{w}}^\alpha(\frac{\alpha}{2}), \beta = l_{\mathbf{w}}^\alpha(\frac{\alpha+1}{2})$ if $\alpha \in \mathbb{T} \setminus \mathbb{P}_*$, and for $\beta = l_{\mathbf{w}}^\alpha(\ddot{\alpha})$ if $\alpha \in \mathbb{P}_*$. This shows the first part of the assertion.

We are interested now in the question of when for given $\alpha \in \mathbb{T}$ and $\mathbf{s} \in \{0,1\}^{\mathbb{N}}$ there exists a $\beta \in \mathbb{T}$ with the property $I^\alpha(\beta) = \mathbf{s}$ $(*)$. As can be seen at the end of Chapter 2.1.3, one has $\mathbb{T}_{\mathbf{s}}^\alpha \neq \emptyset$ for all $\alpha \in \mathbb{T}$ and $\mathbf{s} \in \{0,1\}^{\mathbb{N}}$. In particular, if no σ-iterate of $\mathbf{s}$ is substantially equal to $\hat{\alpha}$, then there exists a $\beta \in \mathbb{T}$ satisfying $(*)$.

In case that $\alpha \notin \mathbb{P}_*$ and $\mathbf{s} = \mathbf{w}0\hat{\alpha}$ ($\mathbf{s} = \mathbf{w}1\hat{\alpha}$) with $\mathbf{w} \in \{0,1\}^*$, one finds a $\beta \in \mathbb{T}$ with $(*)$ if $\mathbb{T}_{\hat{\alpha}}^\alpha$ contains a point γ different from α. Then $(*)$ is valid for $\beta = l_{\mathbf{w}0}^\alpha(\gamma)$ ($\beta = l_{\mathbf{w}1}^\alpha(\gamma)$). (i) and (ii) follow immediately from Theorem 2.19.

Now let $\alpha \in \mathbb{P}_*, \mathbf{v} = \mathbf{v}^\alpha$ and $e = e^\alpha$, and assume $\mathbf{s}$ not to be α-regular. If $\mathbf{s}$ does not end with $\overline{\mathbf{v}e}$ or $\overline{\mathbf{v}(1-e)}$, then by Proposition 2.6 no backward iterate of $\dot{\alpha}$ and $\ddot{\alpha}$ lies in $\mathbb{T}^\alpha_{\mathbf{s}}$. (If there were such backward iterate in $\widetilde{\mathbb{T}^\alpha_{\mathbf{s}}}$, it would be contained in $\widetilde{\mathbb{T}^\alpha_{\mathbf{s}}} \setminus \mathbb{T}^\alpha_{\mathbf{s}}$.) Therefore, each $\beta \in \mathbb{T}^\alpha_{\mathbf{s}}$ satisfies (*).

By Lemma 2.54 and $\mathbb{T}^\alpha_{\overline{\mathbf{v}e}} = \{\alpha\}$ we have $\mathbb{T}^\alpha_{\overline{\mathbf{v}(1-e)}} = \{\beta \in \mathbb{T} \mid I^\alpha(\beta) \cong \hat{\alpha}, \beta$ is periodic with $PER(\beta) = PER(\alpha)\}$. So (iii) and (iv) follow from Theorem 2.44 (see also the proof of that theorem) since $\overline{\alpha} \in \mathbb{T}^\alpha_{\overline{\mathbf{v}(1-e)}}$, and (v) is a consequence of $\widetilde{I^\alpha}(\alpha) = \overline{\mathbf{v}(1-e)}$. ∎

For $\alpha \in \mathbb{P}_*$ and a given $\mathbf{s} \in \{0,1\}^{\mathbb{N}}$ it holds $\mathbb{T}^\alpha_{\mathbf{s}} = \{\beta \in \mathbb{T} \,|\, \widehat{I^\alpha}(\beta) = \mathbf{s}$ or $\widehat{I^\alpha}(\beta) = \mathbf{s}\}$. Thus for different sequences $\mathbf{s}_1, \mathbf{s}_2 \in \{0,1\}^{\mathbb{N}}$, in $\mathbb{T}^\alpha_{\mathbf{s}_1} \cap \mathbb{T}^\alpha_{\mathbf{s}_2}$ there are only points whose orbit contains α. So one of the sequences must end with $\overline{\mathbf{v}^\alpha e^\alpha}$, and the other one with $\overline{\mathbf{v}^\alpha(1-e^\alpha)}$.

On the one hand, before Theorem 2.53 we showed that $\beta \in \mathbb{T}^\alpha_{\mathbf{s}}$ if β has specified itinerary $\mathbf{s}$, and on the other hand, by Proposition 2.56(v) exactly those 0-1-sequences form specified itineraries of a point which do not end with $\overline{\mathbf{v}^\alpha e^\alpha}$. From this one obtains the description of $\mathbb{T}/\approx^\alpha$ announced above.

Theorem 2.57. (Version 2 of Theorem 2.53)
$\mathbb{T}/\approx^\alpha = \{\mathbb{T}^\alpha_{\mathbf{s}} \mid \mathbf{s} \in \{0,1\}^{\mathbb{N}}$, $\mathbf{s}$ does not end with $\overline{\mathbf{v}^\alpha e^\alpha}\}$ *for* $\alpha \in \mathbb{P}_*$. ∎

2.3.2 The set of all Julia equivalences

We want to show now that there are no Julia equivalences other than the equivalence relations $\approx^\alpha, \alpha \in \mathbb{T}$ defined above. For this, we introduce the following concept:

Definition 2.58. (Main point)
$\alpha \in \mathbb{T}$ is said to be a main point *of a $(h, ')$-invariant equivalence relation $\approx \neq \approx^0$ on $\mathbb{T}$ if there exist $\approx$-equivalent points $\beta_1, \beta_2 \in \mathbb{T}$ satisfying $d(\beta_1, \beta_2) = d(\approx)$ and $h(\beta_1) = \alpha$.*

For $\approx = \approx^0$ we declare 0 to be the only main point. (This is convenient since formally 0 and 1 are the farthest identified points.)

Remark. We use the concept of a main point instead of THURSTON's concept of a minor of an invariant lamination, which is the image of a longest chord in an invariant lamination (see [167]). At the end of Chapter 3.1.4 we will say a little about the relation between the two concepts.

By Proposition 2.52 or Corollary 2.55 a point $\alpha \in \mathbb{T}$ forms a main point of $\approx^\alpha$, and for $\alpha \notin \mathbb{P}_*$ the equivalence class $[\alpha]_{\approx^\alpha}$ is no more than the set of all main points of $\approx^\alpha$. The latter is a consequence of the fact that $[\frac{\alpha}{2}]_{\approx^\alpha}$ is '-symmetric.

Let us consider a planar invariant equivalence relation $\approx$ on $(\mathbb{T}, h, ')$ with $0 < d(\approx) < \frac{1}{2}$, and let points $\beta_1, \beta_2 \in \mathbb{T}$ satisfying $\beta_1 < \beta_2$ and $d(\beta_1, \beta_2) = d(\approx)$ be given. Then there cannot exist $\approx$-equivalent points β_3, β_4 such that $\beta_3\beta_4$ crosses one of the chords $\beta_1\beta_2$ and $\beta_1'\beta_2'$. Namely, if both chords were crossed it would hold $\beta_1 \approx \beta_1'$, and if, for example, only $\beta_1\beta_2$ were crossed, all points $\beta_1, \beta_2, \beta_3, \beta_4$ would be $\approx$-equivalent. In both cases there would be two $\approx$-equivalent points of distance greater than $d(\beta_1, \beta_2)$.

By $d(\approx) \geq \frac{1}{3}$, there are no two points β_3, β_4 having distance equal to $d(\approx)$ besides the points β_1, β_2 and the points β_1', β_2'. Otherwise, the corresponding chord $\beta_3\beta_4$ would cross at least one of the chords $\beta_1\beta_2$ and $\beta_1'\beta_2'$. Moreover, $d(h(\beta_1)h(\beta_2)) \leq \frac{1}{3}$ by formula (2.4). Since $h(\beta_1)h(\beta_2)$ does not cross $\beta_1\beta_2$ and $\beta_1'\beta_2'$, the point 0 cannot lie between $h(\beta_1)$ and $h(\beta_2)$.

This shows the existence of at most two main points, and the further statements contained in the following Lemma are obvious.

Lemma 2.59. *If $\approx$ is a planar invariant equivalence relation on $(\mathbb{T}, h, ')$ with $0 < d(\approx) < \frac{1}{2}$, then there exist exactly two main points α and γ for $\approx$.*

If $\alpha < \gamma$, then $\gamma - \alpha \leq \frac{1}{3}$, and two $\approx$-equivalent points are commonly contained in one of the mutually disjoint intervals $]\frac{\alpha}{2}, \frac{\gamma}{2}[, [\frac{\gamma}{2}, \frac{\alpha+1}{2}],]\frac{\alpha+1}{2}, \frac{\gamma+1}{2}[$ and $[\frac{\gamma+1}{2}, \frac{\alpha}{2}]$. In particular, $\hat{\alpha} \cong I^\alpha(\gamma)$ and $\hat{\gamma} \cong I^\gamma(\alpha)$. ■

In connection with Lemma 2.59, we note that α and $\overline{\alpha}$ are the two main points of $\approx^\alpha$ for $\alpha \in \mathbb{P}_* \setminus \{0\}$, but that they do not form a complete equivalence class (compare Example 2.45) in general.

Lemma 2.60. *For two points $\alpha, \gamma \in \mathbb{T}$ satisfying $\alpha < \gamma, \hat{\alpha} \cong I^\alpha(\gamma)$ and $\hat{\gamma} \cong I^\gamma(\alpha)$ the following statements are valid:*

(i) $\frac{\alpha}{2} < \frac{\gamma}{2} \leq \alpha < \gamma \leq \frac{\alpha+1}{2} < \frac{\gamma+1}{2}$.

(ii) $d(\frac{\gamma}{2}, \frac{\alpha+1}{2}) = d(\frac{\alpha}{2}, \frac{\gamma+1}{2}) \geq \frac{1}{3}$ *and* $d(\alpha, \gamma) \leq \frac{1}{3}$.

(iii) The orbits of α and γ are disjoint to the intervals $]\frac{\alpha}{2}, \frac{\gamma}{2}[,]\frac{\alpha+1}{2}, \frac{\gamma+1}{2}[$ and $]\alpha, \gamma[$. Moreover, for all $n \in \mathbb{N}_0$ it holds either $h^n(\alpha), h^n(\gamma) \in [\frac{\gamma}{2}, \frac{\alpha+1}{2}]$ or $h^n(\alpha), h^n(\gamma) \in [\frac{\gamma+1}{2}, \frac{\alpha}{2}]$.

(iv) $\alpha, \gamma \notin \mathbb{P}_\infty$.

(v) If $\alpha \in \mathbb{P}_*$ or $\gamma \in \mathbb{P}_*$, then $\alpha, \gamma \in \mathbb{P}_*$ and $PER(\alpha) = PER(\gamma)$.

(vi) $\hat{\alpha} = \hat{\gamma}$.

Proof: (i) can be obtained by analyzing the formulae $\hat{\alpha} \cong I^\alpha(\gamma)$ and $\hat{\gamma} \cong I^\gamma(\alpha)$ for the first coordinate. It holds $\frac{\gamma}{2} \leq \alpha < \gamma \leq \frac{\alpha+1}{2} < \frac{\gamma+1}{2}$, hence $d(h(\frac{\gamma}{2}), h(\frac{\alpha+1}{2})) \leq d(\frac{\gamma}{2}, \frac{\alpha+1}{2})$. (ii) follows from formula (2.4) and the remark after it.

(iii): $\hat{\alpha} \cong I^\alpha(\gamma)$ and $\hat{\gamma} \cong I^\gamma(\alpha)$ imply that, for all $n \in \mathbb{N}_0$ the chord $h^n(\alpha\gamma)$ does not cross $\frac{\alpha}{2}\frac{\alpha+1}{2}$ or $\frac{\gamma}{2}\frac{\gamma+1}{2}$. Finally, the application of Lemma 2.10 to the chord $S = \frac{\gamma}{2}\frac{\alpha+1}{2}$ leads to (iii).

(iv): Choose a point $\delta \in \mathbb{T} \setminus \mathbb{P}$ between α and γ. Supposing $\alpha \in \mathbb{P}_\infty$ with $PER(\hat{\alpha}) = m$, by (iii) it would hold $I^\delta(\alpha) = I^\delta(h^m(\alpha)) = I^\delta(h^{2m}(\alpha)) = I^\delta(h^{3m}(\alpha)) = \ldots = \hat{\alpha}$. Then $\widetilde{\mathbb{T}^\delta_{\hat{\alpha}}}$ would be an infinite set in contradiction to Corollary 2.22.

(v): Again choose a $\delta \in \mathbb{T} \setminus \mathbb{P}$ between α and γ, and again obtain $I^\delta(\alpha) = I^\delta(\gamma)$ by (iii). We can assume that α is periodic. Then by Corollary 2.23(i) also γ is periodic, and it holds $PER(\alpha) = PER(\gamma)$.

(vi) By use of the statement (iii) one easily gets that $\hat{\alpha} \cong \hat{\gamma}$, and from this and (v) it follows $\hat{\alpha} = \hat{\gamma}$. ■

Let us have a closer look at the case $\alpha, \gamma \in \mathbb{P}_*$ on the assumptions of Lemma 2.60. For this, let $m = PER(\alpha) = PER(\gamma)$. Immediately from Lemma 2.60 (see (iii)) one obtains that $h^{m-1}(\alpha\gamma) = \frac{\alpha}{2}\frac{\gamma+1}{2}$ or $h^{m-1}(\alpha\gamma) = \frac{\alpha+1}{2}\frac{\gamma}{2}$. So we get the following statement:

Lemma 2.61. *Let $\alpha, \gamma \in \mathbb{T}$ be different periodic points with $\hat{\alpha} \cong I^\alpha(\gamma)$ and $\hat{\gamma} \cong I^\gamma(\alpha)$. Then neither $\dot{\alpha} = \frac{\alpha}{2}$ and $\dot{\gamma} = \frac{\gamma}{2}$ nor $\ddot{\alpha} = \frac{\alpha}{2}$ and $\ddot{\gamma} = \frac{\gamma}{2}$.* ■

Now we are able to describe how one can decide whether two periodic points are associated.

Proposition 2.62. (Characterization of the associated point)
For $\alpha \in \mathbb{P}_* \setminus \{0\}$ the following statements are valid:

(i) $\hat{\alpha} = \hat{\overline{\alpha}}$, and between α and $\overline{\alpha}$ there is no $\gamma \in \mathbb{T}$ with $\hat{\gamma} = \hat{\alpha}$.

(ii) $\gamma \in \mathbb{T}$ and $\overline{\alpha}$ coincide iff $\hat{\alpha} \cong I^\alpha(\gamma), \hat{\gamma} \cong I^\gamma(\alpha)$ and $\alpha \neq \gamma$. Furthermore, $\overline{\overline{\alpha}} = \alpha$ and $e^\alpha = e^{\overline{\alpha}}$.

Proof: α and $\overline{\alpha}$ are main points of $\approx^\alpha$, hence $\hat{\alpha} \cong I^\alpha(\overline{\alpha})$ and $\hat{\overline{\alpha}} \cong I^{\overline{\alpha}}(\alpha)$ by Lemma 2.59 and so $\hat{\alpha} = \hat{\overline{\alpha}}$ by Lemma 2.60 (see (v)).

Assuming that $\gamma \in \mathbb{T}$ lies between α and $\overline{\alpha}$ and that $\hat{\alpha} = \hat{\gamma}$, by use of Lemma 2.60 (see (iii)) one deduces $I^\gamma(\overline{\alpha}) = I^\gamma(\alpha) \cong \hat{\alpha} = \hat{\gamma}$, hence by Lemma 2.54 $\alpha, \overline{\alpha}, \gamma \in \mathbb{T}^{\gamma}_{\mathbf{v}^\gamma(1-e^\gamma)}$. Therefore, the points $\alpha, \overline{\alpha}$ and γ define the same orbit (see Corollary 2.23(i)), which contradicts Lemma 2.60 (see (iii)).

Now let γ be a point different from $\alpha, \overline{\alpha}$ and satisfying $\hat{\alpha} \cong I^\alpha(\gamma)$ and $\hat{\gamma} \cong I^\gamma(\alpha)$. Then by Lemma 2.60 (see (v)) and again by Lemma 2.54 $\alpha, \overline{\alpha}, \gamma \in \mathbb{T}^{\alpha}_{\mathbf{v}^\alpha(1-e^\alpha)}$. Finally, Corollary 2.23(i) implies that $\alpha, \overline{\alpha}$ and γ lie on a common orbit.

Consequently, by Lemma 2.60 (see (iii)) we have $\gamma < \alpha < \overline{\alpha}$ or $\overline{\alpha} < \alpha < \gamma$. If now $\frac{\alpha}{2} = \ddot{\alpha}$, then from Lemma 2.61 it follows $\frac{\gamma}{2} = \dot{\gamma}$ and $\frac{\overline{\alpha}}{2} = \dot{\overline{\alpha}}$, and on the other hand, if $\frac{\alpha+1}{2} = \ddot{\alpha}$, then it follows $\frac{\gamma+1}{2} = \dot{\gamma}$ and $\frac{\overline{\alpha}+1}{2} = \dot{\overline{\alpha}}$.

With $m = PER(\alpha)$, in both cases this means $* \neq I^\alpha(\gamma)(m) \neq I^\alpha(\overline{\alpha})(m) \neq *$, in contradiction to $\gamma, \overline{\alpha} \in \mathbb{T}^{\alpha}_{\mathbf{v}^\alpha(1-e^\alpha)}$.

The statement $\hat{\alpha} \cong I^\alpha(\gamma), \hat{\gamma} \cong I^\gamma(\alpha)$ with respect to $\alpha, \gamma \in \mathbb{T}$ is symmetric, such that $\overline{\overline{\alpha}} = \alpha$. By the determination of e^α from the mutual position of $\dot{\alpha}\dot{\overline{\alpha}}$ and $\ddot{\alpha}\ddot{\overline{\alpha}}$ in Proposition 2.41, one obtains $e^\alpha = e^{\overline{\alpha}}$. ■

As a step to the proof of Theorem 1.8 we give a strengthening of Proposition 2.52. At the same time, we show that - as in the periodic case - $\eth\mathcal{B}^\alpha$ can be generated by its longest chord, which is only substantially new if it fails to be a 'diameter'.

Proposition 2.63. (Strengthening of Proposition 2.52)
For $\alpha \in \mathbb{T} \setminus \mathbb{P}_*$, each non-trivial planar and completely invariant equivalence relation on $(\mathbb{T}, h, ')$ with main point α contains $\approx^\alpha$, and $\approx^\alpha$ is the only equivalence relation of that type with finite equivalence classes.

Moreover, if S is a longest chord of the invariant lamination $\mathfrak{B}^\alpha$, then $\mathfrak{B}^\alpha$ is equal to the closure of the set $\{l^\alpha_{\mathbf{w}}(S) | \mathbf{w} \in \{0,1\}^*\} \cup \{l^\alpha_{\mathbf{w}}(S') | \mathbf{w} \in \{0,1\}^*\}$.

Proof: Let $\approx$ be a planar and completely invariant equivalence relation on $(\mathbb{T}, h, ')$ with main point α, and let us verify the first part of the proposition. If $d(\approx) = \frac{1}{2}$, then $\frac{\alpha}{2} \approx \frac{\alpha+1}{2}$. This case is considered in Proposition 2.52. We show that nothing different is possible.

For this, assume that $d(\approx) < \frac{1}{2}$ and that γ is the second main point of $\approx$. By Lemma 2.59 $\hat{\alpha} \cong I^\alpha(\gamma)$ and $\hat{\gamma} \cong I^\gamma(\alpha)$, and by Lemma 2.60 (see (iv),(v)) it holds $\alpha, \gamma \notin \mathbb{P}$. This implies $\alpha \approx^\alpha \gamma$ (see Theorem 2.47).

It holds $\frac{\alpha}{2} \approx \frac{\gamma+1}{2}$ and $\frac{\gamma}{2} \approx \frac{\alpha+1}{2}$, and the set $\{\frac{\alpha}{2}, \frac{\gamma}{2}, \frac{\alpha+1}{2}, \frac{\gamma}{2}\}$ is contained in $[\frac{\alpha}{2}]_{\approx^\alpha}$. By the complete invariance of $\approx$ and by Lemma 2.59 one easily sees that $l^\alpha_{\mathbf{w}}(\frac{\alpha}{2}) \approx l^\alpha_{\mathbf{w}}(\frac{\gamma+1}{2})$ and $l^\alpha_{\mathbf{w}}(\frac{\gamma}{2}) \approx l^\alpha_{\mathbf{w}}(\frac{\alpha+1}{2})$ for all $\mathbf{w} \in \{0,1\}^*$. Moreover, in each case $l^\alpha_{\mathbf{w}}(\{\frac{\alpha}{2}, \frac{\gamma}{2}, \frac{\alpha+1}{2}, \frac{\gamma+1}{2}\})$ forms a subset of an equivalence class of $\approx^\alpha$.

Now one obtains $d(l^\alpha_{\mathbf{w}}(\frac{\alpha}{2}\frac{\gamma}{2})) \leq \frac{1}{2} - d(\approx) < \frac{1}{4}$ for all $\mathbf{w} \in \{0,1\}^*$. Otherwise, there would exist a shortest 0-1-word $w_1 w_2 \ldots w_n$ with $d(l^\alpha_{w_1 w_2 \ldots w_n}(\frac{\alpha}{2}\frac{\gamma}{2})) > \frac{1}{2} - d(\approx)$. Then formula (2.4) would yield the inequality $\frac{1}{2} - d(\approx) \geq d(l^\alpha_{w_2 w_3 \ldots w_n}(\frac{\alpha}{2}\frac{\gamma}{2})) = 1 - 2d(l^\alpha_{w_1 w_2 \ldots w_n}(\frac{\alpha}{2}\frac{\gamma}{2})) > \frac{1}{2} - d(l^\alpha_{w_1 w_2 \ldots w_n}(\frac{\alpha}{2}\frac{\gamma}{2}))$, which implies $d(\frac{\alpha}{2}\frac{\gamma+1}{2}) = d(\approx) \leq d(l^\alpha_{w_1 w_2 \ldots w_n}(\frac{\alpha}{2}\frac{\gamma}{2}))$. Thus $l^\alpha_{w_1 w_2 \ldots w_n}(\frac{\alpha}{2}\frac{\gamma}{2})$ would cross one of the chords $\frac{\alpha}{2}\frac{\gamma}{2}, \frac{\alpha+1}{2}\frac{\gamma+1}{2}, \frac{\alpha}{2}\frac{\gamma+1}{2}$ and $\frac{\gamma}{2}\frac{\alpha+1}{2}$. Then $l^\alpha_{w_1 w_2 \ldots w_n}(\frac{\alpha}{2}) \approx^\alpha \frac{\alpha}{2}$, which is impossible since $\approx^\alpha$ is non-degenerate.

Consider a sequence $(\mathbf{w}_n)_{n=1}^\infty$ such that the chords $S^\alpha_{\mathbf{w}_n} = l_{\mathbf{w}_n}(\frac{\alpha}{2}\frac{\alpha+1}{2})$ converge to a chord $B \in \mathfrak{B}^\alpha$. By Lemma 2.7 and the discussion above, the lengths of the chords $l_{\mathbf{w}_n}(\frac{\alpha}{2}\frac{\gamma}{2})$ converge to 0, thus $(l_{\mathbf{w}_n}(\frac{\alpha}{2}) l_{w_n}(\frac{\gamma+1}{2}))_{n\in\mathbb{N}}$ converges to B. So let us summarize: The endpoints of a chord in $\mathfrak{B}^\alpha$ are $\approx$-equivalent, and by Proposition 2.29 $\approx^\alpha$ must be contained in $\approx$. This contradicts our assumption that $d(\approx) < \frac{1}{2}$.

Now let S be a longest chord of $\mathfrak{B}^\alpha$. We can assume that there is a critical gap bounded by S. The images α, γ of the endpoints of S must be main points of $\approx^\alpha$ and must satisfy $\hat{\alpha} \cong I^\alpha(\gamma)$ and $\hat{\gamma} \cong I^\gamma(\alpha)$. One easily sees that $\{S, S'\} = \frac{\alpha}{2}\frac{\gamma+1}{2}, \frac{\alpha+1}{2}\frac{\gamma}{2}\}$. Now the arguments for the proof of the second part of the proposition are nearly the same as those above. (Replace $d(\approx)$ by $d(S)$.) ■

Proof of Theorem 1.8. We have to show that $\approx = \approx^\alpha$ for each main point α of a Julia equivalence $\approx$, and that for a given planar and completely invariant equivalence relation $\approx$ on $(\mathbb{T}, h, ')$, the following statements are equivalent:

(i) each $\approx$-equivalence class is finite,

(ii) there exists an $\alpha \in \mathbb{T}$ with $\approx\, = \,\approx^\alpha$,

(iii) $\approx$ is a Julia equivalence.

'(i) $\Longrightarrow$ (ii)': Let $\approx$ be a non-trivial, planar and completely invariant equivalence relation on $(\mathbb{T}, h, ')$ possessing only finite equivalence classes. Further, let α be a main point of $\approx$. If $\alpha \in \mathbb{T} \setminus \mathbb{P}_*$, then $\approx = \approx^\alpha$ by Proposition 2.63.

Otherwise, α is periodic and besides α there is a second main point γ. Further by Proposition 2.62, α and γ must be associated periodic points and according to Corollary 2.55 $\approx$ and $\approx^\alpha$ must be equal.

Each equivalence relation $\approx^\alpha; \alpha \in \mathbb{T}$ is a Julia equivalence; thus it remains to be shown the implication '(iii) $\Longrightarrow$ (i)'.

Suppose that (iii) is valid and E is an infinite $\approx$-equivalence class. In the case that $d(\approx) = \frac{1}{2}$ fix an $\alpha \in \mathbb{T}$ with $\frac{\alpha}{2} \approx \frac{\alpha+1}{2}$. There exists an $m \in \mathbb{N}_0$ such that for all $n \geq m$ the iterate $h^n(E)$ is completely contained in one of the intervals $[\frac{\alpha}{2}, \frac{\alpha+1}{2}], [\frac{\alpha+1}{2}, \frac{\alpha}{2}]$. Otherwise, it would hold $[\frac{\alpha}{2}]_\approx = h^{n_1}(E) = h^{n_2}(E)$ for some different $n_1, n_2 \in \mathbb{N}$ and $\approx$ would be degenerate.

In the case that $d(\approx) < \frac{1}{2}$, let α and γ with $\alpha < \gamma$ be the two main points of $\approx$. By Lemma 2.60, each of the equivalence classes $h^n(E); n \in \mathbb{N}_0$ lies in one of the intervals $]\frac{\alpha}{2}, \frac{\gamma}{2}[, [\frac{\gamma}{2}, \frac{\alpha+1}{2}],]\frac{\alpha+1}{2}, \frac{\gamma+1}{2}[$ and $[\frac{\gamma+1}{2}, \frac{\alpha}{2}]$, hence in one of the intervals $[\frac{\alpha}{2}, \frac{\alpha+1}{2}], [\frac{\alpha+1}{2}, \frac{\alpha}{2}]$.

According to Corollary 2.22 in both cases $\alpha \in \mathbb{P}$, and there exist a sequence $\mathbf{s} \in \{0,1\}$ and an $n \in \mathbb{N}_0$ with $\sigma^n(\mathbf{s}) \cong \hat{\alpha}$ and $E \subseteq \widetilde{\mathbb{T}_\mathbf{s}^\alpha}$. By Corollary 2.32, it is impossible that $\alpha \in \mathbb{P}_\infty$, hence we can assume that $\alpha \in \mathbb{P}_*$.

Besides the finitely many elements of $\mathbb{T}_\mathbf{s}^\alpha$, there are only backward iterates of α in $\widetilde{\mathbb{T}_\mathbf{s}^\alpha}$. This is a consequence of Proposition 2.6. The sequence $\mathbf{s}$ possesses only finitely many representations of the form $\mathbf{s} = \mathbf{u}t_1\mathbf{v}t_2\mathbf{v}t_3\mathbf{v}\ldots$ with $\mathbf{v} = \mathbf{v}^\alpha; t_1t_2t_3\ldots \in \{0,1\}^\mathbb{N}$ and α-regular $\mathbf{u}$. By Proposition 2.6 and the fact that E consists of infinitely many points, one can fix such representation and two different $i, j \in \mathbb{N}$ such that $\beta_1 = l^\alpha_{\mathbf{u}t_1\mathbf{v}t_2\mathbf{v}\ldots t_i\mathbf{v}}(\ddot{\alpha})$ and $\beta_2 = l^\alpha_{\mathbf{u}t_1\mathbf{v}t_2\mathbf{v}\ldots t_j\mathbf{v}}(\ddot{\alpha})$ lie in E. The orbit of $\beta_1\beta_2$ contains $\dot{\alpha}\ddot{\alpha}$, i.e. $\approx$ must be non-degenerate. This is contrary to our assumption. ■

Supplementary to Theorem 1.8, from Proposition 2.52 and Corollary 2.55 we get the following statement:

Corollary 2.64. (Strictly monotone diameter)
The diameter d is strictly monotone on the set of all Julia equivalences, i.e. if $\approx_1, \approx_2$ are Julia equivalences and if $\approx_1$ is properly contained in $\approx_2$, then $d(\approx_1) < d(\approx_2)$. ■

It would have been possible to define a Julia equivalence as a planar and completely invariant equivalence relation on $(\mathbb{T}, h, ')$ with only finite equivalence classes. That would have been more elegant, but in direct relationship to quadratic polynomials the way gone is more natural.

2.3.3 Periodic equivalence classes, simple closed curves, and the ramification order

The later discussion will show that much of the structure of quadratic iteration is related to periodic points and to simple closed curves in the (Abstract) Julia sets. Beforehand, we want to give some preparations.

Proposition 2.65. (Characterization of periodic equivalence classes) *For $\alpha \in \mathbb{T}$ let E be a non-trivial periodic $\approx^\alpha$-equivalence class of some period m. Then there exist unique associated periodic points $\gamma, \overline{\gamma} \in \mathbb{P}_*$ and an $n \in \{0, 1, \ldots, m-1\}$ with $E = [h^n(\gamma)]_{\approx^\gamma} = [h^n(\overline{\gamma})]_{\approx^{\overline{\gamma}}}$.*

γ and $\overline{\gamma}$ are the images of the two unique $\approx^\alpha$-equivalent points of maximal distance in $A := \bigcup_{j=0}^{m-1} h^j(E)$, and besides γ and $\overline{\gamma}$ there are no other $\approx^\alpha$-equivalent associated periodic points in A. Moreover, exactly one of the following statements is valid:

(i) γ and $\overline{\gamma}$ lie on different orbits: then A is the disjoint union of the orbits of γ and $\overline{\gamma}$ and $E = h^n(\{\gamma, \overline{\gamma}\})$.

(ii) γ and $\overline{\gamma}$ lie on one orbit: then A is equal to the orbit of γ and $E = \{h^{ik}(\beta) | i = 1, 2, 3, \ldots\}$ for each $\beta \in E$ and $k \in \mathbb{N}$ satisfying $h^k(\gamma) = \overline{\gamma}$.

Proof: Taking into consideration that $\mathbb{T}^\alpha_{\mathbf{s}} = \widetilde{\mathbb{T}^\alpha_{\mathbf{s}}}$ in the case $\alpha \notin \mathbb{P}_*$, from one of the Theorems 2.48, 2.51 and 2.57 one obtains $E = \mathbb{T}^\alpha_{\mathbf{s}}$ for some periodic 0-1-sequence $\mathbf{s}$. Further, by Corollary 2.23(i) all points in E are periodic of the same period, and if $\alpha \in \mathbb{P}_\infty$, then $\mathbf{s}$ does not end with $\hat{\alpha}$.

Fix two $\approx^\alpha$-equivalent points β_1, β_2 in $A = \bigcup_{j=0}^{m-1} h^j(E)$ of maximal possible distance. Since A is h-invariant, one has $d(\beta_1 \beta_2) \geq \frac{1}{3}$, and since $\approx^\alpha$ is $'$-invariant but A does not contain a preperiodic point, β_1, β_2 are the unique points of maximal distance in A.

The chords in the orbit of $h(\beta_1\beta_2)$ cross neither $\beta_1\beta_2$ nor $\beta_1'\beta_2'$, hence neither $\frac{\beta_1}{2}\frac{\beta_1+1}{2}$ nor $\frac{\beta_2}{2}\frac{\beta_2+1}{2}$. Otherwise, $[\beta_1]_{\approx^\alpha} (= [\beta_1']'_{\approx^\alpha})$ contained two points of distance greater than $d(\beta_1, \beta_2)$.

From the symbolic viewpoint, for $\gamma = h(\beta_1)$ and $\delta = h(\beta_2)$ this means $\hat{\gamma} \cong I^\gamma(\delta)$ and $\hat{\delta} \cong I^\delta(\gamma)$, and by virtue of Proposition 2.62 this implies $\delta = \overline{\gamma}$. Now fix some $n \in \{0, 1, 2, \ldots, m-1\}$ satisfying $E = [h^n(\gamma)]_{\approx^\alpha}$.

The chord $\gamma\overline{\gamma}$ bounds both the set $\operatorname{conv}([\gamma]_{\approx^\gamma}) = \mathbb{D}^\gamma_{\overline{\mathbf{v}^\gamma (1-e^\gamma)}}$ and the set $\operatorname{conv}(h^{m-n}(E)) = \mathbb{D}^\alpha_{\sigma^{m-n}(\mathbf{s})}$. Further, by Proposition 2.14, each boundary chord of these two sets lies on the orbit of $\gamma\overline{\gamma}$. Hence the equivalence classes $h^{m-n}(E)$ and $[\gamma]_{\approx^\gamma}$, and so also E and $[h^n(\gamma)]_{\approx^\gamma}$ coincide.

If A would contain $\approx^\alpha$-equivalent associated points $\delta, \overline{\delta}$ different from $\gamma, \overline{\gamma}$, one could show that at least one of its iterates would be equal to β_1 or β_2, hence would lie between $\dot{\delta}$ and $\ddot{\overline{\delta}}$, or between $\dot{\overline{\delta}}$ and $\ddot{\delta}$, which is impossible by Proposition 2.62 and Lemma 2.60.

By Corollary 2.23(i), A is equal to the orbit of γ or the union of the orbits of γ and $\overline{\gamma}$, and in the first case E must consist of two points. So we can finish the proof considering the case (ii). For this assume that $\gamma, \overline{\gamma}$ lie on a common orbit and in E, and that $\overline{\gamma} = h^k(\gamma)$. Then $\operatorname{conv}(E) = \mathbb{D}^{\gamma}_{\overline{\mathbf{v}^{\gamma}(1-e^{\gamma})}}$ is bounded by $\gamma\overline{\gamma}$ and again by Proposition 2.14 all boundary chords of $\operatorname{conv}(E)$ are given by $h^{ik}(\gamma\overline{\gamma})$ with $i = 0, 1, 2, \ldots$. This completes the proof. ■

Note that there exist $\approx^{\alpha}$-equivalence classes consisting of only one point, as for example $\{\frac{2}{3}\}$ for all $\alpha \notin [\frac{1}{3}, \frac{2}{3}]$.

It is not hard to see that to each infinite gap of $\mathcal{B}^{\alpha}$ for $\alpha \in \mathbb{T} \setminus \mathbb{P}_*$ and of $\eth\mathcal{B}^{\alpha}$ for $\alpha \in \mathbb{P}_*$ there corresponds a simple closed curve in the Abstract Julia set $\mathbb{T}/\approx^{\alpha}$. For non-periodic $\hat{\alpha}$ the invariant lamination $\mathcal{B}^{\alpha}$ does not possess infinite gaps (see Theorem 2.16 and Corollary 2.20(ii)). Analyzing Proposition 2.30, Proposition 2.31, and Proposition 2.43, one obtains the following statement:

Proposition 2.66. (Simple closed curves in Abstract Julia sets)
An Abstract Julia set $\mathbb{T}/\approx^{\alpha}$ contains a simple closed curve iff $\alpha \in \mathbb{P}$.

The $h/\approx^{\alpha}$-orbit of each such curve contains $\mathbb{P}^{\alpha}_{\infty}/\approx^{\alpha}$ if $\alpha \in \mathbb{P}_{\infty}$, and $[Gap^{\alpha}_{\mathbf{v}^{\alpha}*} \cap \mathbb{T}]_{\approx^{\alpha}}/\approx^{\alpha}$ if $\alpha \in \mathbb{P}_*$. ■

Clearly, each Abstract Julia set is a *continuum*, i.e. a connected compact metrizable topological space, and since it is locally connected, it forms an arcwise connected continuum. In particular, the above considerations say that $\mathbb{T}/\approx^{\alpha}$ for α with non-periodic kneading sequence is a *dendrite*, i.e. a locally connected continuum which does not contain a simple closed curve.

Let us finish by recalling a simple topological concept for a continuum K. The *ramification order* of a point $x \in K$ is defined to be the number of connectedness components of $K \setminus \{x\}$ (see [90], Vol. 2, p. 274). One easily sees the following: If $\approx$ is a closed equivalence relation on $\mathbb{T}$ with only finite equivalence classes, then the ramification order of each point $E \in \mathbb{T}/\approx$ is finite and equal to the cardinality of the equivalence class E.

We refer to Chapter 4.2.3, which contains statements on the ramification order in Abstract Julia sets and in the Abstract Mandelbrot set.

3. The Abstract Mandelbrot set

In this chapter we investigate the equivalence relation $\sim$ on $\mathbb{T}$, which was defined by $\alpha_1 \sim \alpha_2 \iff \approx^{\alpha_1} = \approx^{\alpha_1}$, and the structure of the corresponding factor space $\mathbb{T}/\sim$, the Abstract Mandelbrot set. Our discussion is connected with a closer look at the Julia equivalences $\approx^\alpha; \alpha \in \mathbb{T}$ from two points of view: When is a Julia equivalence contained in another one, and what are the subsets of $\mathbb{T}$ on which an iterate of h acts like h itself?

3.1 The Abstract Mandelbrot set - an atlas of Abstract Julia sets

3.1.1 Symbolic description of the Abstract Mandelbrot set

Looking for a 'good' topology on the set of all Julia equivalences, one could proceed as follows: Two Julia equivalences are considered to be near if each pair of points equivalent with respect to one of the Julia equivalences is near to a pair of points equivalent with respect to the other one.

This way turns out to be unsuitable in our situation because h is a chaotic map: near pairs of points become far under iteration.

In Chapter 2.3 it was pointed out that the structure of a Julia equivalence is strongly determined by its main points, and so it seems to be natural to measure the distance of two Julia equivalences by the distance of their main points. This leads to the topology induced by the map $\alpha \mapsto \approx^\alpha; \alpha \in \mathbb{T}$, which of course can be identified with the quotient topology on $\mathbb{T}/\sim$.

In the following we want to demonstrate that this topology is very useful for our purposes. We start by listing some statements which were actually proved in Section 2.

Since the main points of a Julia equivalence $\approx$ are $\approx$-equivalent and since for $\alpha \in \mathbb{T}\setminus\mathbb{P}_*$ the equivalence class $[\alpha]_{\approx^\alpha}$ consists of non-periodic main points (see Chapter 2.3.2), one has the following statements:

$$\text{For all } \alpha, \gamma \in \mathbb{T} \text{ it holds: } \approx^\alpha = \approx^\gamma \Longrightarrow \alpha \approx^\alpha \gamma. \tag{3.1}$$

For all $\alpha \in \mathbb{T} \setminus \mathbb{P}_*$ and $\gamma \in \mathbb{T}$ it holds:

$$\alpha \approx^\alpha \gamma \iff (\gamma \notin \mathbb{P}_* \text{ and}) \approx^\alpha = \approx^\gamma . \tag{3.2}$$

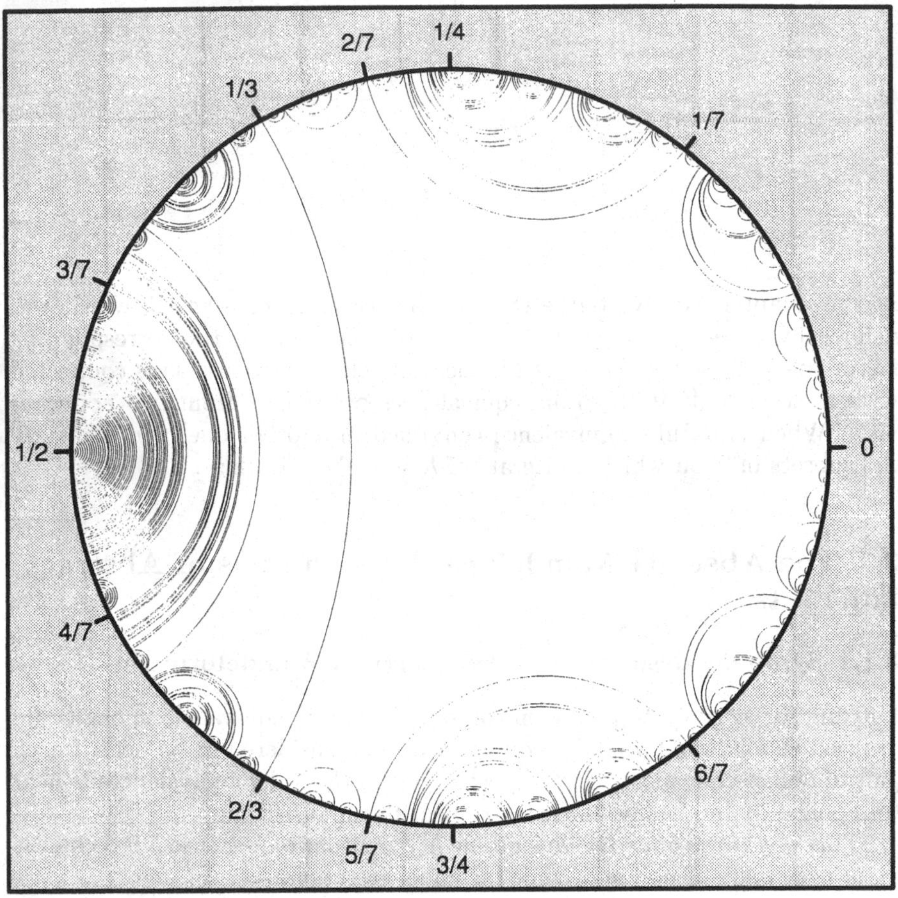

Fig. 3.1. The Abstract Mandelbrot set

Analyzing the symbolic descriptions of Julia equivalences in the Theorems 2.47, 2.50 and 2.53, one obtains the following:

$$\text{For all } \alpha, \gamma \in \mathbb{T} \text{ it holds: } \alpha \approx^{\alpha} \gamma \Longrightarrow \hat{\alpha} \cong I^{\alpha}(\gamma). \tag{3.3}$$

Moreover, in the case $\alpha \in \mathbb{T} \setminus \mathbb{P}$ the fact that $\hat{\alpha}$ is non-periodic and Theorem 2.47 lead to the following stronger statement:

$$\text{For all } \alpha \in \mathbb{T} \setminus \mathbb{P} \text{ and } \gamma \in \mathbb{T} \text{ it holds:}$$
$$\alpha \;\approx^{\alpha}\; \gamma \Longleftrightarrow (\hat{\alpha} \cong I^{\alpha}(\gamma) \Longleftrightarrow)\; \hat{\alpha} = I^{\alpha}(\gamma). \tag{3.4}$$

Now we provide some different descriptions of the equivalence relation $\sim$.

Theorem 3.1. (Symbolic description of $\sim$)
$\sim$ is a closed planar equivalence relation on $\mathbb{T}$, and for all $\alpha, \gamma \in \mathbb{T}$ the following statements are equivalent:

(i) $\alpha \sim \gamma$,

(ii) $\alpha \approx^{\alpha} \gamma$ *and* $\gamma \approx^{\gamma} \alpha$,

(iii) $\hat{\alpha} \cong I^{\alpha}(\gamma)$ *and* $\hat{\gamma} \cong I^{\gamma}(\alpha)$,

(iv) $\hat{\gamma} = \hat{\alpha} \cong I^{\alpha}(\gamma) = I^{\gamma}(\alpha)$,

(v) $I^{\beta}(\alpha) = I^{\beta}(\gamma)$ *for all* β *between* α *and* γ,

(vi) $\alpha \approx^{\beta} \gamma$ *for all* β *between* α *and* γ.

Proof: First of all, let us show that the statements (i), (ii) and (iii) are equivalent and that (iii) implies statement (iv).

The implications '(i) $\Longrightarrow$ (ii)' and '(ii) $\Longrightarrow$ (iii)' are immediate consequences of (3.1) and (3.3), respectively. Finally, if (iii) is valid for $\alpha \neq \gamma$, then by Lemma 2.60 (see (iv),(v) there) either $\alpha, \gamma \in \mathbb{T} \setminus \mathbb{P}$ or $\alpha, \gamma \in \mathbb{P}_*$. In the first case it follows $\hat{\alpha} = I^{\alpha}(\gamma), \hat{\gamma} = I^{\gamma}(\alpha)$, and by (3.2) and (3.4) $\alpha \sim \gamma$.

In the second case, by Proposition 2.62, α and γ are associated. According to the last statement in Theorem 1.8 it follows $\alpha \sim \gamma$, hence $\hat{\alpha} = \hat{\gamma}$ and $e^{\alpha} = e^{\gamma}$ by Proposition 2.62. Taking into consideration that by Proposition 2.41 $1 - I^{\gamma}(\alpha)(m) = e^{\gamma} = e^{\alpha} = 1 - I^{\alpha}(\gamma)(m)$, one gets $I^{\alpha}(\gamma) = I^{\gamma}(\alpha)$.

The validity of '(iii) $\Longrightarrow$ (v)' follows immediately from Lemma 2.60 (see (iii) there). In order to show that also '(v) $\Longrightarrow$ (iii)' is valid, fix two points $\alpha, \gamma \in \mathbb{T}$ with $\alpha < \gamma$. Supposing $I^{\alpha}(\gamma) \ncong \hat{\alpha}$, there exists an $n \in \mathbb{N}_0$ such that $h^n(\alpha)$ and $h^n(\gamma)$ are not both contained in $[\frac{\alpha}{2}, \frac{\alpha+1}{2}]$ and not both $[\frac{\alpha+1}{2}, \frac{\alpha}{2}]$. Then one easily sees that there exists a $\beta \in \,]\alpha, \gamma[$ with the property that $\frac{\beta}{2}\frac{\beta+1}{2}$ separates the points $h^n(\alpha)$ and $h^n(\gamma)$. Thus (v) is impossible. One argues analogously if $I^{\gamma}(\alpha) \ncong \hat{\gamma}$.

Let us prove the equivalence of (v) and (vi). First assume that (v), and so (iii), is valid, and consider a β between α and γ. By use of the symbolic description of $\approx^{\beta}$ in the Theorems 2.47, 2.50 and 2.53, one easily sees that $\alpha \approx^{\beta} \gamma$ could only be false under one of the following conditions:

1. $\beta \in \mathbb{P}_*$ and there exists an $n \in \mathbb{N}_0$ with $\{h^n(\alpha), h^n(\gamma)\} = \{\dot{\beta}, \ddot{\beta}\}$.
2. $\beta \in \mathbb{P}_{\infty}$ and there exists a $\mathbf{w} \in \{0,1\}^{\mathbb{N}}$ with $I^{\beta}(\alpha) = I^{\beta}(\gamma) = \mathbf{w}\hat{\beta}$.

If 1. were valid, then one would find a $\delta \in \mathbb{T}$ between α and γ, such that $h^n(\alpha)$ and $h^n(\gamma)$ are separated by the chord $\frac{\delta}{2}\frac{\delta+1}{2}$, which is contrary to (v).

If 2. were valid, then one would obtain a contradiction as follows: β is an accumulation point of the orbit of α (see Proposition 2.31(ii)), hence one iterate of α must lie between α and γ. This is impossible by (iii) and by Lemma 2.60 (see (iii) there).

Conversely, if (vi) is valid, then by Theorem 2.53 we have $I^\beta(\alpha) \cong I^\beta(\gamma)$ for all $\beta \in \mathbb{P}_*$ between α and γ. Since the periodic points are dense in $\mathbb{T}$, it follows $I^\beta(\alpha) \cong I^\beta(\gamma)$ for all β between α and γ.

Assume that there exist a β between α and γ, an $n \in \mathbb{N}$ and a $t \in \{0,1\}$ with $I^\beta(\alpha)(n) = *$ and $I^\beta(\gamma)(n) = t$, or $I^\beta(\alpha)(n) = t$ and $I^\beta(\gamma)(n) = *$. Then obviously one finds a δ between α and γ with $I^\delta(\alpha)(n) = 1 - t$ and $I^\delta(\gamma)(n) = t$. Therefore, (v) must be valid.

At the end let us verify that $\sim$ is a planar and closed equivalence relation. Let $\alpha < \beta < \gamma < \delta$, and let $\alpha \sim \gamma$ and $\beta \sim \delta$. This means $\gamma \approx^\gamma \alpha$ and $\beta \approx^\beta \delta$ (see (ii)) and $\beta \approx^\gamma \delta$ and $\alpha \approx^\beta \gamma$ (see (vi)). Since $\approx^\beta$ as well as $\approx^\gamma$ are planar, we obtain $\beta \approx^\beta \gamma$ and $\gamma \approx^\gamma \beta$, hence $\beta \sim \gamma$.

Obviously, both statements $\hat{\alpha} \cong I^\alpha(\gamma)$ and $\hat{\gamma} \cong I^\gamma(\alpha)$ are valid iff all iterates of the chord with endpoints α and γ do not cross the chords $\frac{\alpha}{2}\frac{\alpha+1}{2}$ and $\frac{\gamma}{2}\frac{\gamma+1}{2}$. The set of all pairs (α, γ) satisfying this property is closed in $\mathbb{T}\times\mathbb{T}$. Hence $\sim$ forms a closed equivalence relation. ∎

The geometric argument at the end of the proof yields a simple geometric description of $\sim$, which is worth to show off. This is that for all $\alpha, \gamma \in \mathbb{T}$ the following holds:

$$\alpha \sim \gamma \Longleftrightarrow \text{The iterates of } \alpha\gamma \text{ do not cross } \frac{\alpha}{2}\frac{\alpha+1}{2} \text{ and } \frac{\gamma}{2}\frac{\gamma+1}{2}. \tag{3.5}$$

Moreover, by Theorem 3.1 (see (iii)), by the characterization of the point associated with a periodic point given in Proposition 2.62(ii) and by Lemma 2.60 (see (iv) and (v)), for $\alpha, \gamma \in \mathbb{T}$ the following holds:

$$\text{If } \alpha \sim \gamma \text{ and } \hat{\alpha} \text{ is periodic, } \alpha \neq \gamma \text{ implies } \alpha, \gamma \in \mathbb{P}_* \text{ and } \gamma = \overline{\alpha}. \tag{3.6}$$

Whereas the high level of self-similarity in Julia equivalences and the corresponding Abstract Julia sets is caused by the action of h, the very regular structure of $\sim$ is a little surprising. (Compare Figure 3.1, which represents $\sim$ by a lamination: $\sim$-equivalent points are connected by a sequence of chords.)

Perhaps, a feeling for this phenomenon comes up by the following statement which giving an idea how some kind of self-similarity transfers from dynamics to $\sim$. The type of self-similarity explained by such transfer is discussed in detail in Chapter 3.4. The Chapters 3.2 and 3.3 provide another type of self-similarity.

Corollary 3.2. *Let $\alpha, \gamma \in \mathbb{T}$ be associated periodic points, and let $\beta_1, \beta_2 \in \mathbb{T}$ be two $\approx^\alpha$-equivalent points between α and γ. Then the following statements are equivalent:*

(i) $\beta_1 \sim \beta_2$,

(ii) none of the iterates of $\beta_1\beta_2$ separates $\alpha\gamma$ from β_1 or β_2.

Proof: We assume that $\alpha < \gamma$. Since α and γ are the main points of $\approx^\alpha$, it holds $d(\approx^\alpha) = d(\frac{\alpha}{2}\frac{\gamma+1}{2}) = d(\frac{\alpha+1}{2}\frac{\gamma}{2})$. Thus the two endpoints of an iterate of $\beta_1\beta_2$ lie commonly in $]\frac{\alpha}{2},\frac{\gamma}{2}[, [\frac{\gamma}{2},\frac{\alpha+1}{2}],]\frac{\alpha+1}{2}\frac{\gamma+1}{2}[$ or $[\frac{\gamma+1}{2},\frac{\alpha}{2}]$.

Taking into consideration that $h^{-1}([\alpha,\gamma]) = [\frac{\alpha}{2},\frac{\gamma}{2}] \cup [\frac{\alpha+1}{2}\frac{\gamma+1}{2}]$, simple geometric arguments show that for each $n \in \mathbb{N}_0$ the following is valid: $h^n(\beta_1\beta_2)$ crosses one of the chords $\frac{\beta_1}{2}\frac{\beta_1+1}{2}, \frac{\beta_2}{2}\frac{\beta_2+1}{2}$ iff $h^{n+1}(\beta_1\beta_2)$ separates at least one of the points β_1,β_2 from $\alpha\gamma$. According to (3.5) this shows the above statement. ■

3.1.2 Periodic points and Lavaurs' equivalence relation

In Chapter 2.1.1 it was noted that the map $\alpha \mapsto \hat{\alpha}$ is continuous at α iff α does not belong to $\mathbb{P}_*$.

Whereas $\hat{0}^\frown = \hat{0}^\frown = \overline{0}$, for $\alpha \in \mathbb{P}_* \setminus \{0\}$ with $\hat{\alpha} = \overline{\mathbf{w}*}; \mathbf{w} \in \{0,1\}^*$ there exists a symbol $s \in \{0,1\}$ satisfying $\hat{\alpha}^\frown = \overline{\mathbf{w}s}$ and $\hat{\alpha}^\frown = \overline{\mathbf{w}(1-s)}$.

To answer the question whether $s = 0$ or $s = 1$, fix two associated periodic points α,γ with $m = PER(\alpha) = PER(\gamma)$ and $\alpha < \gamma$. By Proposition 2.41, for sufficiently small $\epsilon > 0$ one obtains $I^\alpha(\alpha+\epsilon)(m) = e^\alpha$ and $h^{m-1}(\alpha+\epsilon) = \dot{\alpha} + 2^{m-1}\epsilon \in]\dot{\alpha} + \frac{\epsilon}{2}, \ddot{\alpha} + \frac{\epsilon}{2}[$. This implies $\hat{\alpha}^\frown(m) = e^\alpha$, and since $e^\alpha = e^\gamma$, the following statement is valid:

Lemma 3.3. (Determination of the characteristic symbol II)
Let $\alpha,\gamma \in \mathbb{P}_* \setminus \{0\}$ be associated points, let $\alpha < \gamma$ and $m = PER(\alpha)(= PER(\gamma))$. Then it holds $\hat{\alpha}^\frown(m) = \hat{\gamma}^\frown(m) = 1 - \hat{\alpha}^\frown(m) = 1 - \hat{\gamma}^\frown(m) = e^\alpha = e^\gamma$. ■

Now we are able to give the reason why the characteristic symbol of 0 was fixed to be 0 (compare Definition 2.34). Thinking the circle $\mathbb{T}$ as the unit interval with pasted endpoints 0 and 1, the point $\gamma = 1(= 0)$ can be considered to be associated with $\alpha = 0$. Then $PER(\alpha) = PER(\gamma) = 1$ and $\hat{\alpha}^\frown(0) = \hat{\gamma}^\frown(1) = 0$, and the characteristic symbol 0 for $\alpha = 0$ is compatible with Lemma 3.3.

We want to point out now that the structure of the Abstract Mandelbrot set is determined essentially by the discontinuity of the kneading sequence. For $n = 2,3,4,\ldots$ let $\mathbb{P}_*^n$ be the set of all points $\alpha \in \mathbb{P} \setminus \{0\}$, for which n is a multiple of $PER(\alpha)$. Taking into consideration that for such points $2^n\alpha \equiv \alpha \bmod 1$, hence $(2^n-1)\alpha \bmod 1 = 0$, one easily obtains $\mathbb{P}_*^n = \{\frac{p}{2^n-1} \mid p = 1,2,\ldots,2^n-2\}$ for all $n = 2,3,4,\ldots$.

Exactly at the points of $\mathbb{P}_*^n$ the n-th member of the kneading sequence changes. On one side of such points it is 1, and on the other side 0. This yields the following statement:

Lemma 3.4. *For all $\alpha,\gamma \in \mathbb{T}$ with $0 < \alpha < \gamma$ and all $n \in \mathbb{N} \setminus \{1\}$ it holds $\hat{\alpha}^\frown(n) = \hat{\gamma}^\frown(n)$ iff the number of elements of $]\alpha,\gamma[\cap \mathbb{P}_*^n$ is even.* ■

Finally, applying Lemma 3.4 to points $\alpha, \gamma \in \mathbb{P}_*$ with $0 < \alpha < \gamma$ and with $PER(\alpha) = PER(\gamma)$, by use of Lemma 3.3 one gets

Lemma 3.5. *For $\alpha, \gamma \in \mathbb{P}_*$ with $0 < \alpha < \gamma$ and $PER(\alpha) = PER(\gamma)$ the following statements are equivalent:*

(i) $\hat{\alpha} = \hat{\gamma}$,

(ii) for each $n \in \mathbb{N}$ with $1 < n < PER(\alpha)$, the set $]\alpha, \gamma[\cap \mathbb{P}_*^n$ consists of an even number of points,

(iii) for each $n > 1$ which is not a multiple of $PER(\alpha)$, the number of all periodic points of period n in the interval $]\alpha, \gamma[$ is even. ■

Let us turn to the equivalence relation $\sim$ now. For $\alpha, \gamma \in \mathbb{T}$ with $\alpha \sim \gamma$ it holds $d(\alpha\gamma) \leq \frac{1}{3}$ (and $d(\alpha\gamma) = \frac{1}{3}$ iff $\alpha\gamma = \frac{1}{3}\frac{2}{3}$), and by Lemma 2.60 (see (i), (ii)) the point 0 does not lie between α and γ.

Let us fix the kneading sequence of a periodic point different from 0, and consider the (finite) set $A = \{\alpha_1, \alpha_2, \alpha_3, \ldots\}$ of all points possessing this kneading sequence, where we require that $\alpha_1 < \alpha_2 < \alpha_3 < \ldots$. Since by Proposition 2.62(i) between two associated points of A there is no other element of A, Theorem 3.1 and Proposition 2.62(ii) force that $\alpha_1 \sim \alpha_2, \alpha_3 \sim \alpha_4, \alpha_5 \sim \alpha_6, \ldots$. This leads to the following statement:

Theorem 3.6. ($\sim$ on the periodic points)
The restriction $\equiv$ of $\sim$ to $\mathbb{P}_* \setminus \{0\}$ is characterized by the following two properties:

(i) Each equivalence class consists of two points.

(ii) If $\alpha \equiv \gamma; \alpha, \gamma \in \mathbb{P}_* \setminus \{0\}$ and $\alpha < \gamma$, then $\hat{\alpha} = \hat{\gamma}$ and $\hat{\beta} \neq \hat{\alpha}$ for all $\beta \in \mathbb{P}_* \setminus \{0\}$ with $\alpha < \beta < \gamma$. ■

We want to add another characterization of $\sim$ restricted to $\mathbb{P}_* \setminus \{0\}$. In order to describe the structure of the (concrete) Mandelbrot set, LAVAURS [93] has given an equivalence relation on $\mathbb{P}_* \setminus \{0\}$, which we want to denote by $\sim_L$ for the moment. It is defined as follows.

Call $\alpha \in \mathbb{P}_* \setminus \{0\}$ *combinatorially smaller* than $\beta \in \mathbb{P}_* \setminus \{0\}$ if either $PER(\alpha) < PER(\beta)$, or $PER(\alpha) = PER(\beta)$ and $\alpha < \beta$, and construct chords $S_1, S_2, S_3 \ldots$ with endpoints in $\alpha \in \mathbb{P}_* \setminus \{0\}$ in an inductive way:

First let $B_1 = \frac{1}{3}\frac{2}{3}$. Further, if $B_1, B_2, \ldots, B_n$ are already constructed, the one endpoint α of B_{n+1} is the combinatorially smallest number in $\mathbb{P}_* \setminus (\{0\} \cup B_1 \cup B_2 \cup \ldots \cup B_n)$ and the other one the combinatorially smallest number γ in $\mathbb{P}_* \setminus (\{0, \alpha\} \cup B_1 \cup B_2 \cup \ldots \cup B_n)$ satisfying the property that $B_{n+1} = \alpha\gamma$ does not cut a chord $B_1, B_2, \ldots, B_n$.

For $\alpha, \gamma \in \mathbb{P}_* \setminus \{0\}$ let $\alpha \sim_L \gamma$ if $\alpha\gamma$ coincides with one of the chords $B_1, B_2, B_3, \ldots$.

Theorem 3.7. *The restriction of* $\equiv$ *of* $\sim$ *to* $\mathbb{P}_* \setminus \{0\}$ *is equal to* $\sim_L$.

Proof: We know that $\frac{1}{3} \equiv \frac{2}{3}$, and we assume that $\equiv$ and $\sim_L$ are equal on the set A of all endpoints of $B_1, B_2, \ldots, B_n$ for a given $n \in \mathbb{N}$. Let α be the combinatorially smallest point in $\mathbb{P}_* \setminus (\{0\} \cup A)$ and γ the combinatorially smallest one in $\mathbb{P}_* \setminus (\{0, \alpha\} \cup A)$ such that $\alpha\gamma$ has no point in common with one of the chords $B_1, B_2, \ldots, B_n$. Since chords with associated endpoints do not cross each other (see Proposition 2.62(ii) and Theorem 3.1), $\overline{\alpha}$ is not smaller than γ (in the usual order), and one has $PER(\alpha) = PER(\gamma)$.

By assumption, the interval $]\alpha, \gamma[$ contains with each point of $\mathbb{P}_*^n; n < PER(\alpha)$ also its associated one. Therefore, the cardinality of the set $\mathbb{P}_*^n \cap]\alpha, \gamma[$ is even. From this fact and Lemma 3.5 it follows $\hat{\gamma} = \hat{\alpha}$, and this together with Theorem 3.6 implies $\gamma = \overline{\alpha}$.

One shows by induction that the endpoints of the chords $B_1, B_2, B_3, \ldots$ are mutually associated, which completes the proof. ■

As an immediate consequence of Theorem 3.7 and the definition of $\sim_L$ one gets the following statement known as Lavaurs' lemma (in the language of 'periodic' parameter rays).

Corollary 3.8. (Lavaurs' lemma)
Let $\alpha_1, \alpha_2 \in \mathbb{P}_*$ *of some period* m *be given. If* $\alpha_2\overline{\alpha_2}$ *lies behind* $\alpha_1\overline{\alpha_1}$, *then there exists some* $\gamma \in \mathbb{P}_*$ *of period less than* m, *such that* $\gamma\overline{\gamma}$ *separates* $\alpha_1\overline{\alpha_1}$ *and* $\alpha_2\overline{\alpha_2}$. ■

3.1.3 Kneading sequences, visibility and internal addresses

As suggested by the discussion in the previous subsection, the periodic points play an outstanding role in understanding the structure of the Abstract Mandelbrot set. Those different from 0 are identified pairwisely by $\sim$, and for the corresponding pairs $(\alpha, \overline{\alpha})$ we have $\hat{\alpha}^\frown = \hat{\overline{\alpha}}^\frown$ and $\hat{\alpha}^\frown = \hat{\overline{\alpha}}^\frown$.

For our purposes it is useful to comprise also the point 0. Recall that the point $\overline{0}$ associated with 0 was defined to be 1=0, and that e^0 - the characteristic symbol of 0 - is 0.

Definition 3.9. ('Periodic' parameter chords)

1. ***Often the substitution of*** $\alpha \in \mathbb{P}_*$ ***by*** $\overline{\alpha}$ ***in the notation for some object yields the same object. To indicate this, we also denote the object writing*** $\alpha\overline{\alpha}$ ***instead of*** α ***and*** $\overline{\alpha}$.

In particular, we generalize the period, *the* kneading sequence, *the initial subword of period length decreased by one and the* characteristic symbol *to* $B = \alpha\overline{\alpha}$ *as follows:* $PER(B) := PER(\alpha) = PER(\overline{\alpha}), \hat{B} := \hat{\alpha} = \hat{\overline{\alpha}}, \mathbf{v}^B := \mathbf{v}^\alpha = \mathbf{v}^{\overline{\alpha}}, e^B := e^\alpha = e^{\overline{\alpha}}$. *(Later we will write* $Gap(\alpha\overline{\alpha}), Gap\langle\alpha\overline{\alpha}\rangle, \mathcal{B}_*^{\alpha\overline{\alpha}}, \eth\mathcal{B}^{\alpha\overline{\alpha}}$ *instead of* $Gap(\alpha), Gap\langle\alpha\rangle, \mathcal{B}_*^\alpha, \eth\mathcal{B}^\alpha$.*) Moreover, let* $\dot{B} := \dot{\alpha}\dot{\overline{\alpha}}, \ddot{B} := \ddot{\alpha}\ddot{\overline{\alpha}}$.

(We emphasize that $PER(B)$ is the period of the endpoints of B and not the period in the sense of Definition 1.4, in other words, that $PER(B)$ can exceed some n with $h^n(B) = B$! Then α and $\overline{\alpha}$ lie on a common orbit.)

2. Each subset of the closed unit disk containing a point different from $0 \in \mathbb{T}$ *is defined to be behind the chord* 01, *and we use the following concepts and notations for* $B = \alpha\overline{\alpha}; \alpha \in \mathbb{P}_*$ *with* $\alpha < \overline{\alpha}$:

- $\hat{B}^+ := \hat{\alpha}^\frown = \hat{\overline{\alpha}}^\frown = \overline{\mathbf{v}^B e^B}$ *is called the* kneading sequence just behind *the chord* B.
- $\hat{B}^- := \hat{\alpha}^\smallfrown = \hat{\overline{\alpha}}^\smallfrown = \overline{\mathbf{v}^B(1 - e^B)}$ *is called the* kneading sequence just before B, *where* 01^- *is not defined.*

3. Let $\mathcal{B}_*$ *denote the set of all chords* $\alpha\overline{\alpha}$ *for* $\alpha \in \mathbb{P}_* \setminus \{0\}$. *A chord* $Q \in \mathcal{B}_*$ *behind another one* $B \in \mathcal{B}_* \cup \{01\}$ *is said to be* immediately visible *from* B *if* Q *is not separated from* B *by a chord in* $\mathcal{B}_*$, *and* visible *from* B *if* Q *is not separated from* B *by a chord* $\tilde{B} \in \mathcal{B}_*$ *with* $PER(\tilde{B}) < PER(B)$.

Remark. As emphasized at the end of Chapter 1, the chord system $\mathcal{B}_* \cup \{01\}$ and the system of hyperbolic components of the Mandelbrot set are combinatorially equivalent. Although this fact is not yet proved, the reader should have it in mind. The concepts 'visible' and 'internal address' being discussed now originate in the description of the Mandelbrot set by LAU and SCHLEICHER in [92, 148]. There the concepts were given in the language of hyperbolic components, and most of the statements in the present subsection are only modifications of results in [92, 148].

The chords in $\mathcal{B}_*$ are mutually disjoint. Moreover, by Lavaurs' lemma (Corollary 3.8) two chords $B_1, B_2 \in \mathcal{B}_* \cup \{01\}$ of the same period are separated by a chord in $\mathcal{B}_*$ of less period if one of them lies behind the other one. Therefore in the definition of visibility '$PER(\tilde{B}) < PER(B)$' could be replaced by '$PER(\tilde{B}) \leq PER(B)$', and Lemma 3.4 implies the following

Lemma 3.10. (Separation Lemma)
Let $\alpha, \alpha_1, \alpha_2$ *be non-periodic points, let* B, B_1, B_2 *be chords in* $\mathcal{B}_* \cup \{01\}$, *and let* $n \in \mathbb{N}$. *Further, recall that* $\mathbf{s}|n$ *denotes the initial subword of length* n *of a sequence* $\mathbf{s}$. *Then the following statements are valid:*

(i) α_1 *and* α_2 *are separated by a chord in* $\mathcal{B}_*$ *of period* $m \leq n \iff \widehat{\alpha_1}|n \neq \widehat{\alpha_2}|n$.

(ii) If α *is behind* B, *then it holds:* α *and* B *are separated by a chord in* $\mathcal{B}_*$ *of period* $m \leq n \iff \hat{\alpha}|n \neq \hat{B}^+|n$.

(iii) If α *is not behind* B, *then it holds:* α *and* B *are separated by a chord in* $\mathcal{B}_*$ *of period* $m \leq n \iff \hat{\alpha}|n \neq \hat{B}^-|n$.

(iv) If B_2 *is behind* B_1, *or if* $B_1 = 01$, *then* B_1, B_2 *are separated by a chord in* $\mathcal{B}_*$ *of period* $m \leq n \iff \widehat{B_1}^+|n \neq \widehat{B_2}^-|n$. ■

At this place we want to insert a simple lemma on periodic symbol sequences which will repeatedly be used.

Lemma 3.11. *For each symbol sequence* $\mathbf{s}$ *of period* n *the following holds (where periods are taken with respect to* σ*):*

(i) If $\mathbf{s}(i) = \mathbf{s}(i + mj)$ *for some* $m \in \mathbb{N}$*, and for all* $i = 1, 2, \ldots, m$ *and* $j \in \mathbb{N}_0$*, then* n *divides* m*.*

(ii) Each subword of $\mathbf{s}$ *being equal to* $\mathbf{s}|n-1$ *starts at the* $jn+1$*-th member of* $\mathbf{s}$ *for some* $j \in \mathbb{N}_0$*.*

Proof: (i): Consider $\mathbf{s}$ as a map on the (additive) cyclic group $\{1, 2, \ldots, m = 0\}$, and let k be the greatest common divisor of n and m. The subgroup generated by n has order $\frac{m}{k}$ and consists of the elements $0, k, 2k, \ldots, m-k$. By use of this fact one shows $\mathbf{s}(l) = \mathbf{s}(k+l) = \mathbf{s}(2k+l) = \ldots = \mathbf{s}(m-k+l)$ for all $l = 1, 2, \ldots, k$. Since $k \leq n$ and $n = PER(\mathbf{s})$, this implies $n = k$. Therefore n divides m.

Let e be the n-th symbol of $\mathbf{s}$. If (ii) were false, then $\sigma^k(\mathbf{s})$ would be equal to $(\mathbf{s}|(n-1))d$ for some symbol $d \neq e$ and some k with $0 < k < n$. So one would obtain a contradiction by counting the symbol e in subwords of length n. ■

As an immediately consequence of Lemma 3.10 and Lemma 3.11(ii) let us express visibility and immediate visibility in the frame of kneading sequences.

Corollary 3.12. (Visibility and kneading sequences)
Let $B, Q \in \mathcal{B}_* \cup \{01\}$*, and let* Q *be of period* m *and behind* B*. Then* Q *is visible from* B *iff* $\widehat{B}^+|m = \widehat{Q}^-|m$*.*

Further, Q *is immediately visible from* B *iff* $\widehat{B}^+ = \widehat{Q}^-$*, or equivalently, iff* Q *is visible from* B *and* $PER(B)$ *divides* m*.* ■

One shows by induction that for each non-periodic $\alpha \in \mathbb{T}$ and for each chord $B \in \mathcal{B}_* \cup \{01\}$ there can be constructed a unique finite or infinite sequence $B^{(0)}, B^{(1)}, B^{(2)}, \ldots, B^{(n)}(, \ldots)$ of chords in $\mathcal{B}_* \cup \{01\}$ satisfying the following properties:

1. $B^{(0)} = 01$.
2. For each $i = 1, 2, \ldots, n, (\ldots)$, the chord $B^{(i)}$ is that of minimal period which separates α from $B^{(i-1)}$, or which separates B from $B^{(i-1)}$ or is equal to B but different from $B^{(i-1)}$.
3. The sequence has maximal possible length.

Then $B^{(i)}$ is visible from $B^{(i-1)}$ for each $i = 1, 2, \ldots, n(, \ldots)$, and according to the note after Definition 3.9 the sequence of the $m_i = PER(B^{(i)})$ is properly increasing.

Definition 3.13. (Internal address)
For $\alpha \in \mathbb{T} \setminus \mathbb{P}_*$ *and* $B = \gamma\overline{\gamma} \in \mathcal{B}_* \cup \{01\}$ *the sequence* $m_i = PER(B^{(i)})$ *constructed above is said to be the* internal address *of* α *and* B, γ*. It is denoted by* $m_0 \to m_1 \to \ldots \to m_n(\to \ldots)$

By construction, each $B^{(i)}$ is visible from $B^{(i-1)}$. So by Corollary 3.12 it holds $\widehat{B^{(i-1)}}^{+}|m_i = \widehat{B^{(i)}}^{-}|m_i$ for all $i = 1, 2, \ldots, n, (\ldots)$. One has to distinguish three cases:

a) $\alpha \notin \mathbb{P}$: then the internal address is infinite and from Lemma 3.10 it follows $\hat{\alpha} = \lim_{i\to\infty} \widehat{B^{(i)}}$.

b) $\alpha \in \mathbb{P}_\infty$: then the internal address is finite and $\hat{\alpha}$ coincides with the kneading sequence just behind $B^{(n)}$.

c) If the construction starts from $B \in \mathcal{B}_* \cup \{01\}$, then of course $B = B^{(n)}$.

From this one sees that the internal addresses of $\alpha \in \mathbb{T} \setminus \mathbb{P}_*$ and $B \in \mathcal{B}_* \cup \{01\}$ can be derived from the kneading sequences $\hat{\alpha}, \hat{B} = s_1 s_2 s_3 \ldots$ as follows: If m_{i-1} is known, then m_i is the least number k such that the initial subwords of $s_1 s_2 \ldots s_{k-m_i}$ and $s_{m_i+1} s_{m_i+2} \ldots s_k$ are different. In other words, with Definition 2.18 we have the following

Proposition 3.14. (Determination of the internal address from the kneading sequence)

(i) For $\alpha \in \mathbb{T}\setminus\mathbb{P}$, the internal address of α is $1 \to \chi_{\hat{\alpha}}^{\hat{\alpha}}(1) \to (\chi_{\hat{\alpha}}^{\hat{\alpha}})^2(1) \to \ldots$.

(ii) For $\alpha \in \mathbb{P}_\infty$, the internal address of α is $1 \to \chi_{\hat{\alpha}}^{\hat{\alpha}}(1) \to (\chi_{\hat{\alpha}}^{\hat{\alpha}})^2(1) \to \ldots \to PER(\hat{\alpha})$.

(iii) For $B \in \mathcal{B}_*$, the internal address of B is $1 \to \chi_{\hat{B}}^{\hat{B}}(1) \to (\chi_{\hat{B}}^{\hat{B}})^2(1) \to \ldots \to PER(B)$ and the characteristic symbol only depends on $\mathbf{v}^B$: there exists some n satisfying $PER(B) = (\chi_{\hat{B}}^{\hat{B}})^n(1)$ and $e^B = 1 - \hat{B}(k)$ with $k = PER(B) - (\chi_{\hat{B}}^{\hat{B}})^{n-1}(1)$. ∎

Proposition 3.14 also shows that the kneading sequence of a point $\alpha \in \mathbb{T}$ can be got back from its internal address, with one restriction: in the case $\alpha \in \mathbb{P}$ we need to know whether $\alpha \in \mathbb{P}_*$ or $\alpha \in \mathbb{P}_\infty$.

Example 3.15. $\alpha = \frac{13}{31}$ and $\overline{\alpha} = \frac{18}{31}$ have kneading sequence $\overline{0100*}$. The first symbols of $\sigma(\overline{0100*}) = \overline{100*0}$ and $\overline{0100*}$ are different. Thus the internal address of α (and $\overline{\alpha}$) starts with $1 \to 2$. Further, $\sigma^2(\overline{0100*}) = \overline{00*01}$ and $\overline{0100*}$ firstly differ in the second coordinate, and $\sigma^4(\overline{0100*}) = \overline{*0100}$ and $\overline{0100*}$ have different first symbol. We get that α has finite internal address $1 \to 2 \to 4 \to 5$.

Internal addresses and kneading sequences are two different concepts describing one and the same thing. However, kneading sequences are more convenient from the dynamical point of view, but internal addresses are more compact and contain the geometric information on the position of chords in $\mathcal{B}_*$ (and so of hyperbolic components in the Mandelbrot set) and of points in $\mathbb{T}$ in a direct form.

The knowledge of all 'abstract' internal addresses or kneading sequences occurring for some chord in $\mathcal{B}_*$ or some point in $\mathbb{T}$ - which we call briefly *admissible* - would be a big step to a complete understanding of the combinatorial structure of the Mandelbrot set. Actually, there exist sequences in $\{0,1\}^{\mathbb{N}}$ not being the kneading sequence of a point in $\mathbb{T}$. Sequences starting with 1 are trivial examples, but let us give 'proper' ones:

Example 3.16. (Non-admissible kneading sequences and internal addresses) The determination of the internal address from the kneading sequence provided by Proposition 3.14 does not depend on the point or chord the kneading sequence is taken from. So from each 'abstract' kneading sequence one can deduce an 'abstract' internal address.

$\overline{010}$ has period 3, and the 'abstract' internal address determined by it is $1 \to \chi_{\mathbf{s}}^{\mathbf{s}}(1) \to (\chi_{\mathbf{s}}^{\mathbf{s}})^2(1) \to \ldots = 1 \to 2 \to 4 \to 5 \to 7 \to 8 \to \ldots$. The sequence $\overline{010}$ fails to be admissible. Namely, if a periodic sequence $\mathbf{s} \in \{0,1\}^{\mathbb{N}}$ is admissible, then by Proposition 3.14(ii) the period of $\mathbf{s}$ is the last member of the 'abstract' internal address obtained from $\mathbf{s}$.

However, the latter property is not necessary for a periodic $\mathbf{s} \in \{0,1\}^{\mathbb{N}}$ to form a kneading sequence, in other words, not each 'abstract' finite internal address is admissible. For example, there is no point whose internal address starts with $1 \to 2 \to 4 \to 5 \to 6$ or, equivalently, whose kneading sequence starts with 010011 or 01001$*$. One easily checks this by looking at all points of period not greater than 6.

Remark. The example $\overline{010}$ was described by C. Penrose in [123, 124] (see [124], Example 4.2). Penrose's argument that $\overline{010}$ is not admissible is the following: there cannot be an embedding of the dynamics generated which is orientation-preserving at periodic points. This gives an idea what the general problem is.

The problem of admissibility of internal addresses can be reduced to describing all (finite) internal addresses of chords in $\mathcal{B}_*$. To see this, first fix a point $\alpha \in \mathbb{P}_\infty$. Then, by Proposition 3.14 (ii) $m = PER(\hat{\alpha})$ coincides with the last member m_n of the internal address of α. The construction above Definition 3.13 yields the chord $B = B^{(n)} \in \mathcal{B}_* \cup \{01\}$ being not separated from α by some chord in $\mathcal{B}_*$.

So we have $\hat{\alpha} = \overline{\mathbf{v}^B e^B}$. Moreover, by Proposition 3.14 (iii), e^B can be deduced from the 0-1-word $\mathbf{v}^B$. So one sees that to know all finite internal addresses and periodic kneading sequences means no more than to know all internal addresses of chords in $\mathcal{B}_*$.

The relation between non-periodic admissible kneading sequences and infinite admissible internal addresses is bijective (see Proposition 3.14) and, by the definition of internal addresses, for each admissible infinite internal address $1 \to n_1 \to n_2 \to \ldots$ all internal addresses $1 \to n_1 \to n_2 \to \ldots \to n_j; j \in \mathbb{N}$ are admissible by a chord in $\mathcal{B}_*$

In the other direction, if some $1 \to n_1 \to n_2 \to \ldots \to n_j; j \in \mathbb{N}$ is the internal address of some chord $\alpha_j\overline{\alpha_j} \in \mathcal{B}_*$, then one easily checks that $1 \to n_1 \to n_2 \to \ldots$ is the internal address of each accumulation point of $(\alpha_j)_{j\in\mathbb{N}}$. So the problem to find all admissible internal addresses can be reduced to the following open problem:

Problem. How can the set of all internal addresses of chords in $\mathcal{B}_*$ or, equivalently, the set of all kneading sequences of periodic points in $\mathbb{T}$ be characterized?

The problem is naturally related to asking for all admissible extensions $1 \to n_1 \to n_2 \to \ldots \to n_j = m \to n$ of a given admissible internal address $1 \to n_1 \to n_2 \to \ldots \to n_j = m$. We will give some partial results in this direction in Corollaries 3.85, 3.79 and 3.51.

Remarks. 1. The idea behind internal addresses is not new. It appeared as the concept of the *cutting times* in unimodal real dynamics. Here the admissibility problem is completely solved, and different characterizations of admissible kneading sequences or of the admissible internal addresses, in particular for real quadratic polynomials, were found by COLLET and ECKMANN [27], MILNOR and THURSTON [118], HOFBAUER and G. KELLER [64, 65], and BRUIN [23]. (For a survey, see the latter reference.)

2. The problem of admissible kneading sequences was noted by C. PENROSE [123], and by BANDT and the author (see [6]). The formulation of the problem is on a rather elementary level, and at first glance the relation to complex quadratic iteration is surprising. Perhaps, a 'naive' investigation of the structure of kneading sequences would directly lead to a structure like the set $\mathcal{B}_*$.

3.1.4 Further descriptions of the Abstract Mandelbrot set

We want to point out now that $\sim$ is already determined by the identification on $\mathbb{P}_* \setminus \{0\}$. For this, let $\mathcal{B}$ be the closure of $\mathcal{B}_*$ in $Comp(\mathbb{D})$. The chords in $\mathcal{B}$ have $\sim$-equivalent endpoints, since $\sim$ is closed. Actually, $\mathcal{B}$ is THURSTON's quadratic minor lamination, and the 'non-symbolic' part of the following Theorem 3.17 is due to THURSTON [167] (see Theorem II.6.11.). For some remarks concerning the quadratic minor lamination, we refer to the end of this subsection.

Clearly, a gap G in $\mathcal{B}$ is the convex hull of its intersection with $\mathbb{T}$. Since two points in $G \cap \mathbb{T}$ or two boundary chords of G, or a point in $G \cap \mathbb{T}$ and a boundary chord of G cannot be separated by a chord in $\mathcal{B}_*$, from Lemma 3.10 the following can be deduced: There exists a 0-1-sequence s_G such that for all $\alpha \in G \cap \mathbb{T}$ the kneading sequence $\hat{\alpha}$ is substantially equal to s_G. Precisely, s_G is equal to $\hat{\alpha}^{\frown}$ or $\hat{\alpha}^{\smallfrown}$. (Compare Figure 3.2, where four gaps are labelled by the corresponding sequence.)

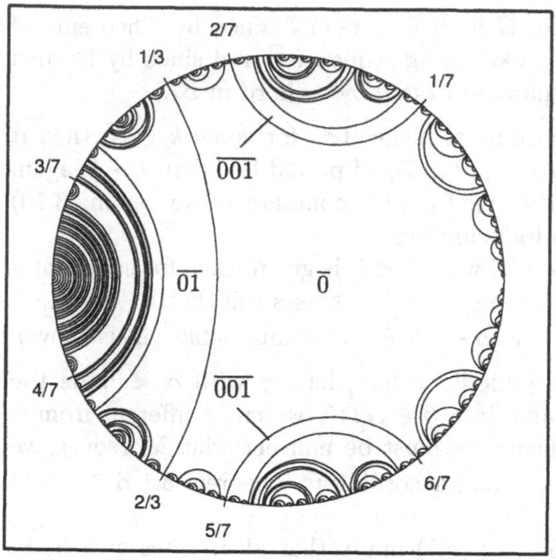

Fig. 3.2. Infinite gaps in $\mathcal{B}$

Theorem 3.17. (Symbolic dynamics and the gaps of $\mathcal{B}$)
For each gap G of $\mathcal{B}$ there exists a unique periodic or preperiodic sequence $\mathbf{s}_G = s_1s_2s_3\ldots \in \{0,1\}^{\mathbb{N}}$ satisfying $\hat{\alpha} \cong \mathbf{s}_G$ for all $\alpha \in G \cap \mathbb{T}$, and the following statements are valid:

(i) If $\mathbf{s}_G$ is preperiodic, then each $\alpha \in G \cap \mathbb{T}$ is preperiodic. Further, G is polygonal and coincides with $\mathbb{D}^{\alpha}_{\hat{\alpha}}$. Conversely, each $\mathbb{D}^{\alpha}_{\hat{\alpha}}$ for a preperiodic $\alpha \in \mathbb{T}$ with $CARD(\mathbb{T}^{\alpha}_{\hat{\alpha}}) > 2$ is a gap of $\mathcal{B}$.

(ii) If $\mathbf{s}_G$ is periodic, then G is an infinite gap with boundary chords given by some $B \in \mathcal{B}_*$ and all chords immediately visible from B. Moreover, $G \cap \mathbb{T}$ is a Cantor set containing a dense set of periodic points. Either
a) G is the 'big' gap with $B = 01$ and $G \cap \mathbb{T} = \{\beta \in \mathbb{T} | \beta \cong \overline{0}\}$, or
b) $\mathbf{s}_G = \overline{\mathbf{v}^B e^B} \neq \overline{0}$.
Conversely, each chord in $\mathcal{B}_*$ is the longest boundary chord of an infinite gap of $\mathcal{B}$.

Proof: Let G be a gap of $\mathcal{B}$. If $G \cap \mathbb{T}$ contains a point $\alpha \in \mathbb{P}_* \setminus \{0\}$, then by the periodicity of α and by $\mathbf{s}_G = \hat{\alpha}^{\frown}$ or $\mathbf{s}_G = \hat{\alpha}^{\frown}$ there exists an $s \in \{0,1\}$ satisfying $\mathbf{s}_G = \overline{\mathbf{v}^{\alpha} s}$. Thus by (3.6) each boundary chord of G must belong to $\mathcal{B}_*$.

If $0 \notin G \cap \mathbb{T}$, then $\mathbf{s}_G$ is equal to the kneading sequence just behind the longest boundary chord B of G, i.e. $\mathbf{s}_G = \overline{\mathbf{v}^B (1 - e^B)}$. Otherwise, $\mathbf{s}_G$ is no more than the kneading sequence just behind the chord 01, which is equal to the sequence $\overline{0}$.

$\mathbb{P}_*$ is dense in $\mathbb{T}$ and $G \cap \mathbb{T}$ arises from $\mathbb{T}$ by taking away infinitely many mutually disjoint open intervals with periodic endpoints. Therefore $G \cap \mathbb{T}$ forms a Cantor set and is the closure of a subset of $\mathbb{P}_*$.

There exists an infinite gap G with $0 \in G \cap \mathbb{T}$ since by Theorem 2.4 infinitely many points in $\mathbb{T}$ have kneading sequence $\overline{0}$ and since by Lemma 3.10 those points cannot be separated from 0 by a chord in $\mathcal{B}_*$.

Further, if $B \in \mathcal{B}_*$ has kneading sequence $\overline{0^k*}$ for some $k \in \mathbb{N}$, then it cannot be separated from 0 by a chord in $\mathcal{B}_*$ of period less than $k+1$, again by Lemma 3.10. Therefore B is visible from 01 (compare above Lemma 3.10) and bounds the gap with 0 on its boundary.

If $B = \alpha\overline{\alpha} \in \mathcal{B}_*$ of period m and with $\alpha < \overline{\alpha}$ is given, then for sufficiently small $\epsilon, \delta > 0$ the chord $h^{m-1}((\alpha+\epsilon)(\overline{\alpha}-\delta))$ crosses the chords $\frac{\alpha+\epsilon}{2}\frac{\overline{\alpha}-\delta+1}{2}$ and $\frac{\alpha+\epsilon+1}{2}\frac{\overline{\alpha}-\delta}{2}$. Therefore, by (3.5) $\alpha+\epsilon \sim \overline{\alpha}-\delta$ is impossible. (ii) is shown.

If $G \cap \mathbb{T}$ does not contain periodic points, let $\alpha\gamma$ with $\alpha < \gamma$ be the longest boundary chord of G and fix a $\beta \in G \cap \mathbb{T}$ which is different from α and γ. Then by (3.6) the sequence $\mathbf{s}_G$ must be non-periodic. Moreover, we have $\hat{\alpha} = \hat{\beta} = \hat{\gamma} = \mathbf{s}_G$ and $\alpha \sim \gamma$, and according to Theorem 3.1 $\hat{\alpha} \cong I^{\alpha}(\gamma)$ and $\hat{\gamma} \cong I^{\gamma}(\alpha)$.

The latter and Lemma 2.60 (see (iii)) imply that all iterates of α lie in $[\frac{\gamma}{2}, \frac{\alpha+1}{2}]$ or $[\frac{\gamma+1}{2}, \frac{\alpha}{2}]$, hence not between $\frac{\alpha}{2}$ and $\frac{\beta}{2}$ or between $\frac{\alpha+1}{2}$ and $\frac{\beta+1}{2}$. By $\hat{\beta} = \hat{\alpha}$ this means $\hat{\beta} = I^{\beta}(\alpha)$, and from (3.2) and (3.4) at the beginning of this subsection it follows $\beta \sim \alpha$.

In particular, $\beta \sim \alpha$ implies $\beta \approx^{\alpha} \alpha$ (see Theorem 3.1). This and again (3.2) imply that $G \cap \mathbb{T}$ is contained in an equivalence class E of $\sim$. In fact $G \cap \mathbb{T}$ and E coincide. Namely, if there existed a point $\delta \in E \setminus (G \cap \mathbb{T})$, then one would find points $\beta_1, \beta_2, \beta_3 \in G \cap \mathbb{T}$ such that $\delta \sim \beta_1$, $\beta_2\beta_3$ bounds G, and $\delta\beta_1$ crosses $\beta_2\beta_3$. Clearly, then $\delta\beta_1$ also crosses a chord in $\mathcal{B}_*$, which contradicts (3.6). By (3.2) and (3.4) one obtains $G \cap \mathbb{T} = \mathbb{T}^{\delta}_{\hat{\delta}} = \mathbb{T}^{\delta}_{\mathbf{s}_G}$ for all $\delta \in G \cap \mathbb{T}$.

Further, the sequence $\mathbf{s}_G = \hat{\delta}$ is non-periodic, and so it must be preperiodic by Corollary 2.20(i). Finally, by Corollary 2.23(ii) all points in $G \cap \mathbb{T}$ are preperiodic.

Conversely, $[\alpha]_{\sim} = [\alpha]_{\approx^{\alpha}} = \mathbb{T}^{\alpha}_{\hat{\alpha}}$ for all preperiodic points $\alpha \in \mathbb{T}$ (again see (3.2) and (3.4)), and the convex hull $\mathbb{D}^{\alpha}_{\hat{\alpha}}$ of $\mathbb{T}^{\alpha}_{\hat{\alpha}}$ is disjoint to all chords in $\mathcal{B}_*$. Otherwise, (3.6) would be false.

If $CARD(\mathbb{T}^{\alpha}_{\hat{\alpha}}) > 2$, then $\mathbb{D}^{\alpha}_{\hat{\alpha}}$ is contained in a gap G in $\mathcal{B}$. Applying the above argumentation to G, it follows $\mathbb{D}^{\alpha}_{\hat{\alpha}} = G$, which completes the proof of the theorem. ■

In dependence on the context and in accordance to Definition 3.9, we denote the gap in $\mathcal{B}_*(\cup\{01\})$ with (formally) longest boundary chord $B = \alpha\overline{\alpha}$ by $Gap(\alpha)$ and $Gap(B)$.

Theorem 2.44 describes when $S = \alpha\overline{\alpha} \in \mathcal{B}_*$ is an accumulation chord of $\eth\mathcal{B}^S := \eth\mathcal{B}^{\alpha}$. We want to give a similar statement relative to $\mathcal{B}$.

Corollary 3.18. *For $S = \alpha\overline{\alpha} \in \mathcal{B}_*, \mathbf{v} = \mathbf{v}^S$ and $e = e^S$ the following statements are equivalent:*

(i) α and $\overline{\alpha}$ lie on a common orbit.
(ii) $PER(\overline{\mathbf{v}(1-e)}) < PER(\alpha)$.
(iii) $S \notin \eth\mathcal{B}$.

Moreover, $PER(\overline{\mathbf{v}(1-e)})$ belongs to the internal address of S.

Proof: Assume that $\alpha < \overline{\alpha}$. If S is not isolated in $\eth\mathcal{B}$, then fix a sequence $(\alpha_i)_{i\in\mathbb{N}}$ of points less then α such that $(\alpha_i\overline{\alpha_i})_{i\in\mathbb{N}}$ converges to $\alpha\overline{\alpha}$. By Theorem 3.1 it holds $\alpha_i \approx^\alpha \overline{\alpha_i}$ for all $i \in \mathbb{N}$, hence $S \in \eth\mathcal{B}^S$.

On the other hand, assume that the chord S is isolated in $\eth\mathcal{B}$. Then it bounds a gap G_1 with longest boundary chord S and $\mathbf{s}_{G_1} = \overline{\mathbf{v}e}$, and a gap G_2 with a longest boundary chord Q different from S and with $s_{G_2} = \widehat{S}^- = \widehat{Q}^+ = \overline{\mathbf{v}(1-e)}$. The period of Q is equal to the period of $\overline{\mathbf{v}(1-e)}$, which by Lemma 3.11(i) must divide $PER(S) = PER(\alpha) = PER(\overline{\mathbf{v}e}) = |\mathbf{v}| + 1$. Since according to Theorem 3.7 $PER(S) = PER(Q)$ is impossible, it follows $PER(\overline{\mathbf{v}(1-e)}) < PER(\alpha)$. Now the equivalence of (i), (ii) and (iii) is an immediate consequence of Theorem 2.44.

To show that $PER(\overline{\mathbf{v}(1-e)})$ belongs to the internal address, we can assume that $PER(\overline{\mathbf{v}(1-e)}) < PER(\alpha)$, hence that (iii) is valid. Then the Q considered above must belong to the chord sequence defining the internal address of S. ■

Now let us relate the structure of the Abstract Mandelbrot set to the structure of the Abstract Julia sets.

Theorem 3.19. (Abstract Mandelbrot set as an atlas of Julia equivalences)

(i) $\sim$ is the smallest closed equivalence relation on $\mathbb{T}$ which identifies associated points, and the only closed planar equivalence relation on $\mathbb{T}$ which identifies two points with periodic kneading sequence iff they are periodic and associated.

(ii) A set $E \subseteq \mathbb{T}$ forms an equivalence class of $\sim$ iff it is equal to the set of main points for some Julia equivalence, i.e. iff $E = \{0\}, E = \{\alpha, \overline{\alpha}\}$ for some $\alpha \in \mathbb{P}_* \setminus \{0\}$ or $E = [\alpha]_{\approx^\alpha}$ for some $\alpha \in \mathbb{T} \setminus \mathbb{P}_*$.

Proof: (ii): By Theorem 3.1 (see (iv)), $[0]_\sim = \{0\}$ since 0 is unique with kneading sequence $\overline{*}$, and by (3.6), $[\alpha]_\sim = \{\alpha, \overline{\alpha}\}$ for $\alpha \in \mathbb{P}_* \setminus \{0\}$. Further, for $\alpha \notin \mathbb{P}$ one has $[\alpha]_\sim = \mathbb{T}^\alpha_{\hat{\alpha}} = [\alpha]_{\approx^\alpha}$ by (3.2) and (3.4). Finally, $[\alpha]_\sim$ for $\alpha \in \mathbb{P}_\infty$ is a single set, again by (3.6).

(i): $\sim$ identifies the endpoints of a chord in $\mathcal{B}$. Let $\alpha \sim \gamma$ with $\alpha \neq \gamma$ and $\alpha\gamma \notin \mathcal{B}$ be given. Then by (3.6) $\alpha\gamma$ is disjoint to all chords in $\mathcal{B}_*$ and completely contained in a polygonal gap (see Theorem 3.17). Thus α, γ are connected by a finite sequence of chords in $\mathcal{B}$ and so equivalent with respect to the smallest equivalence relation which identifies associated points.

Let $\sim'$ be a planar equivalence relation containing $\sim$, and let $\alpha, \gamma \in \mathbb{T}$ with $\alpha \sim' \gamma$ and $\alpha \not\sim \gamma$ be given. Then the chord $\alpha\gamma$ must cross an element of $\mathcal{B}_*$ or it must cut the interior of a gap of $\mathcal{B}$.

By planarity of $\sim'$, in both cases two different points having periodic kneading sequence but not being periodic and associated would be identified by $\sim'$ (see Theorem 3.17). ■

The statement (ii) in Theorem 3.19 is illustrated by Figure 3.3: around the lamination representing the Abstract Mandelbrot set some laminations corresponding to special Julia equivalences are arranged. The convex hulls of the sets of main points are marked both for the Julia equivalences and the Abstract Mandelbrot set.

We want to continue with a statement which underlines the strong relationship between the structure of Julia equivalences and of the Abstract Mandelbrot set.

Proposition 3.20. ('Generation' of $\mathfrak{d}\mathcal{B}^\alpha; \alpha \in \mathbb{T} \setminus \mathbb{P}$ from periodic chords) *If $\alpha \in \mathbb{T}$ lies behind some $S \in \mathcal{B}_*$, then $S \in \mathfrak{d}\mathcal{B}^\alpha$.*

For $\alpha \notin \mathbb{P}$, let $(S_i)_{i \in \mathbb{N}}$ be a sequence of chords in $\mathcal{B}_*$ with the property that α is behind each S_i and no chord in $\mathcal{B}_*$ separates α from each $S_i; i \in \mathbb{N}$. Then $\mathfrak{d}\mathcal{B}^\alpha$ is equal to the closure of the set $\{l^\alpha_{\mathbf{w}}(S_i) | \mathbf{w} \in \{0,1\}^*, i \in \mathbb{N}\}$.

Moreover, $\approx^\alpha$ for $\alpha \notin \mathbb{P}$ is the only Julia equivalence which identifies $\gamma \in \mathbb{P}_*$ with $\overline{\gamma}$ iff α lies behind $\gamma\overline{\gamma}$.

Proof: From Theorem 3.1 it follows $\beta \approx^\alpha \overline{\beta}$, when $\beta, \overline{\beta}$ denote the endpoints of S, hence from Proposition 2.65 one obtains $S \in \mathfrak{d}\mathcal{B}^\alpha$. So let us show the second part.

We can assume that S_{i+1} lies behind S_i for each $i \in \mathbb{N}$. Then $(S_i)_{i \in \mathbb{N}}$ converges to a (possibly degenerate) chord $\gamma\delta$ with $\gamma, \delta \sim \alpha$, hence by Theorem 3.19(ii) $\gamma\delta \in \mathfrak{d}\mathcal{B}^\alpha$, and $\frac{\gamma}{2}\frac{\delta+1}{2}$ is a longest chord of $\mathfrak{d}\mathcal{B}^\alpha$. Now $\mathfrak{d}\mathcal{B}^\alpha = \mathrm{cl}\{l^\alpha_{\mathbf{w}}(S_i) | \mathbf{w} \in \{0,1\}^*, i \in \mathbb{N}\}$ follows from Proposition 2.63.

Finally, fix some $\beta \in \mathbb{T}$ with $\beta \not\sim \alpha$. If β lies behind $\gamma\delta$, then by Theorem 3.17 there exists some chord $S \in \mathcal{B}_*$ with $\approx^\beta$-equivalent endpoints, which separates β from α, δ and γ. Otherwise $\frac{\beta}{2}\frac{\beta+1}{2}$ crosses infinitely many of the $\dot{S}_i$ defined by the above S_i. So according to the description of Julia equivalences in one of the Theorems 2.47, 2.50 and 2.53 the two endpoints of infinitely many S_i cannot be $\approx^\beta$-equivalent. This completes the proof. ■

From Proposition 2.30 we know that for $\alpha \in \mathbb{P}_\infty$ there are chords in $\mathcal{B}^\alpha$, which bound two infinite gaps and have irrational endpoints. Therefore, Propositions 3.20 and 2.42 provide the following

Corollary 3.21. ($\approx^\alpha; \alpha \in \mathbb{T} \setminus \mathbb{P}_*$ is rationally determined)
For $\alpha \in \mathbb{T}$ the following statements are equivalent:

(i) *For all points $\beta, \gamma \in \mathbb{T}$ with $\beta \approx^\alpha \gamma$ there exist sequences $(\beta_i)_{i \in \mathbb{N}}$ and $(\gamma_i)_{i \in \mathbb{N}}$ of rational points in $\mathbb{T}$ converging to β and γ, respectively, and satisfying $\beta_i \approx^\alpha \gamma_i$ for all $i \in \mathbb{N}$.*

(ii) $\alpha \notin \mathbb{P}_\infty$. ∎

$\mathcal{B}_*$ can be considered as the restriction of $\sim$ to $\mathbb{P}_* \setminus \{0\}$, but how can the chords in $\mathcal{B} = \mathrm{cl}\mathcal{B}_*$ be characterized? By the above results, $\mathcal{B}$ is equal to the set of all chords with $\sim$-equivalent endpoints which (of course) do not cut the interior of a polygonal gap of $\mathcal{B}$. Such chords belong to $\mathcal{B}^\alpha$ for some $\alpha \in \mathbb{T}$, so their orbits consist of mutually non-crossing chords.

Indeed, by Theorem 3.17 each polygonal gaps G of $\mathcal{B}$ coincides with $\mathbb{D}^\alpha_{\breve{\alpha}}$ for some preperiodic $\alpha \in \mathbb{T}$, and by Proposition 2.14 a non-degenerate chord with endpoints in such $\mathbb{D}^\alpha_{\breve{\alpha}}$ bounds $\mathbb{D}^\alpha_{\breve{\alpha}}$ iff its iterates do not cut each other.

Further, if a non-degenerate $B = \alpha\gamma \in \mathcal{B} \setminus \mathcal{B}_*$ does not bound a polygonal gap of $\mathcal{B}$, then by Theorem 3.19 $\{\alpha, \gamma\}$ is an equivalence class of $\approx^\alpha$, hence B belongs to $\mathcal{B}^\alpha$. So let us summarize:

$$S = \alpha\gamma \in \mathcal{B} \iff \text{The iterates of } S \text{ do not cross each other and cross neither } \frac{\alpha}{2}\frac{\alpha+1}{2} \text{ nor } \frac{\gamma}{2}\frac{\gamma+1}{2} \tag{3.7}$$

Remarks. 1. In the remark after Definition 2.58 we pointed to THURSTON's central concept of a minor. Minors are images of longest chords in invariant laminations, and since invariant laminations are more general than Julia equivalences, one could think that the concept of a minor is more general than that of a main point. This is false.

Actually, the set of all minors of invariant laminations, by THURSTON called *quadratic minor lamination (QML)*, coincides with $\mathcal{B}$. Indeed, if S and S' are the two longest chords of a non-trivial invariant laminations, then $d(S) \geq \frac{1}{3}$. The point 0 lies between the endpoints of one of these chords since otherwise $h(S)$ would cross one of the chords S, S'. With $\alpha\gamma = h(S)$ it holds $\{\frac{\alpha}{2}\frac{\gamma+1}{2}, \frac{\alpha+1}{2}\frac{\gamma}{2}\} = \{S, S'\}$, and the iterates of the minor $\alpha\gamma$ cross neither S nor S', hence neither $\frac{\alpha}{2}\frac{\alpha+1}{2}$ nor $\frac{\gamma}{2}\frac{\gamma+1}{2}$. By (3.7) it holds $\alpha\gamma \in \mathcal{B}$.

Conversely, let $B = \alpha\gamma$ be an element of $\mathcal{B}$. If $\alpha = \gamma$, then the chord $\frac{\alpha}{2}\frac{\alpha+1}{2}$ lies in $\mathcal{B}^\alpha$, and in the case that α and γ are the only main points of $\approx^\alpha$, the chord B forms a minor of $\mathfrak{d}\mathcal{B}^\alpha$. Otherwise, the union of $\mathfrak{d}\mathcal{B}^\alpha$ with the set of all iterates of $\alpha\gamma$ and all chords $l^\alpha_{\mathbf{w}}(\alpha\gamma)$; $\mathbf{w} \in \{0,1\}^*$ forms an invariant lamination with minor $\alpha\gamma$.

2. A minor S of an invariant lamination has the property that its iterates do not cross each other and are not shorter than S, but this property is not sufficient for some S being a minor. For example, the chord $S = \frac{1}{6}\frac{1}{3}$ satisfies

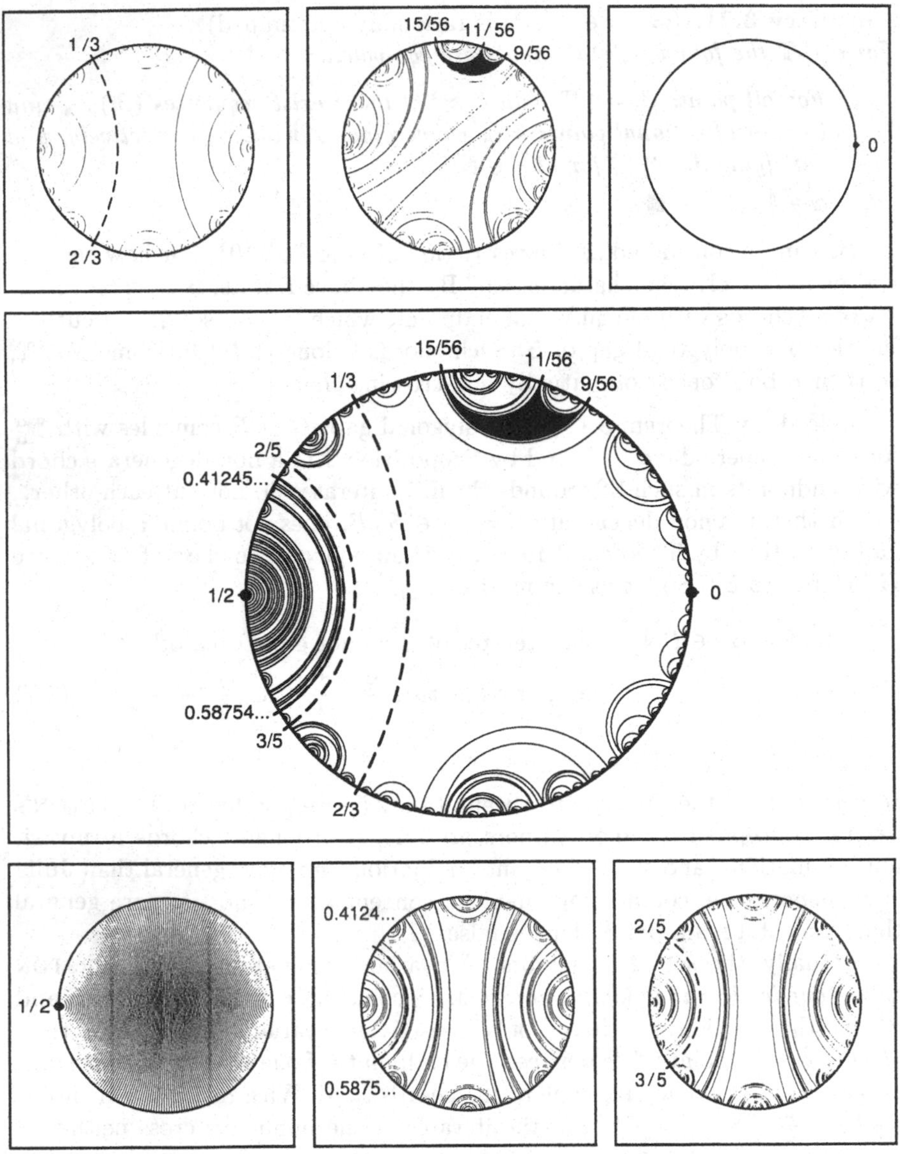

Fig. 3.3. Abstract Mandelbrot set as atlas of Abstract Julia sets

the property, but it fails to be a minor. If S were the minor of an invariant lamination, then one preimage of S would have the endpoint $\frac{1}{6}$ and would be a longest chord of the lamination, but $\frac{1}{6}\frac{7}{12}$ crosses the chord $h(S)$, and $\frac{1}{12}\frac{1}{6}$ is shorter than S. We have $QML = B$, and so minors are characterized by formula (3.7).

3. Julia equivalences can be considered as special invariant laminations. We saw that different invariant laminations can correspond to the same Julia equivalence, and that there exist invariant laminations which do not define a Julia equivalence, but contain a 'sublamination' providing one.

However, each (non-trivial) invariant lamination has a minor, and from that minor there can be constructed an invariant 'sublaminations' being minimal in some sense and providing a Julia equivalence (see [167], Proposition II.6.7, and the first remark above). This will not play any role subsequently, but we want to emphasize again that the concept of a Julia equivalence is useful and highly compact, but that THURSTON's invariant laminations do the most of the work.

4. THURSTON gave a classification of gaps in QML (see [167], II.6). In so far the most parts of Theorem 3.17 are not new. However, the symbolic approach gives some more insight into the structure of QML.

3.2 The ordered Abstract Mandelbrot set

3.2.1 The subtle structure of $\approx^\alpha$ for periodic α

We know exactly when two Julia equivalences $\approx^\alpha$ and $\approx^\gamma$ coincide. Now we are interested in the question when $\approx^\alpha$ is properly contained in $\approx^\gamma$. Let

$$\alpha \prec \gamma :\Longleftrightarrow \approx^\alpha \subset \approx^\gamma$$

for $\alpha, \gamma \in \mathbb{T}$. The relation $\prec$ completed by all pairs $(\alpha, \alpha); \alpha \in \mathbb{T}$ forms a partial order on $\mathbb{T}$. By Corollary 2.64, $\alpha \prec \gamma$ implies $\alpha \in \mathbb{P}_*$, and this is the reason to have a closer look at the structure of the invariant laminations $\eth\mathcal{B}^\alpha$ for $\alpha \in \mathbb{P}_*$.

For such α, let $Gap\langle\alpha\rangle := Gap^\alpha_{\mathbf{v}^\alpha *}$ be the critical value gap of $\eth\mathcal{B}^\alpha$, and set $\mathbb{T}\langle\alpha\rangle := Gap\langle\alpha\rangle \cap \mathbb{T}$. The circle $\mathbb{T} = \mathbb{T}\langle 0\rangle$ is the continuous image of the Cantor discontinuum $\{0,1\}^{\mathbb{N}}$ under the map q^0

$$q^0(\mathbf{b}) = \sum_{i=1}^{\infty} b_i 2^{-i}; \mathbf{b} = b_1 b_2 b_3 \ldots \in \{0,1\}^{\mathbb{N}},$$

the 'inversion' of binary expansion. We want to define an analogous map for each $\alpha \in \mathbb{P}_* \setminus \{0\}$. It will be instructive to keep in mind that each $\approx^\alpha$ is obtained from $\approx^0$ by adding some identification, which of course is trivial.

For $\alpha \in \mathbb{P}_*$, let $\mathbf{v} = \mathbf{v}^\alpha, e = e^\alpha$ and $m = PER(\alpha)$. In case that $\alpha = 0$ we have $\mathbf{v} = \lambda, e = 0$ and $m = 1$, and $\eth\mathcal{B}^\alpha$ only consists of degenerate chords, such that $Gap\langle\alpha\rangle = \mathbb{D}$ is the only gap. It 'arises' from the invariant lamination $\mathcal{B}^\alpha$ by taking away $S^\alpha_{\mathbf{v}*}$ and all chords $S^{\alpha,t}_{\mathbf{v}w_1\mathbf{v}w_2\ldots\mathbf{v}w_n\mathbf{v}*}$ with $w_1 w_2 \ldots w_n \in \{0,1\}^* \setminus \{\lambda\}, t \in \{0,1\}$.

Formally, we assigned to $\alpha = 0$ the associated point $\overline{\alpha} = 1$ (and to 1 the associated point $\overline{1} = 0$). For $\alpha = 0$ and all $\mathbf{b} = b_1 b_2 b_3 \ldots \{0,1\}^{\mathbb{N}}$ it holds

$$
\begin{aligned}
q^{\alpha}(\mathbf{b}) &= \alpha + (l^{\alpha}_{\mathbf{v}^{\alpha}}(\ddot{\alpha}) - \alpha) \sum_{i=1}^{\infty} b_i 2^{-m(i-1)} \\
&= \alpha + (2^m - 1)(\overline{\alpha} - \alpha) \sum_{i=1}^{\infty} b_i 2^{-mi}, \qquad (3.8)
\end{aligned}
$$

where $q^{\alpha}(\overline{0}) = 0$ and $q^{\alpha}(\overline{1}) = 1$ are to be identified. Recall that $l^{\alpha}_{\mathbf{v}^{\alpha}}(\ddot{\alpha}) = \ddot{\alpha} = \frac{1}{2}$ for $\alpha = 0$.

Now for each $\alpha \in \mathbb{P}_* \cup \{1\}$ we define q^{α} according to formula (3.8) and show that this definition makes sense. In the case that $\alpha = 1$, identify $q^{\alpha}(\overline{0}) = 1$ and $q^{\alpha}(\overline{1}) = 0$ and set formally $l^{\alpha}_{\mathbf{v}^{\alpha}} = \frac{1}{2}$.

Obviously, q^{α} is continuous, and one easily verifies that $q^{\alpha}(b_1 b_2 b_3 \ldots) = q^{\overline{\alpha}}((1-b_1)(1-b_2)(1-b_3)\ldots)$ for all $b_1 b_2 b_3 \ldots \in \{0,1\}^{\mathbb{N}}$.

The following abbreviations will be convenient from the technical point of view: $0\langle\alpha\rangle := \mathbf{v}^{\alpha} e^{\alpha}, 1\langle\alpha\rangle := \mathbf{v}^{\alpha}(1-e^{\alpha})$ and $*\langle\alpha\rangle := \mathbf{v}^{\alpha}*$, and more general $\mathbf{t}\langle\alpha\rangle = t_1\langle\alpha\rangle t_2\langle\alpha\rangle t_3\langle\alpha\rangle \ldots$ for a word (sequence) $\mathbf{t} = t_1 t_2 t_3 \ldots$ from $\{0,1,*\}^*$ ($\{0,1,*\}^{\mathbb{N}}$). In particular, $0\langle\alpha\rangle = 0$ and $1\langle\alpha\rangle = 1$ for $\alpha = 0$.

By *sequence tuning* we understand the procedure changing from a sequence $\mathbf{t} \in \{0,1,*\}^{\mathbb{N}}$ to the sequence $\mathbf{t}\langle\alpha\rangle$ for some periodic $\alpha \neq 0$.

According to Proposition 2.43, the gap $Gap\langle\alpha\rangle$ is represented by the α-regular word $\mathbf{w} = \mathbf{v}^{\alpha}$, and it is bounded by the chords $\alpha\overline{\alpha}$ and $l^{\alpha,1-e}_{(\mathbf{u}0)\langle\alpha\rangle}(\alpha\overline{\alpha})$; $\mathbf{u} \in \{0,1\}^*$. In particular, the gaps $h^n(Gap\langle\alpha\rangle); n = 0,1,2,\ldots,m-1$ have mutually disjoint iteriors.

Proposition 3.22. *Let $\alpha \in \mathbb{P}_*$. Then $\mathbb{T}\langle\alpha\rangle = q^{\alpha}(\{0,1\}^{\mathbb{N}})$. Moreover, $\mathbb{T}\langle\alpha\rangle$ forms a Cantor set and arises from the smaller closed interval with the $\approx^{\alpha}$-equivalent endpoints $\alpha, \overline{\alpha}$ by taking away all open intervals with the $\approx^{\alpha}$-equivalent endpoints $q^{\alpha}(\mathbf{w}0\overline{1})$ and $q^{\alpha}(\mathbf{w}1\overline{0})$; $\mathbf{w} \in \{0,1\}^*$.*

In particular, it holds $q^{\alpha}(\overline{0}) = \alpha, q^{\alpha}(\overline{1}) = \overline{\alpha}$, and $q^{\alpha}(\mathbf{w}1\overline{0}) = l^{\alpha,e}_{(\mathbf{w}1)\langle\alpha\rangle}(\alpha) = l^{\alpha,1-e}_{(\mathbf{w}0)\langle\alpha\rangle}(\alpha), q^{\alpha}(\mathbf{w}0\overline{1}) = l^{\alpha,1-e}_{(\mathbf{w}0)\langle\alpha\rangle}(\overline{\alpha})$ for $e = e^{\alpha}$ and all $\mathbf{w} \in \{0,1\}^*$.

Proof: Let us start showing the first part of the proposition. By Theorem 2.15(ii), the mutual position of the triangles $\Delta^{\alpha}_{(\mathbf{w}*)\langle\alpha\rangle}; \mathbf{w} \in \{0,1\}^*$ and chords $S^{\alpha,e}_{(\mathbf{w}*)\langle\alpha\rangle}, S^{\alpha,1-e}_{(\mathbf{w}*)\langle\alpha\rangle}; \mathbf{w} \in \{0,1\}^*$ does not depend on $\alpha \in \mathbb{P}_*$.

Therefore, that which is obvious in the case $\alpha = 0$ remains valid in the general case: If $\mathbf{w} \in \{0,1\}^*$ has the representation $\mathbf{w} = \mathbf{w}_1 1 \mathbf{w}_2 1 \ldots 1 \mathbf{w}_k$ where each word $\mathbf{w}_i; i = 1,2,\ldots,k$ is empty or only consists of symbols 0, then α and $l^{\alpha,e}_{(\mathbf{w}1)\langle\alpha\rangle}(\alpha)$ are connected by a the finite sequence of chords $S^{\alpha,e}_{(\mathbf{w}_1*)\langle\alpha\rangle} = \alpha l^{\alpha,e}_{(\mathbf{w}_1 1)\langle\alpha\rangle}(\alpha), S^{\alpha,e}_{(\mathbf{w}_1 1 \mathbf{w}_2 *)\langle\alpha\rangle} = l^{\alpha,e}_{(\mathbf{w}_1 1)\langle\alpha\rangle}(\alpha) l^{\alpha,e}_{(\mathbf{w}_1 1 \mathbf{w}_2 1)\langle\alpha\rangle}(\alpha)$, $\ldots, S^{\alpha,e}_{(\mathbf{w}_1 1 \mathbf{w}_2 1 \ldots 1 \mathbf{w}_k *)\langle\alpha\rangle} = l^{\alpha,e}_{(\mathbf{w}_1 1 \mathbf{w}_2 1 \ldots 1 \mathbf{w}_{k-1} 1)\langle\alpha\rangle}(\alpha) l^{\alpha,e}_{(\mathbf{w}1)\langle\alpha\rangle}(\alpha)$. Further, it holds

$\alpha < l^{\alpha,e}_{(\mathbf{w}_1 1)\langle\alpha\rangle}(\alpha) < l^{\alpha,e}_{(\mathbf{w}_1 1\mathbf{w}_2 1)\langle\alpha\rangle}(\alpha) < \ldots < l^{\alpha,e}_{(\mathbf{w}1)\langle\alpha\rangle}(\alpha) < \overline{\alpha}$ or $\alpha > l^{\alpha,e}_{(\mathbf{w}_1 1)\langle\alpha\rangle}(\alpha) > l^{\alpha,e}_{(\mathbf{w}_1 1\mathbf{w}_2 1)\langle\alpha\rangle}(\alpha) > \ldots > l^{\alpha,e}_{(\mathbf{w}1)\langle\alpha\rangle}(\alpha) > \overline{\alpha}$.

So $q^\alpha(\mathbf{w}1\overline{0}) = l^{\alpha,e}_{(\mathbf{w}1)\langle\alpha\rangle}(\alpha)$ is obtained by summing up the lengths of the chords above, which are given by the first formula below (2.6) (see Lemma 2.35). For all $k \in \mathbb{N}$ we have $q^\alpha(\mathbf{w}01^k\overline{0}) = l^{\alpha,e}_{(\mathbf{w}01^k)\langle\alpha\rangle}(\alpha) = l^{\alpha,1-e}_{(\mathbf{w}01^{k-1}0)\langle\alpha\rangle}(\alpha)$. Thus by Proposition 2.38(i) $(l^{\alpha,1-e}_{(\mathbf{w}01^{k-1}0)\langle\alpha\rangle}(\alpha))_{k\in\mathbb{N}} = (l^{\alpha}_{(\mathbf{w}01^{k-1})\langle\alpha\rangle\mathbf{v}}(\ddot{\alpha}))_{k\in\mathbb{N}}$ converges to $l^{\alpha,1-e}_{(\mathbf{w}0)\langle\alpha\rangle}(\overline{\alpha})$, hence $q^\alpha(\mathbf{w}0\overline{1}) = l^{\alpha,1-e}_{(\mathbf{w}0)\langle\alpha\rangle}(\overline{\alpha})$.

By Proposition 2.43 we saw that $Gap\langle\alpha\rangle$ is bounded by the mutually different chords $\alpha\overline{\alpha}$ and $l^{\alpha,1-e}_{(\mathbf{w}0)\langle\alpha\rangle}(\alpha\overline{\alpha})$, $\mathbf{w} \in \{0,1\}^*$, and by the already shown, the set of points $q^\alpha(\mathbf{w}1\overline{0})$; $\mathbf{w} \in \{0,1\}^*$ is dense in $\mathbb{T}\langle\alpha\rangle$. Since q^α is continuous and $\{0,1\}^{\mathbb{N}}$ compact, this implies that the set $q^\alpha(\{0,1\}^{\mathbb{N}})$ is closed. So it coincides with $\mathbb{T}\langle\alpha\rangle$. ■

As in the case $\alpha = 0$, the action of h^m with $m = PER(\alpha)$ is related to the shift σ on $\{0,1\}^{\mathbb{N}}$:

Proposition 3.23. *Let $\alpha \in \mathbb{P}_* \setminus \{0\}, e = e^\alpha$ and $m = PER(\alpha)$. Then the following statements are valid:*

(i) $\mathbb{T}\langle\alpha\rangle$ is h^m-invariant.

(ii) The sets $h^n(\mathbb{T}\langle\alpha\rangle \setminus \{\overline{\alpha}\}); n = 0,1,2,\ldots,m-1$ are mutually disjoint.

(iii) For each $n = 0,1,2,\ldots,m-2$, the set $h^n(\mathbb{T}\langle\alpha\rangle)$ lies either behind $\dot{\alpha}\dot{\overline{\alpha}}$ or behind $\ddot{\alpha}\ddot{\overline{\alpha}}$, and there is a point in $h^n(\mathbb{T}\langle\alpha\rangle)$ behind $\alpha\overline{\alpha}$ iff m divides n. Further, h^{n+1} acts injectively and order-preserving on $\mathbb{T}\langle\alpha\rangle$.

(iv) For all $\mathbf{b} \in \{0,1\}^{\mathbb{N}}$ it holds $q^\alpha(\sigma(\mathbf{b})) = h^m(q^\alpha(\mathbf{b}))$.

Proof: (i) and the first part of (iii) are immediate consequences of Proposition 2.43, and the second part of (iii) follows from the first part by the orientation-invariance of h (Lemma 2.9).

(ii): A point β belonging to two different iterates of $\mathbb{T}\langle\alpha\rangle \setminus \{\overline{\alpha}\}$ must be a common endpoint of two chords each bounding one of the corresponding iterates of $Gap\langle\alpha\rangle$. We can assume that $\beta \in (\mathbb{T}\langle\alpha\rangle \setminus \{\overline{\alpha}\}) \cap h^n(\mathbb{T}\langle\alpha\rangle \setminus \{\overline{\alpha}\})$ for some $n = 1,2,\ldots,m-1$. Then by Proposition 2.43 (i) $\beta = \alpha = h^n(\alpha)$, which is obviously false. This shows that the sets $h^n(\mathbb{T}\langle\alpha\rangle \setminus \{\overline{\alpha}\}); n = 0,1,2,\ldots,m-1$ are mutually disjoint.

(iv): If a sequence $\mathbf{b} \in \{0,1\}^{\mathbb{N}}$ ends with $\overline{0}$ or $\overline{1}$, then $q^\alpha(\sigma(\mathbf{b})) = h^m(q^\alpha(\mathbf{b}))$ by Proposition 3.22. (iv) is easily obtained taking into consideration that both q^α, h and σ are continuous. ■

For $\alpha \in \mathbb{P}_* \setminus \{0\}$ let us define a map Q^α from $\mathbb{T}$ into $\mathbb{T}$ by the following prescription:

$$Q^\alpha(q^0(\mathbf{b})) := q^\alpha(\mathbf{b}) \text{ for } \mathbf{b} \in \{0,1\}^{\mathbb{N}} \text{ not ending with } \overline{1}.$$

By the already shown, Q^α maps into $\mathbb{T}\langle\alpha\rangle$, is injective and preserves or reverses the orientation. Let us derive some further properties of Q^α from Propositions 3.22 and 3.23.

Corollary 3.24. *Let $\alpha \in \mathbb{P}_* \setminus \{0\}, e = e^\alpha, m = PER(\alpha)$ and $Q = Q^\alpha$. Then for all $\beta \in \mathbb{T}$ the following holds:*

(i) $Q(h(\beta)) = h^m(Q(\beta))$,

(ii) $\{h^{m-1}(Q(\frac{\beta}{2})), h^{m-1}(Q(\frac{\beta+1}{2}))\} = \{\frac{Q(\beta)}{2}, \frac{Q(\beta)+1}{2}\}$,

(iii) $h^{m-1}(Q(\mathbb{T}_0^\beta \setminus \{\frac{\beta}{2}, \frac{\beta+1}{2}\})) \subseteq \mathbb{T}_e^{Q(\beta)} \setminus \{\frac{Q(\beta)}{2}, \frac{Q(\beta)+1}{2}\}$ and $h^{m-1}(Q(\mathbb{T}_1^\beta \setminus \{\frac{\beta}{2}, \frac{\beta+1}{2}\})) \subseteq \mathbb{T}_{1-e}^{Q(\beta)} \setminus \{\frac{Q(\beta)}{2}, \frac{Q(\beta)+1}{2}\}$.

(iv) $I^{Q(\beta)}(Q(\delta)) = I^\beta(\delta)\langle\alpha\rangle$ for all $\delta \in \mathbb{T}$.

Proof: (i) is no more than a reformulation of (iv) in Proposition 3.23. The statement (ii) can be concluded from (i) as follows: We have $h^m(Q(\frac{\beta}{2})) = Q(\beta)$ and $h^m(Q(\frac{\beta+1}{2})) = Q(\beta)$. Now (ii) follows from the injectivity of $h^{m-1}Q$ (compare Proposition 3.23(iii)).

(iii) is a consequence of (ii) and the fact that $h^{m-1}Q$ preserves or reverses the orientation if the existence of a point $\delta \in l_1^\beta(\mathbb{T} \setminus \{\beta\})$ with $h^{m-1}Q(\delta) \in l_{1-e}^{Q(\beta)}(\mathbb{T} \setminus \{Q(\beta)\})$ can be shown.

In the case $\beta = 0$ such a point is given by $\delta = \frac{3}{4} = q^0(11\overline{0})$ since then $h^{m-1}Q(\delta) = h^{m-1}q^\alpha(11\overline{0}) = h^{m-1}l^{\alpha,e}_{(11)\langle\alpha\rangle}(\alpha) = l^\alpha_{(1-e)\mathrm{v}^\alpha}(\ddot{\alpha})$ and $Q(\beta) = \alpha$ by Proposition 3.22.

For $\beta \neq 0$ let $\delta = 0$. Then $h^{m-1}Q(\delta) = \dot{\alpha}$ and $\dot{\overline{\alpha}} \in l^\alpha_{1-e}(\mathbb{T} \setminus \{\alpha\})$ (since by Proposition 2.41 $e = 1 - I^\alpha(\overline{\alpha})(m)$). Since $\frac{Q(\beta)}{2}\frac{Q(\beta)+1}{2}$ is contained in the critical gap of $\eth\mathcal{B}^\alpha$ and separates the chords $\dot{\alpha}\dot{\overline{\alpha}}$ and $\ddot{\alpha}\ddot{\overline{\alpha}}$, one easily sees that $h^{m-1}Q(\delta) \in l_{1-e}^{Q(\beta)}(\mathbb{T} \setminus \{Q(\beta)\})$.

(iv): By (i), for a given $\delta \in \mathbb{T}$ the map Q assigns $(h^{(n-1)m}(Q(\delta)))_{n\in\mathbb{N}}$ to the sequence $(h^{n-1}(\delta))_{n\in\mathbb{N}}$, and so the statement (iv) follows directly from (ii), (iii) and Proposition 3.23(iii). ∎

Corollary 3.25. *For $\alpha \in \mathbb{P}_* \setminus \{0\}$ and $m = PER(\alpha)$, the point $\beta \in \mathbb{T}$ is periodic with respect to h if $Q^\alpha(\beta)$ is with respect to h^m. If $\beta \in \mathbb{P}_*$, then in particular it holds $PER(Q^\alpha(\beta)) = m\ PER(\beta)$, and in case that $\beta \neq 0$, the following statements are valid:*

(i) $\overline{Q^\alpha(\beta)} = Q^\alpha(\overline{\beta})$,

(ii) $h^{m-1}Q^\alpha(\dot{\beta}) = \dot{Q^\alpha}(\beta)$ and $h^{m-1}Q^\alpha(\ddot{\beta}) = \ddot{Q^\alpha}(\beta)$.

Proof: The first statement immediately follows from Corollary 3.24(i) and the mutual disjointness of the sets $h^n(\mathbb{T}\langle\alpha\rangle \setminus \{\overline{\alpha}\}); n = 0, 1, 2, \ldots, m-1$ (see Proposition 3.23(ii)).

Let $\beta, \delta \in \mathbb{T} \setminus \{0\}$ be associated periodic points and let $Q = Q^\alpha$. Then by Proposition 2.62 $\hat{\beta} \cong I^\beta(\delta)$ and $\hat{\delta} \cong I^\delta(\beta)$, hence according to 3.24(iv) $\widehat{Q(\beta)} \cong$

$I^{Q(\beta)}(Q(\delta))$ and $\widehat{Q(\delta)} \cong I^{Q(\delta)}(Q(\beta))$. Thus $Q(\beta)$ and $Q(\delta)$ are associated and (i) is shown.

In order to get (ii) one uses Corollary 3.24(ii) and the fact that $h^{m-1}(Q(\beta))$ is periodic for periodic β. ■

We have found that the maps q^α for $\alpha \in \mathbb{P}_* \setminus \{0\}$ and q^0 are similar in a certain sense. In fact, the next Theorem shows how $\mathbb{T}\langle\alpha\rangle$ can be considered as a circle in a natural way.

At this place we want to recall the notation $[A]_\equiv$ for the transitive hull of a set A with respect to an equivalence relation $\equiv$. Moreover, let us introduce the following notation which will later be used in a more general context (see Definition 3.43): For $\alpha \in \mathbb{P}_* \setminus \{0\}$ and $m = PER(\alpha)$ let $\mathbb{T}^\alpha_{*_m} = \{\beta \in \mathbb{T} \mid \widetilde{I^\alpha}(\beta) \cong \hat{\alpha}\}$. (The m in the notation is superfluous here since it depends on α, but in the general context it will become substantial.)

Theorem 3.26. ($\mathbb{T}\langle\alpha\rangle$ and sequence tuning of the itinerary)
Let $\alpha \in \mathbb{P}_* \setminus \{0\}$ and $m = PER(\alpha)$. Then the following statements are valid:

(i) q^α forms a conjugacy between the two topological dynamical systems $(\{0,1\}^{\mathbb{N}}, \sigma)$ and $(\mathbb{T}\langle\alpha\rangle, h^m)$.
(ii) $\mathbb{T}^\alpha_{*_m} = [\mathbb{T}\langle\alpha\rangle]_{\approx^\alpha} = [Q^\alpha(\mathbb{T})]_{\approx^\alpha}$.
(iii) The map $\delta \mapsto [Q^\alpha(\delta)]_{\approx^\alpha}; \delta \in \mathbb{T}$ defines a conjugacy between the topological dynamical systems $(\mathbb{T}, h)$ and $(\mathbb{T}^\alpha_{*_m}, h^m)/\approx^\alpha$.
(iv) For all $\delta \in \mathbb{T}$ it holds $I^\alpha(Q^\alpha(\delta)) = I^0(\delta)\langle\alpha\rangle$ and $\widetilde{I^\alpha}(Q^\alpha(\delta)) = \widetilde{I^0}(\delta)\langle\alpha\rangle$.

Proof: By definition q^α is injective and so, as a continuous map on a compact space, it forms a homeomorphism. Thus (i) follows from Proposition 3.23(iv). $\approx^\alpha$ identifies the elements of the equivalence class $[q^\alpha(\overline{1})]_{\approx^\alpha} = [q^\alpha(\overline{0})]_{\approx^\alpha}$ and, for all $\mathbf{w} \in \{0,1\}^*$, the elements of $[q^\alpha(\mathbf{w}0\overline{1})]_{\approx^\alpha} = [q^\alpha(\mathbf{w}1\overline{0})]_{\approx^\alpha}$, Actually, one has no more than the factorization of $\{0,1\}^{\mathbb{N}}$ being induced by binary expansion and yielding the circle. Therefore, it holds $[\mathbb{T}\langle\alpha\rangle]_{\approx^\alpha} = [Q^\alpha(\mathbb{T})]_{\approx^\alpha}$ and (iii).

The first equation in (iv) turns out to be the specification of Corollary 3.24(iv) for $\beta = 0$, and the second one follows from Corollaries 3.24(ii) and 3.25(ii). Finally, by the second part of (iv) and by Proposition 2.56(v) the elements of $\mathbb{T}\langle\alpha\rangle$ provide all itineraries being substantially equal to $\hat{\alpha}$. Thus (ii) is an immediate consequence of Theorem 2.53. ■

q^0 is the 'inversion' of the binary expansion, and also for $\alpha \in \mathbb{P}_* \setminus \{0\}$ the map q^α can be expressed by the binary expansions of the points in $\mathbb{T}\langle\alpha\rangle$. In order to see this, let $PER(\alpha) = m$ and let $\mathbf{b}(\alpha) = \overline{c_1 c_2 \ldots c_m}$ and $\mathbf{b}(\overline{\alpha}) = \overline{d_1 d_2 \ldots d_m}$ be the binary expansion of α and $\overline{\alpha}$, respectively. Then

$$l^\alpha_{\mathbf{v}^\alpha}(\ddot{\alpha}) - \alpha = \frac{2^m - 1}{2^m}(\overline{\alpha} - \alpha)$$

and $\alpha = \frac{2^m}{2^m-1}\sum_{j=1}^m \frac{c_j}{2^j}, \overline{\alpha} = \frac{2^m}{2^m-1}\sum_{j=1}^m \frac{d_j}{2^j}$, which implies

$$\begin{aligned} q^\alpha(\mathbf{b}) &= \alpha + (l^\alpha_{\mathbf{v}^\alpha}(\ddot{\alpha}) - \alpha)\sum_{i=0}^\infty (\frac{b_{i+1}}{2^m})^i \\ &= \sum_{i=0}^\infty (\sum_{j=1}^m \frac{c_j}{2^j})\frac{1}{2^{im}} + (\sum_{j=1}^m \frac{d_j}{2^j} - \sum_{j=1}^m \frac{c_j}{2^j})\sum_{i=0}^\infty (\frac{b_{i+1}}{2^m})^i \\ &= \sum_{i=0}^\infty (\sum_{j=1}^m \frac{c_j}{2^j})\frac{1}{2^{im}} + \sum_{i=0}^\infty b_{i+1}(\sum_{j=1}^m \frac{d_j}{2^j} - \sum_{j=1}^m \frac{c_j}{2^j})\frac{1}{2^{im}} \\ &= \sum_{i=0}^\infty ((1-b_{i+1})\sum_{j=1}^m \frac{c_j}{2^j} + b_{i+1}\sum_{j=1}^m \frac{d_j}{2^j})\frac{1}{2^{im}}, \end{aligned}$$

for all $\mathbf{b} = b_1b_2b_3\ldots \in \{0,1\}^{\mathbb{N}}$, and we get the following statement:

Proposition 3.27. *Let $\alpha \in \mathbb{P}_* \setminus \{0\}$ and $PER(\alpha) = m$. Further, let $\mathbf{b}(\alpha) = \overline{c_1c_2\ldots c_m}$ and $\mathbf{b}(\overline{\alpha}) = \overline{d_1d_2\ldots d_m}$ be the binary expansion of α and $\overline{\alpha}$, respectively. Then the binary expansion of $q^\alpha(\mathbf{a})$ for $\mathbf{a} \in \{0,1\}^{\mathbb{N}}$ is obtained by substituting the symbols in $\mathbf{a}$ as follows: $0 \longrightarrow c_1c_2\ldots c_m, 1 \longrightarrow d_1d_2\ldots d_m$.* ■

Remark. The maps q^α are strongly related to DOUADY and HUBBARD's tuning, which was briefly described in Chapter 1.2.2. In the following we will investigate an abstract setting of tuning. However, with the suitable interpretation of the maps q^α, Proposition 3.27 provides DOUADY's [40] well-known description of tuning for the Mandelbrot set (compare Chapter 4.1 of the present work).

3.2.2 'Small' copies of Abstract Julia sets and of the Abstract Mandelbrot set

The discussion above suggests that from a Julia equivalence $\approx^\alpha$ for $\alpha \in \mathbb{P}_*$ and a second one $\approx$, can be constructed a new Julia equivalence as follows: by virtue of Q^α one transports $\approx$ onto the set $\mathbb{T}\langle\alpha\rangle$ and extends the identification obtained to the whole circle, in a way compatible with the action of h and $'$.

Indeed, this procedure is successful, as the following proposition shows:

Proposition 3.28. *Let $\alpha \in \mathbb{P}_* \setminus \{0\}, Q = Q^\alpha$, and let $\delta \in \mathbb{T}\setminus\{0\}$. Then the following statements are valid:*

(i) $\approx^\alpha \subset \approx^{Q(\delta)}$,

(ii) for all $\beta_1, \beta_2 \in \mathbb{T}$ it holds: $Q(\beta_1) \approx^{Q(\delta)} Q(\beta_2) \Longleftrightarrow \beta_1 \approx^\delta \beta_2$.

Proof: (i): By Proposition 2.42 $\eth\mathcal{B}^\alpha$ coincides with the closure of the set $\mathcal{B}^\alpha_* = \{l^{\alpha,1-e}_{\mathbf{w}}(\alpha\overline{\alpha}) \mid \mathbf{w} \in \{0,1\}^*\}$ with $e = e^\alpha$. All chords contained in this set form boundary chords of infinite gaps of the lamination $\eth\mathcal{B}^\alpha$, and $\frac{Q(\delta)}{2}\frac{Q(\delta)+1}{2}$ lies in the critical gap of this lamination.

Therefore, $\frac{Q(\delta)}{2}\frac{Q(\delta)+1}{2}$ does not cross a chord in the set $\mathcal{B}^\alpha_*$, and if $\frac{Q(\delta)}{2}\frac{Q(\delta)+1}{2}$ has one point in common with such a chord, $Q(\delta)$ must be preperiodic (compare Proposition 2.43).

In dependence on $Q(\delta)$, one of the Theorems 2.47, 2.50 or 2.53 implies that the endpoints of each chord in $\mathcal{B}^\alpha_*$ are $\approx^{Q(\delta)}$-equivalent. Now Proposition 2.42 yields (i). Further, one immediately obtains (ii) from Corollary 3.24(iv) and again from one of the Theorems 2.47, 2.50 and 2.53, where in the case $\delta \in \mathbb{P}_*$ Corollary 3.25(ii) is to be considered. ■

Example 3.29. Let $\alpha = \frac{1}{3}$ and $\delta = \frac{1}{6}$. The binary expansion of α begins with 01 and those of $\overline{\alpha} = \frac{2}{3}$ with 10, and δ has binary expansion $0\overline{01}$. Thus by Proposition 3.27 the binary expansion of $Q^\alpha(\delta)$ is $01\overline{0110}$. So one easily computes that $Q^\alpha(\delta) = \frac{7}{20}$.

Proposition 3.28 is illustrated by Figure 3.4. On the left, one sees the laminations $\eth\mathcal{B}^\alpha$ and $\eth\mathcal{B}^{Q^\alpha(\delta)}$, and on the right the corresponding Julia sets. $\eth\mathcal{B}^{Q^\alpha(\delta)}$ arises from $\eth\mathcal{B}^\alpha$ by adding further chords, and one can imagine that Q^α inserts the invariant lamination given in Figure 1.2 into the gap of $\eth\mathcal{B}^\alpha$.

The Julia set realizing $\approx^\alpha$ must be 'glued' to get the Julia set realizing $\approx^{Q^\alpha(\delta)}$, and the latter contains small copies of J_i, the realization of $\approx^\delta$ (see the small picture at the bottom left).

Section 3.3 includes a comprehensive discussion of the phenomenon that many Abstract Julia sets contain topological copies of other Abstract Julia sets. At this point we want to investigate a certain topological self-similarity of the Abstract Mandelbrot set. As a first step, we show that for $\alpha \in \mathbb{P}_*$ the equivalence relation $\approx^\alpha$ restricted to $\mathbb{T}\langle\alpha\rangle$ is weaker than $\sim$.

Lemma 3.30. *For $\alpha \in \mathbb{P}_* \setminus \{0\}$ the following statements are valid:*

(i) If $\gamma = q^\alpha(\mathbf{w}1\overline{0})$ and $\delta = q^\alpha(\mathbf{w}0\overline{1})$ for some $\mathbf{w} \in \{0,1\}^*$, then it holds $\delta \in [\gamma]_{\approx^\alpha} = [\gamma]_{\approx^\gamma} = [\gamma]_\sim$.

(ii) It holds $[\mathbb{T}\langle\alpha\rangle]_\sim = [Q^\alpha(\mathbb{T})]_\sim$.

(iii) It holds $[\mathbb{T}\langle\alpha\rangle \setminus \{\alpha,\overline{\alpha}\}]_\sim = [\mathbb{T}\langle\alpha\rangle \setminus \{\alpha,\overline{\alpha}\}]_{\approx^\alpha}$, and the restriction of $\approx^\alpha$ to $[\mathbb{T}\langle\alpha\rangle]_\sim$ is contained in the restriction of $\sim$ to $[\mathbb{T}\langle\alpha\rangle]_\sim$.

Proof: (i): First of all, $[\gamma]_{\approx^\gamma} = [\gamma]_\sim$ by Theorem 3.19, and $\delta \in [\gamma]_{\approx^\alpha} \subseteq [\gamma]_{\approx^\gamma}$ by Proposition 3.22 and Proposition 3.28(i). We have to verify that $CARD([\gamma]_{\approx^\gamma}) \leq CARD([\gamma]_{\approx^\alpha})$.

By the above shown, $\alpha = q^\alpha(\overline{0})$ lies on the orbit of γ. On the one hand, this implies $CARD([\gamma]_{\approx^\gamma}) = CARD([\alpha]_{\approx^\gamma})$ since for no $\gamma, \beta \in \mathbb{T}$ an iterate of $[\gamma]_{\approx^\gamma}$ contains both β and β', and on the other hand this implies $CARD([\alpha]_{\approx^\alpha}) \leq CARD([\gamma]_{\approx^\alpha})$. We complete the above inequality by showing $[\alpha]_{\approx^\gamma} \subseteq [\alpha]_{\approx^\alpha}$:

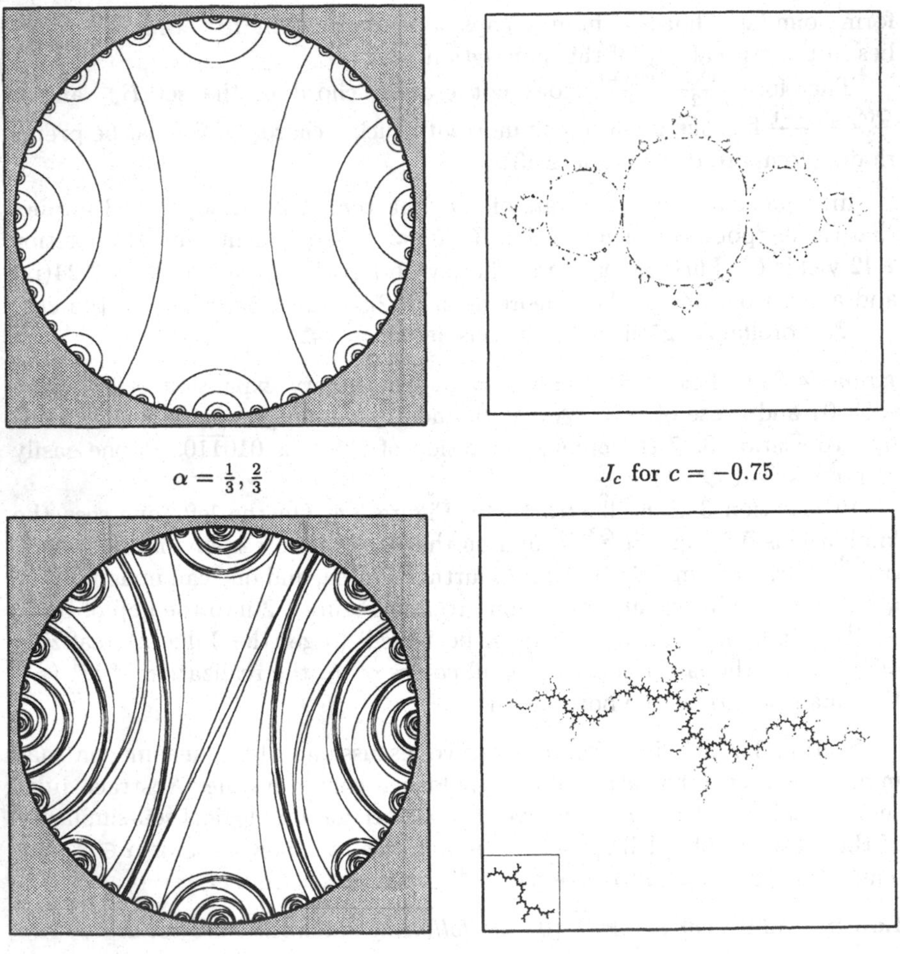

$\alpha = \frac{1}{3}, \frac{2}{3}$ J_c for $c = -0.75$

$Q^\alpha(\delta) = \frac{7}{20}$ J_c for $c = -1,14437\ldots + 0.314719\ldots i$

Fig. 3.4. Extension of a Julia equivalence, and the corresponding Julia sets

By the discussion above Theorem 2.53 and by Theorem 2.57 we have $[\alpha]_{\approx^\alpha} = \mathbb{T}^{\underline{\alpha}}_{\overline{1}\langle\alpha\rangle}$. Since the chord $\alpha\overline{\alpha}$ bounds the gap $Gap\langle\alpha\rangle$, it also bounds $\mathbb{D}^{\underline{\alpha}}_{\overline{1}\langle\alpha\rangle}$. Moreover, by the first statement in Proposition 3.20 $\alpha\overline{\alpha} \in \eth\mathcal{B}^\gamma$, and by Corollary 2.23(i) all boundary chords of $\mathbb{D}^{\underline{\alpha}}_{\overline{1}\langle\alpha\rangle}$ lie in $\eth\mathcal{B}^\gamma$. Therefore, the equivalence class $[\alpha]_{\approx^\gamma}$ lies in $\mathbb{T}^{\underline{\alpha}}_{\overline{1}\langle\alpha\rangle} = [\alpha]_{\approx^\alpha}$.

(ii), (iii): If $\mathbf{w}, \gamma, \delta$ are given as in (i), then $\gamma, \delta \in [\gamma]_{\approx^\alpha} = [\gamma]_{\approx^\gamma} = [\gamma]_\sim$. In case that for all $\mathbf{w} \in \{0,1\}^*$ the point $\gamma \in \mathbb{T}\langle\alpha\rangle \setminus \{\alpha, \overline{\alpha}\}$ is different from $q^\alpha(\mathbf{w}1\overline{0})$ and $q^\alpha(\mathbf{w}0\overline{1})$, it cannot be an endpoint of a boundary chord of $Gap\langle\alpha\rangle$. Then $[\gamma]_{\approx^\alpha}$ must be a single set, and $[\gamma]_{\approx^\alpha} \subseteq [\gamma]_\sim \subset \mathbb{T}\langle\alpha\rangle$. The inclusion on the right follows from the planarity of $\sim$. ■

Now we are able to describe the kind of self-similarity of the Abstract Mandelbrot set which was announced above.

Theorem 3.31. ('Small' copies in the Abstract Mandelbrot set and tuning of the kneading sequence)
For $\alpha \in \mathbb{P}_* \setminus \{0\}$ the following statements are valid:

(i) For all $\mathbf{b} \in \{0,1\}^{\mathbb{N}}$ it holds $\widehat{q^\alpha(\mathbf{b})} = \widehat{q^0(\mathbf{b})}\langle\alpha\rangle$. In particular, $q^\alpha(\mathbf{b})$ is periodic iff $\mathbf{b}$ is, and for different periodic sequences $\mathbf{b},\mathbf{c}$, the points $q^\alpha(\mathbf{b}), q^\alpha(\mathbf{c})$ are associated if $q^0(\mathbf{b}), q^0(\mathbf{c})$ are associated or are equal to $0 (=1)$.

(ii) For all $\mathbf{b}_1, \mathbf{b}_2 \in \{0,1\}^{\mathbb{N}}$ it holds $q^\alpha(\mathbf{b}_1) \sim q^\alpha(\mathbf{b}_2)$ iff $q^0(\mathbf{b}_1) \sim q^0(\mathbf{b}_2)$, and by $[\beta]_\sim \mapsto [Q^\alpha(\beta)]_\sim; \beta \in \mathbb{T}$ there is defined a homeomorphism from $\mathbb{T}/\sim$ onto $[\mathbb{T}\langle\alpha\rangle]_\sim/\sim$.

Proof: (i): One obtains $\widehat{q^\alpha(\mathbf{b})} = \widehat{q^0(\mathbf{b})}\langle\alpha\rangle$ for $\mathbf{b} \in \{0,1\}^{\mathbb{N}}$ setting $\beta = \delta$ in statement (iv) of Corollary 3.24 and using that $q^\alpha(\mathbf{w}0\overline{1})$ and $q^\alpha(\mathbf{w}1\overline{0})$ for all $\mathbf{w} \in \{0,1\}^*$ are $\sim$-equivalent, hence have the same kneading sequences (see Lemma 3.30(i)). The rest of (i) is no more than Corollary 3.25.

Each periodic point in $\mathbb{T}\langle\alpha\rangle$ different from α and $\overline{\alpha}$ is the image of a periodic point in $\mathbb{T}$ with respect to Q^α (see Theorem 3.26). Furthermore, as just shown, $\mathbb{T}\langle\alpha\rangle$ contains with each periodic point its associated one, and by Proposition 2.43 the endpoints of all boundary chords of $Gap\langle\alpha\rangle$ different from $\alpha, \overline{\alpha}$ are preperiodic.

Let us fix the kneading sequence of a periodic point different from 0 and consider the set $\{\delta_1, \delta_2, \delta_3, \ldots\}$ of all points possessing this kneading sequence, where we require that $\delta_1 < \delta_2 < \delta_3 < \ldots$. Then by Theorem 3.6 it follows $\delta_1 \sim \delta_2, \delta_3 \sim \delta_4, \delta_5 \sim \delta_6, \ldots$ and by the already shown $Q^\alpha(\delta_1) \sim Q^\alpha(\delta_2), Q^\alpha(\delta_3) \sim Q^\alpha(\delta_4), Q^\alpha(\delta_5) \sim Q^\alpha(\delta_6), \ldots$. $\sim$ is the smallest closed equivalence relation which identifies associated periodic points (see Theorem 3.19(i)). Now from this, from Theorem 3.26(iii) and from Lemma 3.30(iii) one deduces (ii). ■

In the following, the process which for given $\alpha \in \mathbb{P} \setminus \{0\}$ assigns to $[\beta]_\sim$ the equivalence class $[Q^\alpha(\beta)]_\sim$ is called *tuning.* As can be seen in Theorem 3.31, the result of a tuning is a 'small' copy of the Abstract Mandelbrot set, and the corresponding 'tuned' equivalence relation is defined on the set $[\mathbb{T}\langle\alpha\rangle]_\sim = [Q^\alpha(\mathbb{T})]_\sim$. We want to call this set the *α-tuning* of $\mathbb{T}$.

Theorem 3.31 implies that $Q^\alpha; \alpha \in \mathbb{T}$ maps the set $\mathcal{B}_* \cup \{01\}$ onto all chords in $\mathcal{B}_* \cup \{01\}$ with endpoints in the α-tuning. In particular, by the first statement (i) it multiplies periods with $PER(\alpha)$. This gives the following abstract version of Theorem 1.21.

Corollary 3.32. (Tuning on $\mathcal{B}_*$)
Let α in $\mathbb{T}$ be periodic of period m, and let $\alpha < \overline{\alpha}$. Then $Q^\alpha : \mathbb{T} \mapsto \mathbb{T}$ is orientation-preserving, maps 01 to $\alpha\overline{\alpha}$, and each chord in $\mathcal{B}_*$ of period k to one of period km. Morover, the image of Q^α contains with each two chords of $\mathcal{B}_*$ all chords of $\mathcal{B}_*$ separating them. ■

Let us come back to the beginning of this section and let us characterize $\prec$ in a symbolic way, similarly as $\sim$ in Theorem 3.1.

Theorem 3.33. (Symbolic description of $\prec$, part I; tuning)
For $\alpha, \gamma \in \mathbb{T} \setminus \{0\}$ the following statements are equivalent:

(i) $\alpha \prec \gamma$, i.e. $\approx^\alpha \subset \approx^\gamma$,

(ii) α is periodic, γ lies between α and $\overline{\alpha}$, and $\widetilde{I^\alpha}(\gamma) \cong \hat{\alpha}$.

(iii) α is periodic, γ is different from $\alpha, \overline{\alpha}$ and belongs to the α-tuning of $\mathbb{T}$.

Proof: First of all, recall that $\mathbb{T}^\alpha_{*m} = \{\beta \in \mathbb{T} \mid \widetilde{I^\alpha}(\beta) \cong \hat{\alpha}\} = [\mathbb{T}\langle\alpha\rangle]_{\approx^\alpha}$ by definition and by Theorem 3.26(ii), and that $[Q^\alpha(\mathbb{T}) \setminus \{\alpha, \overline{\alpha}\}]_\sim = [\mathbb{T}\langle\alpha\rangle \setminus \{\alpha, \overline{\alpha}\}]_\sim = [\mathbb{T}\langle\alpha\rangle \setminus \{\alpha, \overline{\alpha}\}]_{\approx^\alpha}$ by Lemma 3.30(ii).

By Proposition 3.28(i), it holds $\approx^\alpha \subset \approx^\gamma$ for $\gamma \in Q^\alpha(\mathbb{T})$, thus the implications '(ii) $\Longrightarrow$ (i)' and '(iii) $\Longrightarrow$ (i)' are obvious now.

Conversely, if $\alpha \prec \gamma$, then $\alpha \in \mathbb{P}_*$ and $\gamma \neq \alpha, \overline{\alpha}$. The point γ must lie between α and $\overline{\alpha}$. Otherwise, the chord $\frac{\gamma}{2}\frac{\gamma+1}{2}$ would separate the periodic $\approx^\gamma$-equivalent points $\dot{\alpha}$ and $\dot{\overline{\alpha}}$, which is impossible by Theorems 2.47, 2.50, 2.53.

The critical gap $h^{-1}(Gap\langle\alpha\rangle)$ of $\eth\mathcal{B}^\alpha$ is bounded by $\dot{\alpha}\dot{\overline{\alpha}}$ and by infinitely many chords with preperiodic endpoints. Moreover, $\approx^\gamma$ does not identify a periodic point with a preperiodic one.

Therefore, if γ is periodic, then the chord $\dot{\gamma}\dot{\overline{\gamma}}$ could cross at most one boundary chord of $h^{-1}(Gap\langle\alpha\rangle)$, namely $\dot{\alpha}\dot{\overline{\alpha}}$. This is impossible since then the points $\dot{\gamma}, \dot{\alpha}, \dot{\overline{\gamma}}, \dot{\overline{\alpha}}$ would be pairwisely $\approx^\gamma$-equivalent, but two of them would have a distance greater than $s(\approx^\gamma)$. Hence γ can only be an element of $\mathbb{T}\langle\alpha\rangle$.

If γ fails to be periodic, it holds $\frac{\gamma}{2} \approx^\gamma \frac{\gamma+1}{2}$, and then, if $\frac{\gamma}{2}\frac{\gamma+1}{2}$ crosses a boundary chord $\delta_1\delta_2$ of $h^{-1}(Gap\langle\alpha\rangle)$, it also crosses $\delta_1'\delta_2'$. By the planarity of $\approx^\gamma$ the point $h(\delta_1) \in \mathbb{T}\langle\alpha\rangle$ (and $h(\delta_2)$) forms a main point of $\approx^\gamma$, such that $\gamma \sim h(\delta_1)$, hence $\gamma \in [\mathbb{T}\langle\alpha\rangle \setminus \{\alpha, \overline{\alpha}\}]_\sim$ (see Theorem 3.19(ii)). ■

In Section 3.3 we will give a further symbolic description of $\prec$ (see Corollary 3.52).

3.2.3 A characterization of points with periodic kneading sequences

In Section 2.1.2 we found all points in $\mathbb{T}$ whose kneading sequence is substantially equal to $\overline{0}$ (see Theorem 2.4). These points are the elements of the set $q^0(Sturm_0 \cup Sturm_0^-) = q^0(Sturm_0) \cup q^1(Sturm_0)$ with $Sturm_0 := \{\mathbf{a(k)} | \mathbf{k} \in \mathbb{N}^{\mathbb{N}}\} \cup \{\overline{\mathbf{a(k)}} | \mathbf{k} \in \mathbb{N}^*\}$. The notation refers to the fact that $Sturm_0$ consists of special Sturmian sequences (compare Section 2.1.2).

We want to show that all points with periodic kneading sequence are obtained by tuning from those with kneading sequence substantially equal to $\overline{0}$.

Theorem 3.34. (Characterization of $\mathbb{P}$ and infinite gaps of $\mathcal{B}$)

(i) *It holds* $\mathbb{P} = \bigcup_{\alpha \in \mathbb{P}_* \cup \{1\}} q^\alpha(Sturm_0)$, *and for different* $\alpha, \gamma \in \mathbb{P}_* \cup \{1\}$ *the set* $q^\alpha(Sturm_0) \cap q^\gamma(Sturm_0)$ *consists at most of* α *or of* γ.

(ii) *For all* $\alpha \in \mathbb{P}_*$ *it holds* $Gap(\alpha) \cap \mathbb{T} = [Gap(\alpha) \cap \mathbb{T}]_\sim = q^\alpha(Sturm_0) \cup q^{\overline{\alpha}}(Sturm_0)$, *and the factor space* $(Gap(\alpha) \cap \mathbb{T})/\sim$ *is a simple closed curve in* $\mathbb{T}/\sim$.

(iii) *For all* $\alpha \in \mathbb{P}_* \setminus \{0\}$, *the correspondence* $[\beta]_\sim \mapsto [Q^\alpha(\beta)]_\sim; \beta \in Gap(0) \cap \mathbb{T}$ *defines a homeomorphism from* $(Gap(0) \cap \mathbb{T})/\sim$ *onto* $(Gap(\alpha) \cap \mathbb{T})/\sim$.

Proof: First note that $Gap(0) \cap \mathbb{T} = q^0(Sturm_0) \cup q^1(Sturm_0)$. This is an immediate consequence of the descriptions of $\{\beta \in \mathbb{T} | \hat{\beta} \cong \overline{0}\}$ in Theorem 2.4 and of Theorem 3.17(ii). The gap $Gap(\alpha)$ has longest boundary chord $\alpha\overline{\alpha}$, and by Theorem 3.31 one has the following: For all $\mathbf{b}_1, \mathbf{b}_2 \in \{0,1\}^{\mathbb{N}}$, in particular for all $\mathbf{b}_1, \mathbf{b}_2 \in Sturm_0 \cup Sturm_0^-$, it holds $q^\alpha(\mathbf{b}_1) \sim q^\alpha(\mathbf{b}_2)$ iff $q^0(\mathbf{b}_1) \sim q^0(\mathbf{b}_2)$.

Therefore, $B \in \mathcal{B}_*$ is a boundary chord of $Gap(0)$ iff $Q^\alpha(B)$ is one of $Gap(\alpha)$ different from $\alpha\overline{\alpha}$. Now (i) and (ii) follow from Theorem 3.17 and (3.6), from the continuity of q^γ, and from $q^{\overline{\gamma}}(Sturm_0) = q^\gamma(Sturm_0^-)$ for all $\gamma \in \mathbb{P}_*$.

Theorem 3.31(i) implies that $\bigcup_{\alpha \in \mathbb{P}_* \cup \{1\}} q^\alpha(Sturm_0) \subseteq \mathbb{P}$. For the other inclusion use that each point $\gamma \in \mathbb{P}_\infty$ lies on the boundary of an infinite gap of $\mathcal{B}$. The reasoning for this is that the point neither can be an endpoint of a chord in $\mathcal{B}$ nor can lie on the boundary of a polygonal gap since by Theorem 3.1 and Lemma 2.60 (see (iv)) the equivalence class $[\gamma]_\sim$ consists only of γ itself. If γ were not a point on the boundary of a gap, then γ would lie between the endpoints of infinitely many chords in $\mathcal{B}_*$ which converge to γ. By Theorem 3.1 (see (vi)), two such endpoints are $\approx^\gamma$-equivalent, contradicting the fact that $\gamma \in \mathbb{P}_\infty^\gamma$ and that $\mathrm{conv}(\mathbb{P}_\infty^\alpha)$ forms an infinite gap of $\mathcal{B}^\gamma$.

This shows $\mathbb{P} = \bigcup_{\alpha \in \mathbb{P}_* \cup \{1\}} q^\alpha(Sturm_0)$. The rest of (i) is obvious since the intersections of two different gaps in $\mathcal{B}_*$ with $\mathbb{T}$ have at most two common points. ■

3.2.4 Sturmian sequences and angle-doubling

In Section 2.1.2 we pointed out for which points in $\mathbb{T}$ the kneading sequence is substantially equal to $\overline{0}$. In this context, we touched the concept of a Sturmian sequence. We want to show now that Sturmian sequences are more strongly related to the dynamics of h than it was indicated.

First of all, let us list some statements on Sturmian sequences which are due to MORSE and HEDLUND [119], but which we present in a way directed to our purposes. By technical reasons, it is convenient also to consider Sturmian bisequences. We recall that the 1-length of a 0-1-word is the number of symbols 1 appearing in it. The 0-length is defined analogously.

Definition 3.35. *A 0-1-sequence or 0-1-bisequence is said to be* Sturmian, *if the 1-lengths (and so the 0-length) of two of its subwords of same lengths differ by at most one.*

A 0-1-sequence $a_1a_2a_3\ldots$ *(0-1-bisequence* $\ldots a_{-2}a_{-1}a_0a_1a_2\ldots$*) is called* recurrent *iff for each index* i *and each* $m > 0$ *there exists a* $j > 0$ *(a* $j > 0$ *and a* $j < 0$*) with* $a_ia_{i+1}\ldots a_{i+m} = a_{i+j}a_{i+j+1}\ldots a_{i+j+m}$.

Let us start our discussion by listing some important properties of Sturmian sequences and bisequences.

1. If $\mathbf{a}$ is a Sturmian sequence or bisequence, then for $n \longrightarrow \infty$ the ratio between the 1-length of a subword of length n and of n converges to a number $\nu(\mathbf{a})$, which is called the 1-*frequency* of $\mathbf{a}$ (see [119], Theorem 2.2.). Besides this number, we consider the number $\mu(\mathbf{a})$ being the limit of ratios between 1-lengths and 0-lengths of subwords, for word-lengths tending to ∞ if $\nu(\mathbf{a}) \leq \frac{1}{2}$, and being the limit of ratios between 0-lengths and 1-lengths if $\nu(\mathbf{a}) \geq \frac{1}{2}$.

Further, if in $\mathbf{a} = (\ldots a_{-2}a_{-1})a_0a_1a_2\ldots$ both 0 and 1 appear, then there is one symbol s with the following property:

$$\text{For all } i \text{ it holds } a_{i+1} \neq s \text{ if } a_i = s. \tag{3.9}$$

First we want to assume that (3.9) is satisfied for $a = 1$, in other words, that $\nu(\mathbf{a}) \leq \frac{1}{2}$. Obviously, then $\mu(\mathbf{a}) = \frac{\nu(\mathbf{a})}{1-\nu(\mathbf{a})}$.

2. Consider a Sturmian bisequence $\mathbf{a}$ different from $\ldots 00000\ldots$ and $\ldots 00100\ldots$, i.e. satisfying $\mu(\mathbf{a}) > 0$. (In the following context, it is not necessary to distinguish two Sturmian bisequences $\ldots a_{-2}a_{-1}a_0a_1a_2\ldots$ and $\ldots a'_{-2}a'_{-1}a'_0a'_1a'_2\ldots = \ldots a_{-2+j}a_{-1+j}a_ja_{1+j}a_{2+j}\ldots$. In fact, one would have to factorize, but there should not be any misunderstanding.)

There exists a minimal number $k = k(\mathbf{a})$ such that 10^k1 is a subword of $\mathbf{a}$. Clearly, no subword of $\mathbf{a}$ only consisting of symbols 0 is longer than $k+1$.

If one substitutes all subwords 0^k1 of $\mathbf{a}$ by $0'$ and then the remaining symbols 0 by $1'$, one gets a new sequence $r(\mathbf{a})$, which turns out to be Sturmian too. (Later we will write 0 and 1 instead of 0' and 1', respectively).

In order to show this, assume that $r(\mathbf{a})$ has two 0'-1'-subwords $\mathbf{v}, \mathbf{w}$ whose 0'-lengths differ by two, of minimal common lengths. Then one can suppose that $\mathbf{v} = 0'\tilde{\mathbf{v}}0'$ and $\mathbf{w} = 1'\tilde{\mathbf{w}}1'$, where $\tilde{\mathbf{v}}$ and $\tilde{\mathbf{w}}$ are 0'-1'-words of common 0'-length. Since obviously a successor of 1' must be 0', the word $\mathbf{w}0'$ is contained in $r(\mathbf{a})$. By the 'backward'-substitutions $0' \longrightarrow 0^k1$ in $\mathbf{v}$ and $1' \longrightarrow 0$ in $\mathbf{w}0'$, and then, by deleting the first k symbols and the last one symbol in the first and second result, respectively, one obtains two 0-1-subwords of $\mathbf{a}$ of common length. Their 0-length differs by two, but this is impossible by assumption.

It is easy to see that $\mu(r(\mathbf{a})) = \frac{1}{\mu(\mathbf{a})} \bmod 1$ and that $k(\mathbf{a})$ is the integer part of $\frac{1}{\mu(\mathbf{a})}$.

3. One can form $r(\mathbf{a}), r^2(\mathbf{a}) = r(r(\mathbf{a})), \ldots, r^n(\mathbf{a})$ as long as $r^{n-1}(\mathbf{a})$ does not coincide with $\ldots 00000\ldots$ or $\ldots 00100\ldots$. If $r^n(\mathbf{a})$ is defined, then the last substitution yields the number $k_n = k(r^{n-1}(\mathbf{a})) \in \mathbb{N}$. One distinguishes three cases:

a) If the substitution process leads to the bisequence $\ldots 00000\ldots$, then the bisequence $\mathbf{a}$ is periodic. Moreover, $\mu(\mathbf{a})$ is rational, and the k_i form the (regular) continued fraction of $\mu(\mathbf{a})$:

$$\mu(\mathbf{a}) = [; k_1, k_2, k_3, \ldots, k_n] = \cfrac{1}{k_1 + \cfrac{1}{k_2 + \cfrac{1}{k_3 + \ldots \frac{1}{k_n}}}} .$$

By construction in each case $k_n > 1$. Nevertheless, we want to refer to the well-known fact that continued fraction expansion is not unique anymore, if one allows expansions ending by 1: from the above expansion of $\mu(\mathbf{a})$ one gets a second one by $[; k_1, k_2, k_3, \ldots, k_n - 1, 1]$. This will play a certain role later.

b) If the substitution process leads to the bisequence $\ldots 00100\ldots$ - in [119], Sturmian bisequences of this type were called skew -, the sequence $\mathbf{a}$ is not recurrent. $\mu(\mathbf{a})$ is represented by the same formula as above. Sturmian sequences of that type will be irrelevant for our purposes.

c) If the substitution process does not end, then the bisequence $\mathbf{a}$ is recurrent, but non-periodic, and $\mu(\mathbf{a})$ is irrational with

$$\mu(\mathbf{a}) = [; k_1, k_2, k_3, \ldots] = \cfrac{1}{k_1 + \cfrac{1}{k_2 + \cfrac{1}{k_3 + \cfrac{1}{\ldots}}}} .$$

4. By the assumption $\nu(\mathbf{a}) \le \frac{1}{2}$, we have $\mu(\mathbf{a}) = \frac{\nu(\mathbf{a})}{1-\nu(\mathbf{a})} = (\frac{1}{1/\nu(\mathbf{a})-1})^{-1}$, and so the continued fraction expansion of $\nu(\mathbf{a})$ is equal to $[; k_1+1, k_2, k_3, \ldots]$.

Each Sturmian bisequences (sequence) with 1-frequency between $\frac{1}{2}$ and 1 arises from one with 1-frequency not greater than $\frac{1}{2}$, substituting 1 by 0 and 0 by 1. Therefore, the case $\nu(\mathbf{a}) \geq \frac{1}{2}$ can be dealt with as the case $\nu(\mathbf{a}) \leq \frac{1}{2}$. One only needs to use $\mu(\mathbf{a}) = \frac{1-\nu(\mathbf{a})}{\nu(\mathbf{a})}$ and to change the symbols as indicated. Then one gets the same results as above excepting the difference that $\nu(\mathbf{a}) = \frac{1}{1+\mu(\mathbf{a})}$. Thus, if the number $\nu(\mathbf{a})$ satisfies $\frac{1}{2} \leq \nu(\mathbf{a}) < 1$, it has continued fraction expansion $[;1, k_1, k_2, \ldots, k_n(,\ldots)]$.

Finally, note that each 1-frequency ν_0 between 0 and 1 is realized by a Sturmian bisequence. For example, if $\nu \leq \frac{1}{2}$ is given, then fix a point $\gamma \in \mathbb{T}$, and define $a_i = 1$ if $x+i\nu$ mod 1 lies in the interval $[0, \nu[$, and $a_i = 0$ else. The bisequence $\ldots a_{-2}a_{-1}a_0a_1a_2\ldots$ constructed in this way is Sturmian and has 1-frequency ν. Bisequences constructed in this way were called mechanical bisequences and coincide with the recurrent Sturmian sequences (see [119]).

Each Sturmian sequence can be extended to a Sturmian bisequence as shown in [119] (see Theorems 6.2., 6.3., 6.4.), and so to deal with (recurrent) Sturmian sequences is no more than to deal with the 'right' parts of Sturmian bisequences.

Subsequently, let *Sturm* be the set of all recurrent Sturmian sequences, and for $\nu \in [0,1[$ let $Sturm^\nu$ be the set of all recurrent Sturmian sequences with 1-frequency ν. By the above discussion, recurrent sequences for rational ν must be periodic.

Let us have a closer look at the Cantor sets $\mathbb{P}_\infty^\alpha$ which played an important role for the description of $\approx^\alpha$ for $\alpha \in \mathbb{P}_\infty$. For this, according to Theorem 3.34(ii) we assign to each $\alpha \in \mathbb{P}_\infty$ a periodic point and a Sturmian sequence as follows:

$$\alpha^* \in \mathbb{P}_* \text{ and } \mathbf{a}(\alpha) \in Sturm_0 \text{ with } \alpha = q^{\alpha^*}(\mathbf{a}(\alpha)). \tag{3.10}$$

α^* and $\mathbf{a}(\alpha)$ are uniquely defined and can be determined by use of $m = PER(\hat{\alpha})$ since $\alpha \in Gap(\alpha^*) \cap \mathbb{T}$: According to Theorem 3.34 and Proposition 3.27 the binary expansion of α possesses a decomposition $(\mathbf{b}_1, \mathbf{b}_2, \mathbf{b}_3, \ldots)$ into subwords of length m of two types. The binary expansion of α^* is given by $\overline{\mathbf{b}_1}$, and that of $\mathbf{a}(\alpha)$ is obtained by the substitution $\mathbf{b}_i \longrightarrow 0$ if $\mathbf{b}_i = \mathbf{b}_1$ and $\mathbf{b}_i \longrightarrow 1$ if $\mathbf{b}_i \neq \mathbf{b}_1$.

Now we are able to complete the description of $\mathbb{P}_\infty^\alpha$ in Proposition 2.31.

Theorem 3.36. (Properties of $\mathbb{P}_\infty^\alpha$, part II)
Let $\alpha \in \mathbb{P}_\infty, m = PER(\hat{\alpha})$ and let ν be the 1-frequency of $\mathbf{a}(\alpha)$. Then $\mathbb{P}_\infty^\alpha = q^{\alpha^*}(Sturm^\nu)$, and $(\mathbb{P}_\infty^\alpha/\approx^\alpha, (h/\approx^\alpha)^m)$ is conjugate to an irrational rotation on the unit circle with rotation number ν (relative to 2π.)

Proof: Let $k_1k_2k_3\ldots \in \mathbb{N}^\mathbb{N}$, let $\mathbf{a} = \mathbf{a}(k_1k_2k_3\ldots)$, and let ν be the 1-frequency of $\mathbf{a}$. Clearly, the set $Sturm^\nu$ is closed in $\{0,1\}^\mathbb{N}$, hence it is compact.

The orbit of $\mathbf{a}$ with respect to σ is dense in $Sturm^\nu$ since each initial subword of a sequence in $Sturm^\nu$ forms a subword of $\mathbf{a}$. This follows from the recurrence of the elements in $Sturm^\nu$ and the considerations on substitution of Sturmian sequences. For $\gamma \in \mathbb{P}_*$ and $m = PER(\gamma)$, the orbit of $\alpha = q^\gamma(\mathbf{a})$ with respect to h^m is dense in $\mathbb{P}^\alpha_\infty$ (see Proposition 2.31(ii)). Therefore, Proposition 3.23(iv) implies $\mathbb{P}^\alpha_\infty = q^\gamma(Sturm^\nu)$.

Finally, Q^α conjugates the dynamical systems $(\mathbb{P}^\alpha_\infty = q^\gamma(Sturm^\nu), h^m)$ and $(\mathbb{P}^\delta_\infty = q^0(Sturm^\nu), h)$, where $\delta = q^0(\mathbf{a})$. (For all maps under consideration we mean the corresponding restrictions.) In particular, Q^α preserves the order or reverses it. Therefore, by Proposition 2.31(iii) the topological dynamical systems $(\mathbb{P}^\alpha_\infty, h^m)/\approx^\alpha$ and $(\mathbb{P}^\delta_\infty, h)/\approx^\delta$ are conjugate too.

Now one easily sees that the 'relative' number of symbols 1 in the binary expansion of δ is equal to the average number of rotations of $h/\approx^\delta$ on $\mathbb{P}^\delta_\infty/\approx^\delta$. Thus $h/\approx^\delta$ defines a 'rotation' with rotation number ν, which completes the proof. ■

Our discussion on the irrational rotations did not include the orientation. Thus we had only rotation numbers less or equal to $\frac{1}{2}$ (relative to 2π). This was meaningful in view of the topological classification of Abstract Julia sets in Chapter 4.2.2, but in view of concrete quadratic dynamics it is convenient to indicate orientation.

For this, fix some $\alpha \in \mathbb{P}_*$ with $\alpha < \overline{\alpha}$ and define $\mathbb{T}(\alpha)$ to be the intersection of the gap $Gap(\alpha)$ and $\mathbb{T}$. Then we have $\mathbb{T}(\alpha) = q^\alpha(Sturm_0) \cup q^{\overline{\alpha}}(Sturm_0) = q^\alpha(Sturm_0 \cup Sturm_0^-)$ and $q^\alpha(\mathbf{a}) = q^{\overline{\alpha}}(\mathbf{a}^-)$.

To each $\gamma \in \mathbb{T}(\alpha)$ we assign a point $\nu^\alpha(\gamma) = \nu^{\overline{\alpha}}(\gamma)$ in the following way:

$$\nu^\alpha(q^\alpha(\mathbf{a})) = \begin{cases} \nu(\mathbf{a}) & \text{if } \mathbf{a} \in Sturm_0 \\ 1 - \nu(\mathbf{a}^-) & \text{if } \mathbf{a} \in Sturm_0^- \end{cases} . \tag{3.11}$$

In this way we get a map $\nu^\alpha = \nu^{\overline{\alpha}}$ from $\mathbb{T}(\alpha)$ onto $\mathbb{T}$, which we want to characterize in more detail now. We begin with the map ν defined on $Sturm_0$.

Lemma 3.37. *With respect to the lexicographic order, the 1-frequency ν is monotonically non-decreasing on $Sturm_0$, and ν is equal for two different sequences in $Sturm_0$ iff these sequences are equal to $\mathbf{a}(k_1k_2\ldots k_n1)$ and $\mathbf{a}(k_1k_2\ldots k_n+1)$ for some non-empty $k_1k_2\ldots k_n \in \mathbb{N}^*$. The latter sequences lie one a common orbit with respect to σ.*

Proof: For $k_1k_2\ldots k_n \in \mathbb{N}^*$, it is convenient to write $\mathbf{a}(k_1k_2\ldots k_n\infty(= k_{n+1}))$ instead of $\mathbf{a}(k_1k_2\ldots k_n)$. This enables us to describe the lexicographic order on $Sturm_0$ by use of the finite or infinite sequences $k_1k_2k_3\ldots$.

According to the discussion of Sturmian sequences in 2., one easily obtains the following:

The substitutions $0^{k_1}1 \to 0$ and $0 \to 1$ transform $\mathbf{a}(k_1k_2k_3\ldots)$ into $\mathbf{a}(k_2k_3\ldots)$, and $\mathbf{a}(k_2k_3\ldots) > \mathbf{a}(k_2'k_3'\ldots)$ if $\mathbf{a}(k_1k_2k_3\ldots) < \mathbf{a}(k_1k_2'k_3'\ldots)$.

Thus, if for given $k_1k_2k_3\ldots, k_1'k_2'k_3'\ldots$, the index i is the minimal one with $k_i \neq k_i'$ then $\mathbf{a}(k_1k_2k_3\ldots) < \mathbf{a}(k_1'k_2'k_3'\ldots)$ iff i is odd and $k_i > k_i'$, or i is even and $k_i < k_i'$.

According to 4. $\nu(\mathbf{a}(k_1k_2k_3\ldots))$ has continued fraction expansion $[;k_1+1,k_2,k_3,\ldots]$. So ν maps $Sturm_0$ onto $[0,\frac{1}{2}]$, and the monotonicity follows from simple continued fraction theory (e.g., compare [83]).

As mentioned above, $[;k_1,k_2,\ldots,k_n,1]$ and $[;k_1,k_2,\ldots,k_n+1]$ represent one and the same rational number. This can also be seen in the frame of the 1-frequencies of the sequences $\overline{\mathbf{a}(k_1k_2\ldots k_n1)}$ and $\overline{\mathbf{a}(k_1k_2\ldots k_n+1)}$. The first sequence is equal to $\overline{\mathbf{a}(k_1k_2\ldots k_{n-1})^{k_n}\mathbf{a}(k_1k_2\ldots k_{n-2})\mathbf{a}(k_1k_2\ldots k_{n-1})}$ and the second one to $\overline{\mathbf{a}(k_1k_2\ldots k_{n-1})^{k_n+1}\mathbf{a}(k_1k_2\ldots k_{n-2})}$, such that they lie on a common orbit with respect to σ. ■

Now let $\nu\in\,]0,\frac{1}{2}]$ have continued fraction expansion $[;\kappa_1,\kappa_2,\ldots,\kappa_l(,\ldots)]$ and let $\frac{p_n}{q_n}=[;\kappa_1,\kappa_2,\ldots,\kappa_n];n=1,2,\ldots,l(,\ldots)$ be the corresponding n-th convergents. It is well-known that $q_n=\kappa_nq_{n-1}+q_{n-2}$ for all $n=1,2,\ldots,l(,\ldots)$, where $q_{-1}=0$ and $q_0=1$ (e.g., compare [83]).

Setting $k_1k_2k_3k_4\ldots=\kappa_1-1\kappa_2\kappa_3\kappa_4\ldots$, one easily checks that q_n is no more than the length of $\mathbf{a}(k_1k_2\ldots k_n)$ and ν is the 1-frequency of $\mathbf{a}(k_1k_2k_3\ldots)$. In particular, the denominator of the (reduced) fraction corresponding to some $\nu(\overline{\mathbf{a}(k_1k_2\ldots k_n)})$ is equal to $|\mathbf{a}(k_1k_2\ldots k_n)|$.

Proposition 3.38. (Parameterization of $\mathbb{T}(\alpha)$ for $\alpha\in\mathbb{P}_*$)
Let $\alpha\in\mathbb{P}_*$. Then ν^α maps $\mathbb{T}(\alpha)=Gap(\alpha)\cap\mathbb{T}=q^\alpha(Sturm_0)\cup q^{\overline{\alpha}}(Sturm_0)$ continuously onto $\mathbb{T}$ and satisfies the following properties:

(i) ν^α is monotonically non-decreasing and invertible on the set of irrational numbers.

(ii) If $\gamma\overline{\gamma}$ is immediately visible from $\alpha\overline{\alpha}$, then $\nu^\alpha(\gamma)$ is rational, it holds $\nu^\alpha(\gamma)=\nu^\alpha(\overline{\gamma})$, and the denominator of the fraction corresponding to $\nu^\alpha(\gamma)$ is equal to $\frac{PER(\gamma)}{PER(\alpha)}$.

Proof: By Lemma 3.37, ν^α maps $q^\alpha(Sturm_0)$ monotonically non-decreasing onto $[0,\frac{1}{2}]$ and $q^{\overline{\alpha}}(Sturm_0)$ monotonically non-decreasing onto $[\frac{1}{2},1]$. Further, by Theorem 3.26(i), a sequence $\mathbf{s}$ in $Sturm_0$ is periodic with respect to σ iff $q^\alpha(\mathbf{s})$ (and $q^{\overline{\alpha}}(\mathbf{s})$) is. This together with Theorem 3.34 shows (i), and the continuity of ν^α is an immediate consequence.

In order to obtain (ii), by the discussion preceding the proposition, one has only to show that for each $k_1k_2\ldots k_n$, the period of $\overline{\mathbf{a}(k_1k_2\ldots k_n)}$ is equal to the length of $q=\mathbf{a}(k_1k_2\ldots k_n)$.

If there existed some $k_1k_2\ldots k_n\in\mathbb{N}^*$ not satisfying this property, we could assume that the length of $k_1k_2\ldots k_n\in\mathbb{N}^*$ is minimal with this property. According to Lemma 3.11(i), there would exist some $l>1$ and some $\mathbf{b}\in\{0,1\}^*$ with $\mathbf{a}(k_1k_2\ldots k_n)=\mathbf{b}^l$. Obviously, $\mathbf{b}$ would begin with $0^{k_1}1$ and end with $0^{k_1}1$ or $0^{k_1}10$ (see 2.)

Substituting in the 0-1-word $\mathbf{a}(k_1k_2\ldots k_n)$ by $0^{k_1}1\to 0,0\to 1$ as described in 2., 3., one would get $\mathbf{a}(k_2k_3\ldots k_n)$. Since the substitution is within the words **b**, it would hold $PER(\overline{\mathbf{a}(k_1k_2\dots k_{n-1})}) < |\mathbf{a}(k_2k_3\dots k_n)|$, contradicting our assumption. ■

3.3 Renormalization

3.3.1 A symbolic concept for renormalization

In Chapter 3.2, we observed two interesting phenomena:

a) There exist Cantor sets A on which some iterate h^n of h acts like h on the circle $\mathbb{T}$. More precisely, A is h^m-invariant, and $(\mathbb{T}, h)$ is conjugate to the topological dynamical system (A, h^n) factorized by the equivalence relation which identifies endpoints of neighboring (interval) components of $\mathbb{T} \setminus A$ pairwisely.
Moreover, for some special Julia equivalences $\approx$ the topological dynamical system $(A, h^n)/\approx$ is conjugate to $(\mathbb{T}, h)/\tilde{\approx}$ for some other Julia equivalence $\tilde{\approx}$.
b) There exist points $\gamma \in \mathbb{T}$, whose kneading sequences become periodic after substituting each n-th symbol by $*$, but possibly fail to be periodic themselves.

Both phenomena are strongly related, which we want to show now. Our systematic studies start with the symbolic side of the medal.

Definition 3.39. (Renormalizable sequences)
Let $\mathbf{s} \in \{0, 1, *\}^{\mathbb{N}}$ *and set* $PER(\mathbf{s}) := \infty$ *if* $\mathbf{s}$ *is non-periodic. For* $n \in \mathbb{N}$, *let* $\mathbf{s}^{*_n}$ *be the sequence obtained from* $\mathbf{s}$ *by substituting each* n*-th symbol by* $*$.

$\mathbf{s}$ *is said to be* n-renormalizable *for* $n \in \mathbb{N} \setminus \{1\}$ *if* $n \leq PER(\mathbf{s})$, *and if the sequence* $\mathbf{s}^{*_n}$ *is periodic, but its initial subword of length* $n-1$ *does not contain* $*$.

If $\mathbf{s} \in \{0, 1, *\}$ is n-renormalizable, then by definition $PER(\mathbf{s}^{*_n}) = n$. Further, we have the following

Lemma 3.40. *Let* $\mathbf{s} \in \{0, 1, *\}^{\mathbb{N}}$ *be* n*-renormalizable for some* $n \in \mathbb{N} \setminus \{1\}$. *Then* n *divides* $PER(\mathbf{s})$.

Proof: Let $m = PER(\mathbf{s}) < \infty$. If n were not a divisor of m, then the greatest common divisor k of n and m would be less than n. Consider $\mathbf{s}$ as a map on the (additive) cyclic group $\{1, 2, \dots, m = 0\}$. Its subgroup generated by n has order $\frac{m}{k}$ and consists of the elements $0, k, 2k, \dots, m-k$.

Since $\mathbf{s}^{*_n}$ is periodic with $PER(\mathbf{s}^{*_n}) = n$, one would obtain $\mathbf{s}(l) = \mathbf{s}(k + l) = \mathbf{s}(2k+l) = \dots = \mathbf{s}(m-k+l)$ for all $l = 1, 2, \dots, k$. This would contradict $k < m = PER(\mathbf{s})$. ■

Clearly, equality of sequences in $\{0, 1, *\}^{\mathbb{N}}$ is stronger than their substantial equality, but we know that Julia equivalences can very well be described by use of the concept of substantial equality. So we want to discuss the question of which iterates of the sequence $\mathbf{s}^{*_n}$ with respect to σ are substantially equal to $\mathbf{s}^{*_n}$.

Lemma 3.41. *Let* $\mathbf{s} \in \{0,1,*\}^{\mathbb{N}}$ *be* n*-renormalizable for* $n \in \mathbb{N} \setminus \{1\}$ *and let* $\sigma^k(\mathbf{s}^{*_n}) \cong \sigma^l(\mathbf{s}^{*_n})$ *for* $k,l \in \{0,1,2,\ldots,n-1\}$ *with* $k < l$. *Then there exists a unique 0-1-sequence* $\mathbf{t} \in \{0,1\}^{\mathbb{N}}$ *satisfying* $\mathbf{t} \cong \sigma^k(\mathbf{s}^{*_n}), \sigma^l(\mathbf{s}^{*_n})$. *The sequence* $\mathbf{t}$ *is periodic with* $PER(\mathbf{t}) < n$, *and* $PER(\mathbf{t})$ *divides* n.

Proof: The existence, the uniqueness and the periodicity of $\mathbf{t}$ are obvious. So let $\mathbf{s}_1 = \sigma^k(\mathbf{s}^{*_n}), \mathbf{s}_2 = \sigma^l(\mathbf{s}^{*_n})$, and consider $\mathbf{s}_1, \mathbf{s}_2$ and $\mathbf{t}$ as maps on the (additive) cyclic group $\{1,2,\ldots,n=0\}$. We can assume that $\mathbf{s}_1(0) = \mathbf{s}_2(l-k) = *$ and that $\mathbf{s}_1, \mathbf{s}_2 \cong \mathbf{t}$.

If r is the order of the subgroup generated by $l-k < n$, then $* = \mathbf{s}_1(0) \cong \mathbf{s}_2(0) = \mathbf{s}_1(k-l) = \mathbf{s}_2(k-l) = \ldots = \mathbf{s}_1((r-2)(k-l)) = \mathbf{s}_2((r-2)(k-l)) = \mathbf{s}_1((r-1)(k-l)) \cong (\mathbf{s}_2((r-1)(k-l)) =)\mathbf{s}_2(l-k) = *$. Hence $\mathbf{t}$ is constant on the considered subgroup.

In a similar way one shows that $\mathbf{t}(i) = \mathbf{t}(i + \frac{n}{r})$ for all i not lying in this subgroup, and by Lemma 3.11(i) it follows that $PER(\mathbf{t})$ is a proper divisor of n. ■

Example 3.42. In a special case the sequence $\mathbf{t}$ given in Lemma 3.41 is well known for us: If $\alpha \in \mathbb{P}_*$ and $n = PER(\alpha)$, then $\hat{\alpha}^{*_n} = \hat{\alpha}$, and on the assumptions of Lemma 3.41 with $\mathbf{s} = \hat{\alpha}$ and $0 = k < l < n$, one has $\mathbf{t} = \mathbf{v}^\alpha(1 - e^\alpha)$ by Theorem 2.44.

We come to the central concepts of this section.

Definition 3.43. (n-renormalizable angles and renormalization domains) *An angle* $\gamma \in \mathbb{T}$ *is called* n-renormalizable *for* $n \in \mathbb{N} \setminus \{1\}$ *if* $\hat{\gamma}^{*_n}$ *is.*

Then we call $\mathbb{T}^\gamma_{*_n} = \{\beta \in \mathbb{T} \mid I^\gamma(\beta) \cong \hat{\gamma}^{*_n}\}$ *for* $\gamma \in \mathbb{T} \setminus \mathbb{P}_*$ *and* $\mathbb{T}^\gamma_{*_n} = \{\beta \in \mathbb{T} \mid \widetilde{I^\gamma}(\beta) \cong \hat{\gamma}^{*_n}\}$ *for* $\gamma \in \mathbb{P}_*$ *the* n-renormalization domain *of* γ.

Note that in the case of a periodic $\gamma \in \mathbb{T}$ of period n the set $\mathbb{T}^\gamma_{*_n}$ was already defined (see above Theorem 3.26). As in this special case it is obviously h^n-invariant.

In the following, we want to investigate the geometrical structure of n-renormalization domains and the mutual position of their finitely many iterates. As a first step we discuss under which conditions two such iterates have a non-empty intersection.

Proposition 3.44. *If* $\alpha \in \mathbb{T}$ *is* n*-renormalizable for* $n \in \mathbb{N} \setminus \{1\}$, *then for* $k,l \in \{0,1,2,\ldots,n-1\}$ *it holds* $\sigma^k(\hat{\gamma}^{*_n}) \cong \sigma^l(\hat{\gamma}^{*_n})$ *iff* $h^k(\mathbb{T}^\gamma_{*_n}) \cap h^l(\mathbb{T}^\gamma_{*_n}) \neq \emptyset$.

Proof: Let $k,l \in \{0,1,2,\ldots,n-1\}$ be different. If $\sigma^k(\hat{\gamma}^{*_n}) \cong \sigma^l(\hat{\gamma}^{*_n})$, then let $\mathbf{t}$ be the 0-1-sequence with $\mathbf{t} \cong \sigma^k(\hat{\gamma}^{*_n}), \sigma^l(\hat{\gamma}^{*_n})$, which is uniquely determined by Lemma 3.41. We show the existence of a $\beta \in \mathbb{T}$ with $I^\gamma(\beta) = \mathbf{t}$ for $\gamma \notin \mathbb{P}_*$ and $\widetilde{I^\gamma}(\beta) = \mathbf{t}$ for $\gamma \in \mathbb{P}_*$, from which $h^k(\mathbb{T}^\gamma_{*_n}) \cap h^l(\mathbb{T}^\gamma_{*_n}) \neq \emptyset$ follows immediately. For this, look at Proposition 2.56, in particular at the statements (i), (ii) and (v).

In case that $\gamma \notin \mathbb{P}_*$ and that $CARD^\gamma(\hat{\gamma}) > 1$, the existence of a point β with the above properties is obvious by (i). In case that $\gamma \notin \mathbb{P}_*$ and that $CARD^\gamma(\hat{\gamma}) = 1$, by (ii) such point β does not exist if $\mathbf{t}$ ends with $0\hat{\gamma}$ or $1\hat{\gamma}$. Then it would hold $PER(\hat{\gamma}) = PER(\mathbf{t})$, which is impossible by Lemma 3.40 and Lemma 3.41. Finally, in case that $\gamma \in \mathbb{P}_*$, by (v) there is no such β if $\mathbf{t}$ ends with $\overline{\mathbf{v}^\gamma e^\gamma}$. Theorem 2.44 yields the following contradiction: $PER(\hat{\gamma}) = PER(\overline{\mathbf{v}^\gamma e^\gamma}) = PER(\mathbf{t})$.

Conversely, let $\beta \in h^k(\mathbb{T}^\gamma_{*_n}) \cap h^l(\mathbb{T}^\gamma_{*_n})$, in other words, let $I^\gamma(\beta) \cong \sigma^k(\hat{\gamma}^{*_n}), \sigma^l(\hat{\gamma}^{*_n})$ if $\gamma \in \mathbb{T} \setminus \mathbb{P}_*$ and $\widetilde{I^\gamma}(\beta) \cong \sigma^k(\hat{\gamma}^{*_n}), \sigma^l(\hat{\gamma}^{*_n})$ if $\gamma \in \mathbb{P}_*$.

In case that $I^\gamma(\beta)$ contains at most one $*$, one has $\sigma^k(\hat{\gamma}^{*_n}) \cong \sigma^l(\hat{\gamma}^{*_n})$ by the periodicity of $\sigma^k(\hat{\gamma}^{*_n})$ and $\sigma^l(\hat{\gamma}^{*_n})$.

Otherwise, γ is periodic and it holds $\widetilde{I^\gamma}(\beta) \cong \sigma^k(\hat{\gamma}^{*_n}) \cong \sigma^l(\hat{\gamma}^{*_n})$. So by Lemma 3.41 $\widetilde{I^\gamma}(\beta)$ is periodic, and β lies in the orbit of γ. In this case let $m = PER(\gamma)$ and $\mathbf{v}^\gamma = v_1 v_2 \ldots v_{m-1}$, and fix the $j \in \{0, 1, 2, \ldots, m-1\}$ with $\beta = h^j(\gamma)$.

Now we have $I^\gamma(\beta) = \sigma^j(\hat{\gamma}) = \overline{v_{j+1} v_{j+2} \ldots v_{m-1} * v_1 v_2 \ldots v_j} \cong \sigma^k(\hat{\gamma}^{*_n}), \sigma^l(\hat{\gamma}^{*_n})$, and we use that by Lemma 3.40 n divides m.

If $n < m$, then the $m - j + n$-th symbols of the sequences $\sigma^k(\hat{\gamma}^{*_n})$ and $\sigma^l(\hat{\gamma}^{*_n})$, hence also their $m - j$-th symbols, coincide, or in each case one of the symbols is equal to $*$. This implies $\sigma^k(\hat{\gamma}^{*_n}) \cong \sigma^l(\hat{\gamma}^{*_n})$.

For $n = m$ one obtains $I^\gamma(\beta) = \sigma^j(\hat{\gamma}) \cong \sigma^k(\hat{\gamma}^{*_n}) = \sigma^k(\hat{\gamma})$ and $I^\gamma(\beta) = \sigma^j(\hat{\gamma}) \cong \sigma^l(\hat{\gamma}^{*_n}) = \sigma^l(\hat{\gamma})$. If $j = k$ or $j = l$, this immediately implies $\sigma^k(\hat{\gamma}^{*_n}) \cong \sigma^l(\hat{\gamma}^{*_n})$.

Otherwise, one applies Lemma 3.41 to k and j as well as to l and j. This results in sequences $\mathbf{t}_1, \mathbf{t}_2 \in \{0,1\}^{\mathbb{N}}$ with $\mathbf{t}_1 \cong \sigma^k(\hat{\gamma}^{*_n}), \sigma^j(\hat{\gamma}^{*_n})$ and $\mathbf{t}_2 \cong \sigma^l(\hat{\gamma}^{*_n}), \sigma^j(\hat{\gamma}^{*_n})$. According to Example 3.42 both sequences must be iterates of $\overline{\mathbf{v}^\gamma(1 - e^\gamma)}$ and so coincide, showing $\sigma^k(\hat{\gamma}^{*_n}) \cong \sigma^l(\hat{\gamma}^{*_n})$. ■

For a given n-renormalizable $\gamma \in \mathbb{T}$ we want to fix the minimal $k > 0$ satisfying $h^l(\mathbb{T}^\gamma_{*_n}) \cap \mathbb{T}^\gamma_{*_n} \neq \emptyset$. This leads to the following

Definition 3.45. *If $\gamma \in \mathbb{T}$ is n-renormalizable for $n \in \mathbb{N} \setminus \{1\}$, let $\mathbf{v}$ be the initial subword of $\hat{\gamma}$ of length $n-1$. Let $SUBPER_n(\gamma)$ be the minimum of $PER(\overline{\mathbf{v}0})$ and $PER(\overline{\mathbf{v}1})$. In case that $SUBPER_n(\gamma) < n$, let $s \in \{0,1\}$ be the unique symbol with $PER(\overline{\mathbf{v}s}) = SUBPER_n(\gamma)$ and let $\mathbf{t}^\gamma_{*_n} := \overline{\mathbf{v}s}$.*

By the above considerations, for $l > k \geq 0$ the sets $h^k(\mathbb{T}^\gamma_{*_n})$ and $h^l(\mathbb{T}^\gamma_{*_n})$ have non-empty intersection iff $SUBPER_n(\gamma)$ divides $l - k$. Moreover, if $l = SUBPER_n(\gamma) < n$, the unique sequence $\mathbf{t}$ with $\mathbf{t} \cong \hat{\gamma}^{*_n}, \sigma^l(\hat{\gamma}^{*_n})$ (see Lemma 3.41) is no more than $\mathbf{t}^\gamma_{*_n}$.

We proceed with two lemmata in preparation of the main result of this subsection.

Lemma 3.46. *Let $\gamma \in \mathbb{T}$ and $s \in \{0,1\}$, and let $\beta_1, \beta_2 \in \mathbb{T}_s^\gamma$ with $\beta_1\beta_2 \neq \frac{\gamma}{2}\frac{\gamma+1}{2}$ and $\delta_1, \delta_2 \in \mathbb{T}$ be given such that the chord $\delta_1\delta_2$ separates the points $h(\beta_1)$ and $h(\beta_2)$. Then there exist preimages $\widetilde{\delta_1}, \widetilde{\delta_2}$ of δ_1, δ_2 lying in $\mathbb{T}_s^\gamma$ and forming the endpoints of a chord which separates β_1 and β_2.*

Proof: If $\delta_1, \delta_2 \neq \gamma$, then let $\widetilde{\delta_1} = l_s^\gamma(\delta_1)$ and $\widetilde{\delta_2} = l_s^\gamma(\delta_2)$. Otherwise, let $\widetilde{\delta_1} = l_s^\gamma(\delta_1)$ and $\widetilde{\delta_2} = \frac{\gamma}{2}$ or $\widetilde{\delta_2} = \frac{\gamma+1}{2}$. Namely, we can assume that $\delta_2 = \gamma$, and have $\beta_1, \beta_2 \notin \{\frac{\gamma}{2}, \frac{\gamma+1}{2}\}$. So there is an open semi-circle containing $\beta_1, \beta_2, \widetilde{\delta_1}, \widetilde{\delta_2}$, and by the orientation invariance described in Lemma 2.9, the chord $\widetilde{\delta_1}\widetilde{\delta_2}$ must separate β_1 and β_2. ■

Lemma 3.47. *Let $\gamma \in \mathbb{P}_\infty$ be n-renormalizable for $n \in \mathbb{N} \setminus \{1\}$, and let β be a point in $\mathbb{T}_{*_n}^\gamma$ whose orbit contains $\frac{\gamma}{2}$ or $\frac{\gamma+1}{2}$. Then in each 'clockwise' and each 'counter-clockwise' neighborhood of β there exists at least one element of $\mathbb{T}_{*_n}^\gamma$.*

Proof: If $SUBPER_n(\alpha) = n$, then the sets $h^i(\mathbb{T}_{*_n}^\gamma); i = 0, 1, 2, \ldots, n-1$ are mutually disjoint. So the points $\frac{\gamma}{2}, \frac{\gamma+1}{2}$ lie in exactly one of these sets, namely in $h^{n-1}(\mathbb{T}_{*_n}^\gamma)$. The latter is also valid if $SUBPER_n(\gamma) < n$, but in this case we have to argue in another way: If it were false, there would exist a j with $0 < j < n$ and $\gamma \in \mathbb{T}_{*_n}^\gamma \cap h^j(\mathbb{T}_{*_n}^\gamma)$. From this one could easily deduce that $PER(\hat{\gamma}) < n$, which would contradict the definition of n-renormalizable.

So if β is given as above, we find a multiple k of n such that the k-th symbol of $I^\gamma(\beta)$ coincides with $*$. Let $\mathbf{u} = \hat{\gamma}|k-1$. Then β lies in the Cantor sets $\mathbb{T}_{\mathbf{u}0\hat{\gamma}}^\gamma \subset \mathbb{T}_{*_n}^\gamma$ and $\mathbb{T}_{\mathbf{u}1\hat{\gamma}}^\gamma \subset \mathbb{T}_{*_n}^\gamma$ (see Proposition 2.30 and Proposition 2.31). β is an accumulation point for both sets and we are finished. ■

Now we are able to provide the announced geometric description of n-renormalization domains and the mutual position of their iterates. For this, let $\mathbb{D}_{*_n}^\gamma$ be the convex hull of $\mathbb{T}_{*_n}^\gamma$.

Theorem 3.48. (Renormalization domain and its orbit)
Let $\gamma \in \mathbb{T}$ be n-renormalizable for $n \in \mathbb{N} \setminus \{1\}$. Then the sets $h^j(\mathbb{T}_{*_n}^\gamma); j = 0, 1, 2, \ldots, n-1$ are closed, h^n-invariant and transitively closed with respect to $\approx^\gamma$. Moreover, the endpoints of each chord bounding one of the sets $h^j(\mathbb{D}_{*_n}^\gamma); j = 0, 1, 2, \ldots, n-1$ are $\approx^\gamma$-equivalent.

For $k, l \in \{0, 1, \ldots, n-1\}$ with $k < l$, both the sets $h^k(\mathbb{T}_{*_n}^\gamma), h^l(\mathbb{T}_{*_n}^\gamma)$ and the sets $h^k(\mathbb{D}_{*_n}^\gamma), h^l(\mathbb{D}_{*_n}^\gamma)$ are disjoint iff $SUBPER_n(\gamma)$ does not divide $l-k$. Otherwise, $h^k(\mathbb{T}_{*_n}^\gamma) \cap h^l(\mathbb{T}_{*_n}^\gamma) = \mathbb{T}_{\sigma^r(\mathbf{t})}^\gamma$ and $h^k(\mathbb{D}_{*_n}^\gamma) \cap h^l(\mathbb{D}_{*_n}^\gamma) = \mathbb{D}_{\sigma^r(\mathbf{t})}^\gamma$ with $\mathbf{t} = \mathbf{t}_{*_n}^\gamma$ and $r = k \bmod SUBPER_n(\gamma)$, and $\mathbb{T}_{\sigma^r(\mathbf{t})}^\gamma$ is a $\approx^\gamma$-equivalence class.

Proof: We start our proof by showing the following: If $\beta_1, \beta_2 \in \mathbb{T}_{*_n}^\gamma$ are given such that one of the open intervals with endpoints β_1 and β_2 is disjoint to $\mathbb{T}_{*_n}^\gamma$, then it holds $\beta_1 \approx^\gamma \beta_2$. We first show $I^\gamma(\beta_1) \cong I^\gamma(\beta_2)$.

If this were false, then there would exist a least coordinate k with $\{I^\gamma(\beta_1)(k), I^\gamma(\beta_2)(k)\} = \{0,1\}$. Clearly, k would be a multiple of n, and the chord $\frac{\gamma}{2}\frac{\gamma+1}{2}$ would separate $h^{k-1}(\beta_1)$ and $h^{k-1}(\beta_2)$, and it would coincide with no $h^i(\beta_1\beta_2); i = 0, 1, \ldots, k-2$. By successive application of Lemma 3.46 one could show the existence of a chord with endpoints in $\mathbb{T}^\gamma_{*_n}$ which separates β_1 and β_2. This would contradict the above assumptions.

If $\gamma \in \mathbb{T} \setminus \mathbb{P}$, then $I^\gamma(\beta_1) \cong I^\gamma(\beta_2)$ implies $\beta_1 \approx^\gamma \beta_2$ by Theorem 2.47. In the case $\gamma \in \mathbb{P}_\infty$, by Lemma 3.47 the symbol $*$ is not contained in $I^\gamma(\beta_1)$ or $I^\gamma(\beta_2)$, and it holds $I^\gamma(\beta_1) = I^\gamma(\beta_2)$. By Theorem 2.50 one has only to exclude that $I^\gamma(\beta_1) = \mathbf{w}\hat{\gamma}$ for some $\mathbf{w} \in \{0,1\}^*$.

If there existed such $\mathbf{w}$, it would hold $\beta_1, \beta_2 \in \mathbb{T}^\gamma_{\mathbf{w}\hat{\gamma}} \subseteq \mathbb{T}^\gamma_{*_n}$. The boundary chords of $\mathbb{D}^\gamma_{\mathbf{w}\hat{\gamma}}$ are contained in the backward orbit of $\frac{\gamma}{2}\frac{\gamma+1}{2}$, and $\mathbb{T}^\gamma_{\mathbf{w}\hat{\gamma}}$ is a Cantor set (see Proposition 2.30 and Proposition 2.31). Thus β_1 and β_2 would be two-side accumulation points of $\mathbb{T}^\gamma_{*_n}$, in contradiction to the assumption.

In the case $\gamma \in \mathbb{P}_*$ one shows $\widetilde{I^\gamma}(\beta_1) = \widetilde{I^\gamma}(\beta_2)$ and is finished by Theorem 2.53. For this, let k be the least coordinate where $\widetilde{I^\gamma}(\beta_1)$ and $\widetilde{I^\gamma}(\beta_2)$ are different. k is a multiple of n, and by $I^\gamma(\beta_1) \cong I^\gamma(\beta_2)$ we can assume that $h^{k-1}(\beta_1) \in \{\dot{\gamma}, \ddot{\gamma}\}$. The points $\dot{\gamma}$ and $\ddot{\gamma}$ are elements of the Cantor set $h^{-1}(\mathbb{T}\langle\gamma\rangle)$, which lies between $\dot{\gamma}\dot{\overline{\gamma}}$ and $\ddot{\gamma}\ddot{\overline{\gamma}}$. Therefore $h^{k-1}(\beta_1)$ and $h^{k-1}(\beta_2)$ can be separated by a chord with endpoints in $h^{-1}(\mathbb{T}\langle\gamma\rangle) \subseteq \mathbb{T}^\gamma_{*_n}$, not lying at the orbit of γ. As above, Lemma 3.46 yields a chord with endpoints in $\mathbb{T}^\gamma_{*_n}$ separating β_1 and β_2, which is impossible again.

With the exception of transitive closeness, the first part of the theorem is obvious now. So assume that $k, l \in \{0, 1, \ldots, n-1\}$ with $k \neq l$ and $h^k(\mathbb{T}^\gamma_{*_n}) \cap h^l(\mathbb{T}^\gamma_{*_n}) \neq \emptyset$ are given. Then let $\mathbf{t}$ be the unique sequence with $\mathbf{t} \cong \sigma^k(\hat{\gamma}^{*_n}), \sigma^l(\hat{\gamma}^{*_n})$. Let us show that $h^k(\mathbb{T}^\gamma_{*_n}) \cap h^l(\mathbb{T}^\gamma_{*_n}) = \mathbb{T}^\gamma_{\mathbf{t}}$, and that $\mathbb{T}^\gamma_{\mathbf{t}}$ forms an equivalence class with respect to $\approx^\gamma$.

If $\gamma \notin \mathbb{P}_*$, then obviously $h^k(\mathbb{T}^\gamma_{*_n}) \cap h^l(\mathbb{T}^\gamma_{*_n}) = \widetilde{\mathbb{T}^\gamma_{\mathbf{t}}} = \mathbb{T}^\gamma_{\mathbf{t}}$. The latter equality follows from Proposition 2.6. By Theorem 2.48 and Theorem 2.51, $\mathbb{T}^\gamma_{\mathbf{t}}$ is a $\approx^\gamma$-equivalence class. In the case $\gamma \in \mathbb{P}_\infty$, consider that $PER(\mathbf{t}) < PER(\hat{\gamma})$. If $\gamma \in \mathbb{P}_*$, then $h^k(\mathbb{T}^\gamma_{*_n}) \cap h^l(\mathbb{T}^\gamma_{*_n}) = \{\beta \in \mathbb{T} \mid \widetilde{I^\gamma}(\beta) = \mathbf{t}\} = \mathbb{T}^\gamma_{\mathbf{t}}$ according to the discussion preceding Theorem 2.53, and by Theorem 2.57 $\mathbb{T}^\gamma_{\mathbf{t}}$ forms a $\approx^\gamma$-equivalence class.

Now come back to the general case, and show that $\mathbb{T}^\gamma_{*_n}$ is transitively closed with respect to $\approx^\gamma$. For this, fix elements β_1, β_2 of $\mathbb{T}$ satisfying $\beta_1 \in \mathbb{T}^\gamma_{*_n}$ and $\beta_2 \approx^\gamma \beta_1$.

In the case $\gamma \in \mathbb{P}$, the statement $\beta_2 \in \mathbb{T}^\gamma_{*_n}$ follows directly from Theorem 2.50 or Theorem 2.53. If $\gamma \in \mathbb{T} \setminus \mathbb{P}$ and $I^\gamma(\beta_1)$ does not end with $0\hat{\gamma}, 1\hat{\gamma}$ or $*\hat{\gamma}$, then $\beta_2 \in \mathbb{T}^\gamma_{*_n}$ is an immediate consequence of Theorem 2.47.

Otherwise, let $k \in \mathbb{N}$ be given with $\sigma^k(I^\gamma(\beta_1)) = I^\gamma(h^k(\beta_1)) = \hat{\gamma}$. The point $h^k(\beta_1)$ lies in $\mathbb{T}^\gamma_{*_n}$ but not in $h^j(\mathbb{T}^\gamma_{*_n})$ for all $j = 1, 2, 3, \ldots, n-1$. The latter follows since by the above shown in each case $\mathbb{T}^\gamma_{*_n} \cap h^j(\mathbb{T}^\gamma_{*_n})$ is empty or a $\approx^\gamma$-equivalence class which by Corollary 2.23(i) contains only periodic points. Thus k is a multiple of n, and it holds $\beta_2 \in \mathbb{T}^\gamma_{*_n}$ by Theorem 2.47.

So we have shown that $\mathbb{T}^{\gamma}_{*_n}$ is transitively closed with respect to $\approx^{\gamma}$. Since images of $\approx^{\gamma}$-equivalence classes are equivalence classes, also the sets $h^j(\mathbb{T}^{\gamma}_{*_n}); j \in \mathbb{N}$ are transitively closed. ■

3.3.2 Renormalization, simple renormalization and tuning

Now we investigate the relation between (sequence) tuning and renormalization. In order to motivate the following definition we have a closer look at the results of sequence tuning.

The kneading sequence of a point coincides with $\overline{*}$, or its first symbol is 0. Hence sequence tuning with respect to some $\alpha \in \mathbb{P}_* \setminus \{0\}$ yields a sequence starting with $\mathbf{v}^{\alpha}*$ or $\mathbf{v}^{\alpha}e^{\alpha}$. This sequence is n-renormalizable for $n = PER(\alpha)$, and by Proposition 3.14 there exists a $k \in \mathbb{N}$ satisfying $(\chi^{\hat{\alpha}}_{\hat{\alpha}})^k(1) = n$. (Recall that $\chi^{\hat{\alpha}}_{\hat{\alpha}}(i)$ for given $\alpha \in \mathbb{T}$ and $i \in \mathbb{N}$ denotes the smallest $j > i$ for which $\sigma^i(\hat{\alpha})$ and $\hat{\alpha}$ have different initial subwords of length $j - i$). The following definition is devoted to this fact.

Definition 3.49. *A point $\gamma \in \mathbb{T}$ is said to be* simply n-renormalizable *for $n \in \mathbb{N} \setminus \{1\}$ if it is n-renormalizable and n belongs to the internal address of γ, i.e. $(\chi^{\hat{\gamma}}_{\hat{\gamma}})^k(1) = n$ for some $k \in \mathbb{N}$.*

So, if γ lies in $[\mathbb{T}\langle\alpha\rangle]_{\sim} = [Q^{\alpha}(\mathbb{T})]_{\sim}$ - the α-tuning of $\mathbb{T}$ - for some $\alpha \in \mathbb{P}_* \setminus \{0\}$, then γ is simply $PER(\alpha)$-renormalizable. In particular, the points in $\mathbb{T}(\alpha) = q^{\alpha}(Sturm_0) \cup q^{\overline{\alpha}}(Sturm_0)$ are simply n-renormalizable for $n = PER(\alpha)$ (see Theorem 3.34). Also note the obvious fact that $\gamma \in \mathbb{T}$ is simply n-renormalizable if its internal address is of the form

$$1 \to \ldots \to n \to k_1 n \to k_2 n \to \ldots \to k_l n \ (\to \ldots)$$

with $1 < k_1 < k_2 < \ldots < k_l \ (< \ldots)$.

The aim of this subsection is to show that simple renormalization can be considered as the inversion of tuning.

Theorem 3.50. (Characterization of simple renormalization and tuning) *For $\gamma \in \mathbb{T}$ and $n \in \mathbb{N} \setminus \{1\}$ the following statements are equivalent:*

(i) γ is simply n-renormalizable.

(ii) There exists an $\alpha \in \mathbb{P}_*$ satisfying $n = PER(\alpha)$ and $\mathbb{T}^{\gamma}_{*_n} = \mathbb{T}^{\alpha}_{*_n}$.

(iii) There exists an $\alpha \in \mathbb{P}_*$ with $n = PER(\alpha)$ and γ in the α-tuning of $\mathbb{T}$.

If $\gamma \in \mathbb{T}$ is n-renormalizable with $SUBPER_n(\gamma) = n$, then γ is simply n-renormalizable.

Proof: '(i) $\Longrightarrow$ (ii)': Let γ be simply n-renormalizable with internal address

$m_0 = 1 \to m_1 \to \ldots \to m_k = n > 1$. Further, let $B^{(m_k)} = \alpha\overline{\alpha}$ (see before Definition 3.13). We can assume that γ lies behind $B^{(m_k)}$, and so by Theorem 3.6(ii) that γ is not periodic with $PER(\gamma) = n$.

$\frac{\gamma}{2}\frac{\gamma+1}{2}$ separates the chords $\dot{\alpha}\dot{\overline{\alpha}}$ and $\ddot{\alpha}\ddot{\overline{\alpha}}$, and it holds $\mathbf{v}^\alpha = \hat{\gamma}|n-1$ (see below Definition 3.13). Moreover, we have $\hat{\alpha} \cong I^\gamma(\alpha) = I^\gamma(\overline{\alpha})$, and $\hat{\alpha} \cong \widetilde{I^\gamma}(\alpha) = \widetilde{I^\gamma}(\overline{\alpha})$ if $\gamma \in \mathbb{P}_*$, since each iterate of $\alpha\overline{\alpha}$ lies behind $\dot{\alpha}\dot{\overline{\alpha}}$ or $\ddot{\alpha}\ddot{\overline{\alpha}}$.

By Proposition 3.23(iii) each set $h^i(\mathbb{T}\langle\alpha\rangle); i = 0,1,2,\ldots,n-2$, hence also each set $h^i([\mathbb{T}\langle\alpha\rangle]_{\approx^\alpha}); i = 0,1,2,\ldots,n-2$ lies behind $\dot{\alpha}\dot{\overline{\alpha}}$ or behind $\ddot{\alpha}\ddot{\overline{\alpha}}$. So by $\mathbb{T}^\alpha_{*_n} = [\mathbb{T}\langle\alpha\rangle]_{\approx^\alpha}$ (see Theorem 3.26(ii)) we have $\mathbb{T}^\alpha_{*_n} \subseteq \mathbb{T}^\gamma_{*_n}$.

In order to show $\mathbb{T}^\gamma_{*_n} \subseteq \mathbb{T}^\alpha_{*_n}$ use that $\mathbb{T}^\gamma_{*_n} = \bigcup\{\mathbb{T}^\gamma_{\mathbf{s}} \mid \mathbf{s} \in \{0,1\}^{\mathbb{N}}, \mathbf{s} \cong \overline{\mathbf{v}*}\}$, where $\mathbf{v} = \mathbf{v}^\alpha$. In the case $\gamma \in \mathbb{P}_*$ this equality follows from the fact that a point with specified itinerary $\mathbf{s}$ belongs to the equivalence class $\mathbb{T}^\gamma_{\mathbf{s}}$ (see Theorem 2.57 and the discussion before Theorem 2.53). Otherwise, we have $\mathbb{T}^\gamma_{\mathbf{s}} = \widetilde{\mathbb{T}^\gamma_{\mathbf{s}}}$ for each 0-1-sequence $\mathbf{s}$ (see Proposition 2.6), which immediately implies the above equality.

By Proposition 3.23(iii), $l^\gamma_{\mathbf{v}}(\eta) = l^\alpha_{\mathbf{v}}(\eta)$ for all $\eta \in h^{n-1}(\mathbb{T}\langle\alpha\rangle)$ whose backward iterates are different from α and γ. So for some fixed $\beta \in \mathbb{T}\langle\alpha\rangle$ not belonging to the orbit of α or γ it holds $l^\gamma_0(\beta), l^\gamma_1(\beta) \in h^{n-1}(\mathbb{T}\langle\alpha\rangle)$, hence $l^\gamma_{\mathbf{v}0}(\beta), l^\gamma_{\mathbf{v}1}(\beta) \in \mathbb{T}\langle\alpha\rangle$.

The sequence $l^\gamma_{\mathbf{v}t_1}(\beta), l^\gamma_{\mathbf{v}t_1\mathbf{v}t_2}(\beta), l^\gamma_{\mathbf{v}t_1\mathbf{v}t_2\mathbf{v}t_3}(\beta), \ldots$ assigned to some given 0-1-sequence $t_1t_2t_3\ldots$ accumulates in $\mathbb{T}^\gamma_{\mathbf{v}t_1\mathbf{v}t_2\mathbf{v}t_3\ldots}$. Therefore, by the closeness of $\mathbb{T}\langle\alpha\rangle$ we can deduce that $\mathbb{T}^\gamma_{\mathbf{s}} \cap \mathbb{T}\langle\alpha\rangle \neq \emptyset$ for all 0-1-sequences $\mathbf{s}$ satisfying $\mathbf{s} \cong \overline{\mathbf{v}*}$. By Theorem 3.1 (see (vi)) it holds $\alpha \approx^\gamma \overline{\alpha}$, and by Proposition 2.65 this implies $[\alpha]_{\approx^\gamma} = [\alpha]_{\approx^\alpha}$, hence also $[\dot{\alpha}]_{\approx^\gamma} = [\dot{\alpha}]_{\approx^\alpha}$ and $[\ddot{\alpha}]_{\approx^\gamma} = [\ddot{\alpha}]_{\approx^\alpha}$.

Now let $\mathbf{s} \cong \overline{\mathbf{v}*}$. If $\alpha \in \mathbb{T}^\gamma_{\mathbf{s}}$, it follows $\mathbb{T}^\gamma_{\mathbf{s}} = [\alpha]_{\approx^\gamma} = [\alpha]_{\approx^\alpha} \subseteq \mathbb{T}^\alpha_{*_n}$, again by Proposition 2.65. If $\alpha \notin \mathbb{T}^\gamma_{\mathbf{s}}$, one obtains $\mathbb{T}^\gamma_{\mathbf{s}} \subseteq \mathbb{T}^\alpha_{*_n}$ as follows: In the case $\gamma \in \mathbb{P}_*$, the set $\mathbb{T}^\gamma_{\mathbf{s}}$ forms a $\approx^\gamma$-equivalence class, otherwise the set $\mathbb{D}^\gamma_{\mathbf{s}}$ is at least a gap or an element of the invariant lamination $\mathcal{B}^\gamma$ or consists of only one point. If an iterate of a point in $\mathbb{T}^\gamma_{\mathbf{s}} \cap \mathbb{T}\langle\alpha\rangle$ lies behind $\dot{\alpha}\dot{\overline{\alpha}}$ or $\ddot{\alpha}\ddot{\overline{\alpha}}$, then the iterate of the whole set $\mathbb{T}^\gamma_{\mathbf{s}}$ lies there. This follows from the planarity of $\approx^\gamma$.

'(ii) $\Longrightarrow$ (iii)': Let $\alpha \in \mathbb{P}_*$ and $\gamma \notin \{\alpha, \overline{\alpha}\}$ with $\mathbb{T}^\gamma_{*_n} = \mathbb{T}^\alpha_{*_n}$ be given. By Lemma 3.30(iii), we only have to show that not $\gamma \approx^\alpha \alpha$. Assuming the contrary, by Proposition 2.65 the point γ must be periodic with $PER(\gamma) = PER(\alpha)$. Further, since by Proposition 3.23(ii) and Theorem 3.26(iii) the set $\mathbb{T}\langle\gamma\rangle \setminus \{\gamma, \overline{\gamma}\}$ does not contain a periodic point of period $PER(\gamma)$, one obtains $\alpha \approx^\gamma \gamma$. The latter and $\gamma \approx^\alpha \alpha$ imply $\gamma = \overline{\alpha}$ (see Theorem 3.1 and (3.6)) contradicting our assumption.

The statement (iii) was the motivation to define simple renormalization, and so the proof of equivalence of (i), (ii), and (iii) is complete.

Now let $\gamma \in \mathbb{T}$ be n-renormalizable and let $SUBPER_n(\gamma) = n$. Then by Theorem 3.48 the sets $h^j(\mathbb{D}^\gamma_{*_n}); j = 0,1,2,\ldots,n-1$ are mutually disjoint. If S

is a longest in the set of all boundary chords of these $h^j(\mathbb{D}^\gamma_{*n})$, then it belongs to $h^{n-1}(\mathbb{D}^\gamma_{*n})$ and is contained in $\mathbb{D}^\gamma_0$ or $\mathbb{D}^\gamma_1$ since it holds $\frac{\gamma}{2}, \frac{\gamma+1}{2} \in h^{n-1}(\mathbb{T}^\gamma_{*n})$.

By the orientation-invariance of h (Lemma 2.9), one easily sees that $h^n(S)$ is a boundary chord of $h^{n-1}(\mathbb{D}^\gamma_{*n})$, and from Lemma 2.10 it follows $h^n(S) = S$ or $h^n(S) = S'$. Since $h^{n-1}(\mathbb{D}^\gamma_{*n})$ is $'$-symmetric, we can assume that $h^n(S) = S$. Moreover, by the orientation-invariance of h (Lemma 2.9) the endpoints of S are fixed points for h^n.

According to Proposition 2.65 there is an $\alpha \in \mathbb{P}_*$ with $h(S) = \dot{\alpha}\ddot{\overline{\alpha}}$. Obviously, $d(S) \geq \frac{1}{3}$, such that γ lies between α and $\overline{\alpha}$. So one easily sees that $\hat{\gamma}|n-1 = \mathbf{v}^\alpha$. By Lemma 3.10(ii), $\alpha\overline{\alpha}$ and γ cannot be separated by a chord in $\mathcal{B}_*$ of period less than n. So γ must be simply n-renormalizable. ■

Statement (ii) in Theorem 3.50 illustrates the character of simple renormalization: the n-renormalization domains for periodic points α are universal: they also support n-renormalization for Julia equivalences with main points obtained by tuning with respect to α. Moreover, the characterization of simple n-renormalizability by internal addresses above Theorem 3.50 turns into the following statement:

Corollary 3.51. (Admissible internal addresses, tuning, and simple renormalization)
If an internal address $1 \to \ldots \to m$ *is admissible and* $q_1, q_2, \ldots, q_l\ (,\ldots) \in \mathbb{N}$, *then the following statements are equivalent:*

(i) $1 \to q_1 \to q_2 \to \ldots \to q_l\ (\to \ldots)$ *is admissible,*

(ii) $1 \to \ldots \to m \to q_1 m \to q_2 m \to \ldots \to q_l m\ (\to \ldots)$ *is admissible.*

■

Using Theorem 3.50, we are able to express the order $\prec$ in terms of simple renormalization.

Corollary 3.52. (Symbolic description of $\prec$, part II; simple renormalization)
For $\alpha, \gamma \in \mathbb{T} \setminus \{0\}$ *the following statements are equivalent:*

(i) $\alpha \prec \gamma$, *i.e.* $\approx^\alpha \subset \approx^\gamma$,

(ii) α *lies in* $\mathbb{P}_*$ *and* γ *between* α *and* $\overline{\alpha}$, γ *is simply* n*-renormalizable with* $n = PER(\alpha)$, *and it holds* $\hat{\gamma}^{*n} = \hat{\alpha}$.

Proof: Obviously, (ii) follows from (i) by Theorem 3.33, Theorem 3.50 and Theorem 3.31(i).

If (ii) is valid, the by Theorem 3.50 there exists a $\delta \in \mathbb{P}_*$ satisfying $PER(\delta) = n$ such that γ lies in the δ-tuning of $\mathbb{T}$. Therefore $\hat{\delta} = \hat{\alpha}$, and γ lies between the points δ and $\overline{\delta}$ or is equal to one of these points. So by Theorem 3.6(ii) it holds $\alpha = \delta$ or $\alpha = \overline{\delta}$, and the statement (i) follows from Theorem 3.33. ■

3.3.3 Further 'small' copies of Abstract Julia sets

Now we want to justify the concept of a renormalization domain. First of all, look again at the simple case. For this, fix some $\alpha \in \mathbb{P}_*$ of period n and some $\delta \in \mathbb{T}$, and let $\gamma \sim Q^\alpha(\delta)$, which means that γ is simply n-renormalizable (see Theorem 3.50).

According to Theorem 3.50 it holds $\mathbb{T}^\gamma_{*_n} = \mathbb{T}^\alpha_{*_n}$ and by Corollary 3.52 the restriction $\approx^\alpha$ to $\mathbb{T}^\gamma_{*_n}$ - which we denote by $\approx$ - is contained in (the corresponding restriction of) $\approx^\gamma$. Further, the map

$$\theta : [Q^\alpha(\beta)]_\approx \mapsto \beta; \beta \in \mathbb{T}$$

satisfies the following properties ($\approx^\gamma$ is considered as the restriction to $\mathbb{T}^\gamma_{*_n}$):

1. θ conjugates $(\mathbb{T}^\gamma_{*_n}, h^n)/\approx$ to $(\mathbb{T}, h)$,
2. $\theta([\gamma]_\approx) = \delta$,
3. $\theta(\approx^\gamma / \approx) = \approx^\delta$.

The notation used in 3. is the following: If $\equiv_1$ and $\equiv_2$ are equivalence relations on a set X and $\equiv_1$ is contained in $\equiv_2$ (as a subset of $X \times X$), then $\equiv_2 / \equiv_1$ denotes that equivalence relation on $X/\equiv_1$ which identifies $[x_1]_{\equiv_1}$ and $[x_2]_{\equiv_1}$ for $x_1, x_2 \in X$ if $x_1 \equiv_2 x_2$ (see Appendix A.1, V.).

Property 2. is obvious and 1. is a consequence of Theorem 3.26(iii). Property 3. following from the fact that $\approx^\alpha \subseteq \approx^\gamma$ (see Proposition 3.28) shows that the n-renormalization domain $\mathbb{T}^\gamma_{*_n}$ supports a 'small' Abstract Julia set contained in $\mathbb{T}/\approx^\gamma$ and homeomorphic to $\mathbb{T}/\approx^\delta$. This phenomenon was already touched on in Chapter 3.2.2, and it is the aim of the present subsection to investigate it in its generality.

If $\gamma \in \mathbb{T}$ is n-renormalizable but not simply n-renormalizable, then $\approx^\gamma$ does not contain a non-trivial Julia equivalence different from $\approx^\gamma$, and $\approx$ cannot be defined from a Julia equivalence $\approx^\alpha$ for some periodic α. Nevertheless, we want to have a similar equivalence relation on the set $\mathbb{T}^\gamma_{*_n}$ and also on its iterates $h^j(\mathbb{T}^\gamma_{*_n}); j = 1, 2, \ldots, n-1$. We use the same notation for all n equivalence relations. This avoids a further increase of number of notations, and there should not be any confusion.

Definition 3.53. *For n-renormalizable $\gamma \in \mathbb{T}$ and $j = 0, 1, 2, \ldots, n-1$ let $\approx^\gamma_{*_n}$ be the smallest equivalence relation on $h^j(\mathbb{T}^\gamma_{*_n})$ identifying $\beta_1, \beta_2 \in \mathbb{T}$ if one of the open arcs with endpoints $\beta_1, \beta_2 \in \mathbb{T}$ is disjoint to $h^j(\mathbb{T}^\gamma_{*_n})$ (in other words, the smallest equivalence relation identifying endpoints of boundary chords of $h^j(\mathbb{D}^\gamma_{*_n})$).*

If γ is simply renormalizable, then $\approx^\gamma_{*_n}$ is equal to the equivalence relation $\approx$ defined above, but before we can find a conjugacy θ for the general case, we must have a closer look at the mutual position of the iterates of $\mathbb{T}^\gamma_{*_n}$. Of course, in the case of simple renormalization we do not get anything new.

Lemma 3.54. *Let $\gamma \in \mathbb{T}$ be n-renormalizable for some $n \in \mathbb{N} \setminus \{1\}$, let $r = SUBPER_n(\gamma) < n$ and $\mathbf{t} = \mathbf{t}^{\gamma}_{*_n}$.*

Then there exists an $\alpha \in \mathbb{P}_*$ satisfying $\mathbb{T}^{\gamma}_{*_n} \cap h^r(\mathbb{T}^{\gamma}_{*_n}) = \mathbb{T}^{\gamma}_{\mathbf{t}} = [\alpha]_{\approx^\alpha}$. The point α is unique except for its associated one, and the following is valid:

- *(i) γ lies between the points α and $\overline{\alpha}$ or coincides with one of them.*
- *(ii) r divides $PER(\alpha)$ and belongs to the internal address of γ, and so γ is simply r-renormalizable if $r > 1$.*
- *(iii) $\hat{\gamma}^{*_n} = \hat{\alpha}^{*_n}$.*
- *(iv) If γ is simply n-renormalizable, then it holds $PER(\alpha) = n$, and $\mathbb{D}^{\gamma}_{*_n} \setminus \mathbb{D}^{\gamma}_{\mathbf{t}}$ is connected.*
- *(v) If γ is not simply n-renormalizable, then $PER(\alpha) > n$, and $\mathbb{D}^{\gamma}_{*_n} \setminus \mathbb{D}^{\gamma}_{\mathbf{t}}$ is disconnected.*

Proof: According to Theorem 3.48, $E := \mathbb{T}^{\gamma}_{*_n} \cap h^r(\mathbb{T}^{\gamma}_{*_n})$ is a periodic $\approx^\gamma$-equivalence class. So we can apply Proposition 2.65:

The two $\approx^\gamma$-equivalent points β_1, β_2 of maximal possible distance in $\bigcup_{j=0}^{\infty} h^j(E)$ are contained in $h^{n-1}(\mathbb{T}^{\gamma}_{*_n})$, the unique $'$-symmetric set on the orbit of $\mathbb{T}^{\gamma}_{*_n}$. Otherwise, some iterate of $\beta_1\beta_2$ would lie (a first time) between $\beta_1\beta_2$ and $\beta_1'\beta_2'$, which is impossible by Lemma 2.10. Therefore, with $\{\alpha, \overline{\alpha}\} = \{h(\beta_1), h(\beta_2)\}$ it holds $E = [\alpha]_{\approx^\alpha} = \mathbb{T}^{\alpha}_{\mathbf{v}^\alpha(1-e^\alpha)}$ (see Theorem 2.57), and r divides the period of α.

$\frac{\gamma}{2}\frac{\gamma+1}{2}$ cannot cut $\beta_1\beta_2$ and $\beta_1'\beta_2'$. In the case $\gamma \in \mathbb{P}$ this is impossible by Theorems 2.53 and 2.50. Otherwise, we would have $\frac{\gamma}{2} \approx^\gamma \frac{\gamma+1}{2}$ and planarity of $\approx^\gamma$ would imply $\approx^\gamma$-equivalence of the periodic β_1 and the non-periodic β_1', which is impossible. Thus (i) is shown.

The chord $\alpha\overline{\alpha}$ is contained in $\mathbb{D}^{\gamma}_{*_n}$ (as γ itself) and its iterates are never between $\beta_1\beta_2 = \dot{\alpha}\dot{\overline{\alpha}}$ and $\beta_1'\beta_2' = \ddot{\alpha}\ddot{\overline{\alpha}}$. This provides $\mathbf{t} = \overline{\mathbf{v}^\alpha(1-e^\alpha)}$ (see Example 3.42), in particular (iii). Now we have $SUBPER_n(\gamma) = SUBPER_n(\alpha) = PER(\mathbf{t})$. Moreover, $PER(\mathbf{t})$ belongs to the internal address of α (see Corollary 3.18). Since the initial subwords of $\mathbf{t}$ and $\hat{\gamma}$ of length $n-1$ coincide, (ii) is an immediate consequence of the prescription for determining internal addresses from kneading sequences, given by Proposition 3.14.

(iv): If γ is simply n-renormalizable, then there exists a $\delta \in \mathbb{P}_*$ satisfying $PER(\delta) = n$ and $\mathbb{T}^{\gamma}_{*_n} = \mathbb{T}^{\delta}_{*_n}$. By use of Proposition 3.23(ii) one obtains $\mathbb{T}^{\gamma}_{*_n} \cap h^r(\mathbb{T}^{\gamma}_{*_n}) = \mathbb{T}^{\delta}_{*_n} \cap h^r(\mathbb{T}^{\delta}_{*_n}) = [\delta]_{\approx^\delta}$. Further, by the above shown it holds $\delta \in \{\alpha, \overline{\alpha}\}$, and we have $\mathbb{D}^{\gamma}_{\mathbf{t}} = \mathbb{D}^{\gamma}_{*_n} \cap h^r(\mathbb{D}^{\gamma}_{*_n}) = \mathbb{D}^{\alpha}_{*_n} \cap h^r(\mathbb{D}^{\alpha}_{*_n}) = \mathbb{D}^{\alpha}_{\overline{\mathbf{v}^\alpha(1-e^\alpha)}}$. The gap $\mathbb{D}^{\alpha}_{\overline{\mathbf{v}^\alpha(1-e^\alpha)}}$ has a boundary chord in common with $Gap\langle\alpha\rangle$, and so from $\mathbb{T}^{\alpha}_{*_n} = [\mathbb{T}\langle\alpha\rangle]_{\approx^\alpha}$ it follows connectivity of $\mathbb{D}^{\gamma}_{*_n} \setminus \mathbb{D}^{\gamma}_{\mathbf{t}}$.

We turn to the non-simple case, for which $PER(\alpha) > n > r$ by Theorem 3.50. Since $r = PER(\mathbf{t})$ divides n, the set $\mathbb{D}^{\gamma}_{\mathbf{t}} = \operatorname{conv}([\alpha]_{\approx^\alpha})$ is a gap of $\eth\mathcal{B}^\gamma$. We are finished if no boundary chord of $\mathbb{D}^{\gamma}_{\mathbf{t}}$ bounds $\mathbb{D}^{\gamma}_{*_n}$.

Otherwise, the boundary of $\mathbb{D}^\gamma_{\mathfrak{t}}$ consists of boundary chords of $h^{jr}(\mathbb{D}^\gamma_{*_n})$; $j = 0, 1, \ldots, \frac{n}{r} - 1$. By Proposition 2.14 these chords are contained in an orbit of length $PER(\alpha)$, and since $PER(\alpha) > n$, more than one of the chords bound $\mathbb{D}^\gamma_{*_n}$.

On the other hand, no two chords bounding $\mathbb{D}^\gamma_{\mathfrak{t}}$ and $\mathbb{D}^\gamma_{*_n}$ have a common endpoint. Otherwise, all boundary chords of $\mathbb{D}^\gamma_{\mathfrak{t}}$ would bound $\mathbb{D}^\gamma_{*_n}$, which is impossible. Hence $\mathbb{D}^\gamma_{*_n} \setminus \mathbb{D}^\gamma_{\mathfrak{t}}$ must be disconnected. This shows (v). ■

The following statements are devoted to the h^n-invariant equivalence classes being contained in n-renormalization domains.

Proposition 3.55. (Fixed classes in the renormalization domain)
Let $\gamma \in \mathbb{T}$ be n-renormalizable for $n \in \mathbb{N} \setminus \{1\}$ and let $\mathbf{v} = \hat{\gamma}|n-1$.

In the case that $\gamma \in \mathbb{P}$ and $n = PER(\hat{\gamma})$, the only h^n-invariant $\approx^\gamma$-equivalence class contained in $\mathbb{T}^\gamma_{*_n}$ is $\mathbb{T}^\gamma_{\overline{\mathbf{v}(1-e)}}$, where $e = e^\gamma$ if $\gamma \in \mathbb{P}_*$ and $e = \hat{\gamma}(n)$ else. Otherwise, $\mathbb{T}^\gamma_{\overline{\mathbf{v}0}}$ and $\mathbb{T}^\gamma_{\overline{\mathbf{v}1}}$ are the only equivalence classes of that kind.

Proof: Clearly, by Theorems 2.48, 2.51 and 2.57 (and by Proposition 2.6) in $\mathbb{T}^\gamma_{*_n}$ at most the sets $\mathbb{T}^\gamma_{\overline{\mathbf{v}0}}$ and $\mathbb{T}^\gamma_{\overline{\mathbf{v}1}}$ are h^n-invariant equivalence classes, and in case that $\gamma \notin \mathbb{P}$ or $n \neq PER(\hat{\gamma})$ both are such equivalence classes indeed.

If $\gamma \in \mathbb{P}_*, n = PER(\gamma)$, then $\mathbb{T}^\gamma_{\overline{\mathbf{v}e}} = \{\alpha\}$ is contained in $\mathbb{T}^\gamma_{\overline{\mathbf{v}(1-e)}}$ (see the discussion above Definition 2.37), and by Theorem 2.57 $\mathbb{T}^\gamma_{\overline{\mathbf{v}(1-e)}}$ forms an equivalence class. If $\gamma \in \mathbb{P}_\infty, n = PER(\hat{\gamma})$ and $e = \hat{\gamma}(n)$, then by Proposition 2.6 and Corollary 2.22 $\mathbb{T}^\gamma_{\overline{\mathbf{v}e}}$ is infinite and by Theorem 2.51 $\mathbb{T}^\gamma_{\overline{\mathbf{v}(1-e)}}$ forms an equivalence class. ■.

In the non-simple case one of the considered h^n-invariant equivalence classes consists of only one point. The precise statement is the following

Lemma 3.56. *Let $\gamma \in \mathbb{T}$ be n-renormalizible but not simply n-renormalizible for some $n \in \mathbb{N} \setminus \{1\}$. Further, let $\mathbf{v} = \hat{\gamma}|n-1$ and let e be the unique symbol with $\mathfrak{t}^\gamma_{*_n} = \overline{\mathbf{v}e}$. Then $\mathbb{T}^\gamma_{\overline{\mathbf{v}(1-e)}}$ is a single set.*

Proof: Look at the γ structure graph of the sequence $\overline{\mathbf{v}(1-e)}$, whose period is n. Obviously, $\{n, 2n, 3n, \ldots\}$ generates a component K containing $K + n$. (Clearly, $\hat{\gamma}$ starts with $\mathbf{v}e$.) By Corollary 2.24 (see case (i)) $\mathbb{T}^\gamma_{\overline{\mathbf{v}(1-e)}}$ consists of at most two points, of period n. We have only to exclude that $\mathbb{T}^\gamma_{\overline{\mathbf{v}(1-e)}}$ is a two-point set.

Assume that $\mathbb{T}^\gamma_{\overline{\mathbf{v}(1-e)}} = \{\gamma_1, \gamma_2\}$. Then, again by Corollary 2.24(i), it holds $h^n(\gamma_1) = \gamma_1$ and $h^n(\gamma_2) = \gamma_2$. Moreover, the chord $\gamma_1\gamma_2$ bounds $\mathbb{D}^\gamma_{*_n}$ as $S = h^{n-1}(\gamma_1\gamma_2)$ forms one of $h^{n-1}(\mathbb{D}^\gamma_{*_n})$.

Otherwise, S would have common points with the interior of $h^{n-1}(\mathbb{D}^\gamma_{*_n})$. Then the set A being the intersection of $h^{n-1}(\mathbb{T}^\gamma_{*_n})$ with the whole closed interval behind $h^{n-1}(\gamma_1)h^{n-1}(\gamma_2)$ would be h^n-invariant. This follows from the orientation-invariance of h (Lemma 2.9). In a small neighborhood of γ there would exist a point $\delta \in \mathbb{T} \setminus \mathbb{P}$ with the following: All points of A have the same itinerary with respect to δ, which is impossible.

S must be the longest chord in the orbit of $\gamma_1\gamma_2$. Otherwise, by Lemma 2.10, the longest chord would separate S and S' and so would bound an iterate of $\mathbb{D}^{\gamma}_{*_n}$ not coinciding with $h^{n-1}(\mathbb{D}^{\gamma}_{*_n})$. This is impossible since a point lying in different iterates of $\mathbb{T}^{\gamma}_{*_n}$ belongs to an iterate of $\mathbb{T}^{\gamma}_{\mathbf{ve}}$ (see Theorem 3.48). In particular, one sees that S and S' are the longest boundary chords of the set $h^{n-1}(\mathbb{D}^{\gamma}_{*_n})$.

By Proposition 2.65 there exists a point $\alpha \in \mathbb{P}_*$ with $\gamma_1\gamma_2 = \alpha\overline{\alpha}$, hence with $S = \dot{\alpha}\dot{\overline{\alpha}}$. Each of the points $h^i(\alpha); i = 0, 1, \ldots, n-2$ lies behind $\dot{\alpha}\dot{\overline{\alpha}}$ or $\ddot{\alpha}\ddot{\overline{\alpha}}$, hence also the whole sets $h^i(\mathbb{T}^{\gamma}_{*_n})$ for $i = 0, 1, \ldots, n-2$. Otherwise, by planarity $\dot{\alpha}$ or $\ddot{\alpha}$ would belong to different iterates of $\mathbb{T}^{\gamma}_{*_n}$. This is impossible, again by Theorem 3.48. Moreover, each $h^i(\mathbb{T}^{\gamma}_{*_n}); i = 0, 1, \ldots, n-2$ is disjoint to the set $h^{n-1}(\mathbb{T}^{\gamma}_{*_n})$, in contradiction to the fact that γ is not simply n-renormalizable. ■

The following says that the boundary of the convex hull of an n-renormalization domain is 'generated' by one chord.

Corollary 3.57. *Let $\gamma \in \mathbb{T}$ be n-renormalizable for $n \in \mathbb{N} \setminus \{1\}$. Further, let $\alpha \in \mathbb{P}_*$ be given such that $\mathbb{T}^{\gamma}_{*_n} = \mathbb{T}^{\alpha}_{*_n}$ in the simple case and $\mathbb{T}^{\gamma}_{\mathbf{t}} = [\alpha]_{\approx^\alpha}$ with $\mathbf{t} = \mathbf{t}^{\gamma}_{*_n}$ in the non-simple one. Then $\alpha\overline{\alpha}$ lies in the orbit of all boundary chords of $\mathbb{D}^{\gamma}_{*_n}$.*

Proof: In the simple renormalizable case, where $\mathbb{T}^{\gamma}_{*_n} = \mathbb{T}^{\alpha}_{*_n} = [\mathbb{T}\langle\alpha\rangle)]_{\approx^\alpha}$, the fact is already known. Each boundary chord of *Gap*$\langle\alpha\rangle$ is a backward iterate of $\alpha\overline{\alpha}$ and the rest follows from Proposition 2.14.

In the non-simple case, let S be a longest periodic chord on the orbit of a fixed boundary chord of $\mathbb{D}^{\gamma}_{*_n}$. By Lemma 2.10 S lies in $h^{n-1}(\mathbb{D}^{\gamma}_{*_n})$.

If S is a boundary chord of $h^{n-1}(\mathbb{D}^{\gamma}_{*_n})$, then by Lemma 2.10 it holds $h^n(S) = S$. Further, by Lemma 3.56 $h(S)$ is a boundary chord of $\mathbb{D}^{\gamma}_{\mathbf{t}}$. We are done since $\mathbb{T}^{\gamma}_{\mathbf{t}} = [\alpha]_{\approx^\alpha}$.

If S has common points with the interior of $h^{n-1}(\mathbb{D}^{\alpha}_{*_n})$, then also $h(S)$ intersects the interior of $\mathbb{D}^{\alpha}_{*_n}$. Then $h(S)$ forms a boundary chord of $\mathbb{D}^{\gamma}_{\mathbf{ve}}$ by Theorem 3.48. The last statement of the following Corollary is an immediate statement of the orientation-invariance of h (Lemma 2.9) and Proposition 2.65 (see (i) and (ii)). ■

As announced at the beginning of this subsection, let us characterize n-renormalization domains as supports of small Abstract Julia sets. In fact, (analogously to renormalization in the complex plane) there will be n conjugate Abstract Julia sets, each of which is supported on an iterate of the n-renormalization domain. If $\gamma \in \mathbb{T}$ is n-renormalizable, then h maps $h^j(\mathbb{T}^{\gamma}_{*_n})$

onto $h^{j+1}(\mathbb{T}^{\gamma}_{*_n})$ in a homeomorphic way for $j = 0, 1, \ldots, n-2$ and acts as a two to one local homeomorphism for $j = n-1$:

$$\mathbb{T}^{\gamma}_{*_n} \xrightarrow{h} h(\mathbb{T}^{\gamma}_{*_n}) \xrightarrow{h} \ldots \xrightarrow{h} h^{n-1}(\mathbb{T}^{\gamma}_{*_n}) \xrightarrow{h} \mathbb{T}^{\gamma}_{*_n} \tag{3.12}$$

By factorizing all sets and all h in (3.12) with respect to $\approx^{\gamma}_{*_n}$, one gets n simple closed curves interchanged by the $h/\approx^{\gamma}_{*_n}$, and the factorization by $\approx^{\gamma}$ provides n small Abstract Julia sets interchanged by $h/\approx^{\gamma}$. This is what we want to show now.

Theorem 3.58. (Renormalization and 'small' copies of Abstract Julia sets) *Let $\gamma \in \mathbb{T}$ be n-renormalizable for $n \in \mathbb{N} \setminus \{1\}$. Then there exist a $\delta \in \mathbb{T}$ and a map θ from $\mathbb{T}^{\gamma}_{*_n}/\approx^{\gamma}_{*_n}$ onto $\mathbb{T}$ satisfying the following properties:*

(i) θ conjugates $(\mathbb{T}^{\gamma}_{*_n}, h^n)/\approx^{\gamma}_{*_n}$ to $(\mathbb{T}, h)$.

(i') $\theta \circ H^{-1}$ conjugates $(h^{-1}(\mathbb{T}^{\gamma}_{*_n}), h^n, ')/\approx^{\gamma}_{*_n}$ to $(\mathbb{T}, h, ')$, where H denotes the restriction of h^{n-1} to $\mathbb{T}^{\gamma}_{*_n}$ factorized by $\approx^{\gamma}_{*_n}$.

(ii) $\theta([\gamma]_{\approx^{\gamma}_{*_n}}) = \delta$.

(iii) $\theta(\approx^{\gamma} / \approx^{\gamma}_{*_n}) = \approx^{\delta}$.

(iv) With $e = e^{\gamma}$ if $\gamma \in \mathbb{P}_*$ and $PER(\gamma) = n$, and with $e = \hat{\gamma}(n)$ otherwise, $\hat{\delta}$ is obtained from the sequence $\hat{\gamma}(n)\hat{\gamma}(2n)\hat{\gamma}(3n)\ldots$ by the substitutions $e \longrightarrow 0, 1-e \longrightarrow 1$.

For all $j = 0, 1, 2, \ldots n-1$, the topological dynamical system $(h^j(\mathbb{T}^{\gamma}_{*_n}), h^n)/\approx^{\gamma}$ is conjugate to $(\mathbb{T}, h)/\approx^{\delta}$, in particular, $(h^{-1}(\mathbb{T}^{\gamma}_{*_n}), h^n, ')/\approx^{\gamma}$ is conjugate to $(\mathbb{T}, h, ')/\approx^{\delta}$.

Proof: If γ is simply renormalizable, let α, δ and θ be given as at the beginning of this subsection. Then (i), (ii) and (iii) are given by 1., 2. and 3., and (iv) is no more than tuning of the kneading sequence (see Theorem 3.31(i)).

(i') results from (i) and the following statement due to Corollary 3.24(ii): $\{H(\theta^{-1}(\frac{\beta}{2})), H(\theta^{-1}(\frac{\beta+1}{2}))\} = \{[h^{n-1}(Q^{\alpha}(\frac{\beta}{2}))]_{\approx^{\alpha}}, [h^{n-1}(Q^{\alpha}(\frac{\beta+1}{2}))]_{\approx^{\alpha}}\} = \{[\frac{Q^{\alpha}(\beta)}{2}]_{\approx^{\alpha}}, [\frac{Q^{\alpha}(\beta)+1}{2}]_{\approx^{\alpha}}\}$.

In the non-simple case let us first establish a conjugacy $\tilde{\theta}$ from the topological dynamical system $(h^{-1}(\mathbb{T}^{\gamma}_{*_n}), h^n, ')/\approx^{\gamma}_{*_n}$ to $(\mathbb{T}, h, ')$. According to Lemma 3.56 there is a periodic point η of period n satisfying $\{\eta\} = [\eta]_{\approx^{\gamma}} = \mathbb{T}^{\gamma}_{\overline{(1-e)\mathbf{v}}}$ with $e = \hat{\gamma}(n)$ and $\mathbf{v} = \hat{\gamma}|n-1$.

The first part of Theorem 3.48 shows that the restriction of $\approx^{\gamma}$ to $h^{-1}(\mathbb{T}^{\gamma}_{*_n})$ contains $\approx^{\gamma}_{*_n}$. Thus $\{\eta\}$ also forms a $\approx^{\gamma}_{*_n}$-equivalence class. We will see that $\{\eta\}$ plays the same role in $(h^{-1}(\mathbb{T}^{\gamma}_{*_n}), h^n, ')/\approx^{\gamma}_{*_n}$ as 0 in $(\mathbb{T}, h, ')$, and we will get a kind of binary expansion for the elements of $h^{-1}(\mathbb{T}^{\gamma}_{*_n})/\approx^{\gamma}_{*_n}$.

First of all deduce that the backward orbit of $\{\eta\}$ with respect to $h^n/\approx^{\gamma}_{*_n}$ is dense in $h^{-1}(\mathbb{T}^{\gamma}_{*_n})/\approx^{\gamma}_{*_n}$. For this, we show that each infinite set A being the intersection of an open interval with $h^{-1}(\mathbb{T}^{\gamma}_{*_n})$ contains a backward iterate of η or of η' with respect to h, Since obviously $h^n(\eta') = \eta$ and since $\eta, \eta' \notin h^j(\mathbb{T}^{\gamma}_{*_n})$ for $j = 0, 1, 2, \ldots, n-2$, the set A contains a backward iterate of η also with respect to h^n.

Assume that neither η nor η' lies in the orbit of A. Then by the orientation-invariance of h (Lemma 2.9) one shows that $h^i(A) \subset]\eta, \eta'[$ or $h^i(A) \subset]\eta', \eta[$ for all $i \in \mathbb{N}_0$ and so that infinitely many points have a common specified itinerary with respect to $h(\eta)$. The latter is impossible.

We assign a 0-1-sequence $I(\beta) := \widetilde{I^{h(\eta)}}(\beta)(1)\widetilde{I^{h(\eta)}}(\beta)(n+1)\widetilde{I^{h(\eta)}}(\beta)(2n+1)\ldots$ to each $\beta \in h^{-1}(\mathbb{T}^{\gamma}_{*_n})$. Obviously, the map I obtained is continuous at all points whose orbit does not contain η and is constant on each $\approx^{\alpha}_{*_n}$-equivalence class. We use this fact to define $\tilde{\theta}$. If E is a $\approx^{\gamma}_{*_n}$-equivalence class, then let $\tilde{\theta}(E)$ be the point in $\mathbb{T}$ with binary expansion $I(\beta)$ for all $\beta \in E$. (Here binary expansion ending with $\overline{0}$ is allowed.)

$\tilde{\theta}$ maps the backward orbit of $\{\eta\}$ with respect to h^n to the set of points with binary expansion ending by $\overline{0}$ ($\overline{1}$), and one easily sees that I is injective on the backward orbit of $\{\eta\}$.

By definition of the equivalence relation $\approx^{\gamma}_{*_n}$ each equivalence class contains a left-side and a right-side accumulation point of $h^{-1}(\mathbb{T}^{\gamma}_{*_n})$. Therefore $h^{-1}(\mathbb{T}^{\gamma}_{*_n})/\approx^{\gamma}_{*_n}$ forms a simple closed curve. Although I fails to be continuous at the points of the backward orbit of $\{\eta\}$ with respect to h^n, for each such point the left-side and the right-side limit exist, and it is not hard to see that these limits are given by $\overline{0}$ and $\overline{1}$, or by $\mathbf{w}1\overline{0}$ and $\mathbf{w}0\overline{1}$ for some 0-1-word $\mathbf{w}$.

So $\tilde{\theta}$ must be continuous. The statement that $\tilde{\theta}$ forms a conjugacy from $(h^{-1}(\mathbb{T}^{\gamma}_{*_n}), h^n)/\approx^{\gamma}_{*_n})$ onto $(\mathbb{T}, h)$ can be verified using standard arguments and the properties of I described above.

For $\beta \in h^{-1}(\mathbb{T}^{\gamma}_{*_n})$ the sequences $I(\beta)$ and $I(\beta')$ are different only in the first symbol, and in the same way $'$ changes only the first symbol of binary expansion. So if $\theta := \tilde{\theta} \circ H$, we have $\tilde{\theta} = \theta \circ H^{-1}$, and (i'), (i) are obvious.

Both the glueings of $h^{-1}(\mathbb{T}^{\gamma}_{*_n})/\approx^{\gamma}_{*_n}$ to $h^{-1}(\mathbb{T}^{\gamma}_{*_n})/\approx^{\gamma}$ and of $\mathbb{T}^{\gamma}_{*_n}/\approx^{\gamma}_{*_n}$ to $\mathbb{T}^{\gamma}_{*_n}/\approx^{\gamma}$ can be considered as homotopic processes. Hence $\approx = \theta(\approx^{\gamma} / \approx^{\gamma}_{*_n}) = \tilde{\theta}(\approx^{\gamma} / \approx^{\gamma}_{*_n})$ must be planar, when the first $\approx^{\gamma} / \approx^{\gamma}_{*_n}$ is meant to be supported on $h^{-1}(\mathbb{T}^{\gamma}_{*_n})/\approx^{\gamma}_{*_n}$ and the second one on $\mathbb{T}^{\gamma}_{*_n}/\approx^{\gamma}_{*_n}$. It follows from the properties of $\tilde{\theta}$ shown above that $\approx$ is a Julia equivalence.

Let us show that $\delta = \theta([\gamma]_{\approx^{\gamma}_{*_n}})) = \tilde{\theta}([h^n(\frac{\gamma}{2})]_{\approx^{\gamma}_{*_n}}) = \tilde{\theta}([h^n(\frac{\gamma+1}{2})]_{\approx^{\gamma}_{*_n}}) = h(\tilde{\theta}([\frac{\gamma}{2}]_{\approx^{\gamma}_{*_n}})) = h(\tilde{\theta}([\frac{\gamma+1}{2}]_{\approx^{\gamma}_{*_n}}))$ is a main point of $\approx$, and so that (iii) is valid. If $\gamma \in \mathbb{T} \setminus \mathbb{P}_*$, this is obvious since $\tilde{\theta}([\frac{\gamma}{2}]_{\approx^{\gamma}_{*_n}})$ and $\tilde{\theta}([\frac{\gamma+1}{2}]_{\approx^{\gamma}_{*_n}}$ are $\approx$-equivalent and interchanged by $'$.

In the case $\gamma \in \mathbb{P}_*$ let $\beta_1 = \tilde{\theta}([\dot{\gamma}]_{\approx^{\gamma}_{*_n}})$ and $\beta_2 = \tilde{\theta}([\dot{\overline{\gamma}}]_{\approx^{\gamma}_{*_n}})$ and $S = \beta_1\beta_2$. Then $S' = \tilde{\theta}([\ddot{\gamma}]_{\approx^{\gamma}_{*_n}})\tilde{\theta}([\ddot{\overline{\gamma}}]_{\approx^{\gamma}_{*_n}})$. Since obviously $\tilde{\theta}$ preserves or reverses the orientation, between the two whole open parts of the circle between S and

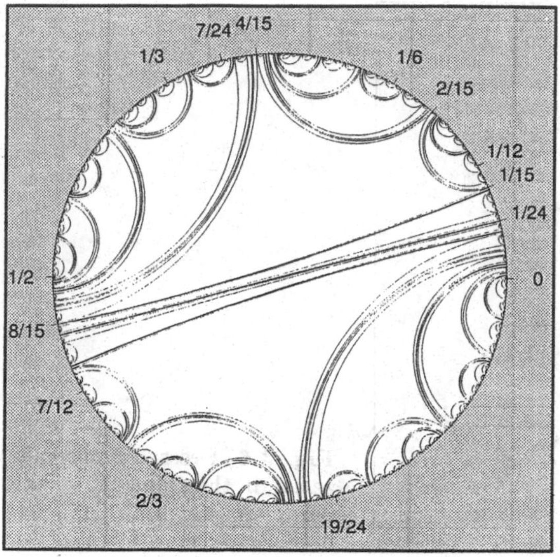

Fig. 3.5. 'Crossed' renormalization: $\approx^\alpha$ for $\alpha = \frac{1}{12}$

S' there is no identification with respect to $\approx$, hence by Proposition 2.65, the endpoints of $h(S)$ must be associated. Thus $h(\tilde{\theta}([\frac{\gamma}{2}]_{\approx^\gamma_{*n}})), h(\tilde{\theta}([\frac{\gamma+1}{2}]_{\approx^\gamma_{*n}}))$ form main points.

We have $\{\tilde{\theta}([\frac{\gamma}{2}]_{\approx^\gamma_{*n}}), \tilde{\theta}([\frac{\gamma+1}{2}]_{\approx^\gamma_{*n}})\} = \{\frac{\delta}{2}, \frac{\delta+1}{2}\}$, and since θ preserves or reverses the orientation, (iv) is obvious now. ∎

Let us finish our discussion on renormalization with two examples for renormalization, one for the non-simple and one for the simple case.

Example 3.59. $\gamma = \frac{1}{12}$ has kneading sequence $\hat{\gamma} = 00\overline{01}$, hence it is 2-renormalizable, but the renormalization fails to be simple. ($\approx^\gamma$ represented in Figure 3.5 corresponds to the Julia set which is represented by Figure 3.6.)

It holds $\mathbf{t}^\gamma_{*_2} = \overline{0}$, and the two sets $\mathbb{T}^\gamma_{*_2}$ and $h(\mathbb{T}^\gamma_{*_2})$ have the h-invariant, hence also h^2-invariant, equivalence class $\mathbb{T}^\gamma_{\overline{0}} = \{\frac{1}{15}, \frac{2}{15}, \frac{4}{15}, \frac{8}{15}\}$ in common. The second h^2-invariant equivalence class is given by $\mathbb{T}^\gamma_{\overline{01}} = \{\frac{1}{3}\}$.

A δ as given in Theorem 3.58 must have kneading sequence $0\overline{1}$. Moreover, by Theorem 2.19 the set $\mathbb{T}^\delta_{\overline{1}}$ consists of exactly one point, which implies $\mathbb{T}^\delta_{\overline{1}} = \{0\}$ and $\delta = \frac{1}{2}$. According to Example 2.49, the 'small' Abstract Julia sets $\mathbb{T}^{1/12}_{*_2}/\approx^{1/12}$ and $h(\mathbb{T}^{1/12}_{*_2})/\approx^{1/12}$ form arcs, as seen in Figure 3.6. There is one long arc in vertical direction and a second one lying in the upper part of the Julia set in horizontal direction. The two arcs cross each other. Later we will show that crossed renormalization in the plane and non-simple renormalization in the abstract setting coincide in the general.

An example for a simply 2-renormalizable γ has already been given by $\gamma = \frac{7}{20}$ (see Example 3.29 and Figure 3.4). Here $\hat{\gamma} = 01\overline{0100}$ and $\mathbf{t}^\gamma_{*_2} = \overline{0}$,

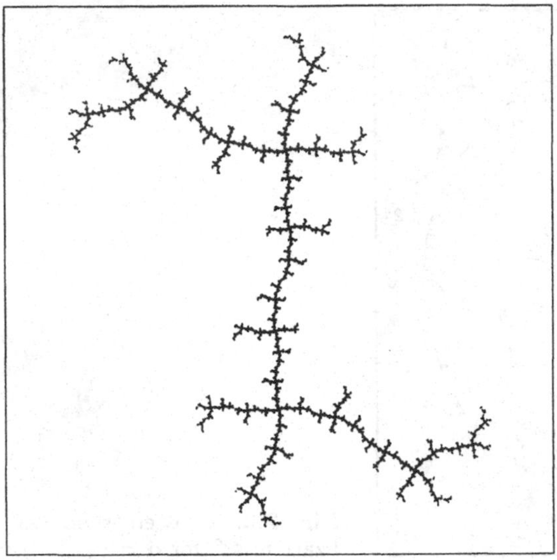

Fig. 3.6. Crossed renormalization in the plane: J_c for the parameter $c = 0,419643\ldots + 0,606291\ldots i$

and the 'small' Abstract Julia sets $\mathbb{T}_{*_2}^{7/20}/\approx^{7/20}$ and $h(\mathbb{T}_{*_2}^{7/20})/\approx^{7/20}$ have the common equivalence class $\mathbb{T}_{\overline{0}}^{\gamma} = \{\frac{1}{3}, \frac{2}{3}\}$, but they do not 'cross' each other.

Let us finish our considerations on renormalization by a remark. We saw that simple renormalizable points in $\mathbb{T}$ are 'organized' in form of the α-tunings. There is a statement by RIEDL and SCHLEICHER [141] describing the loci of non-simple renormalization in a similar way. (In our language) it could be derived from the above considerations. Let us only note the following fact, which is a consequence of Lemma 3.54(ii): If γ is n-renormalizable, but the renormalization is not simple, then either $r = SUBPER_n(\gamma) = 1$, or for $k = \frac{n}{r}$ there exist some non-simply k-renormalizable $\delta \in \mathbb{T}$ with $SUBPER_k(\gamma) = 1$ and some periodic α of period r such that $\gamma = Q^{\alpha}(\delta)$.

3.4 Correspondence and Translation Principles

3.4.1 Parts of the Abstract Mandelbrot set related to special Julia equivalences

Given some $\alpha \in \mathbb{P}_* \setminus \{0\}$, there is much similarity between $\approx^{\alpha}$ behind $\alpha\overline{\alpha}$ and $\sim$ behind $\alpha\overline{\alpha}$. This fact, which was already suggested in Chapter 3.1.1 and which provides certain symmetries in the Mandelbrot set, is the focus of our interest now.

We will present statements in the frame of laminations, which is convenient from the technical point of view. The starting point of our considerations is the dichotomy of $S = \alpha\overline{\alpha}$ for some $\alpha \in \mathbb{P}_* \setminus \{0\}$ as a parameter and a dynamic chord. In this subsection, let the **chord** S be **fixed** of period m, and let $\mathbf{v} := \mathbf{v}^S = v_1 v_2 \dots v_{m-1}, e := e^S$. If we speak of visibility, then we mean visibility from this S.

Correspondence: a) $S \in \mathcal{B}_* \mapsto S \in \mathcal{B}_*^S$. The chord S belongs both to $\mathcal{B}_*$ and to $\mathcal{B}_*^S \subset \eth\mathcal{B}^S$, where $\mathcal{B}_*^S := \mathcal{B}_*^\alpha = \{l_{\mathbf{w}}^{\alpha,1-e}(\alpha\overline{\alpha}) | \mathbf{w} \in \{0,1\}^*\}$ is dense in $\eth\mathcal{B}^S$ (see Proposition 2.42).

Further, S is the longest boundary chord of the gaps $Gap(S)$ in $\mathcal{B}$ and $Gap\langle S\rangle := Gap\langle\alpha\rangle$ in $\eth\mathcal{B}^S$, and $Gap(S)$ is contained in $Gap\langle S\rangle$.

Dynamic pairs. Let us introduce a concept of visibility in $\eth\mathcal{B}^S$. For this, recall the fact that for each 0-1-word $\mathbf{w}$ not ending with $\mathbf{v}$, the chords $l_{\mathbf{w}0}^{\alpha,1-e}(\alpha\overline{\alpha})$ and $l_{\mathbf{w}1}^{\alpha,1-e}(\alpha\overline{\alpha})$ are the longest boundary chords of the gap $Gap_{\mathbf{w}*}^\alpha$ in $\eth\mathcal{B}^S = \eth\mathcal{B}^\alpha$, which was provided by Proposition 2.43. Instead of $Gap_{\mathbf{w}*}^\alpha$, we want to work with these longest chords.

Definition 3.60. (Dynamic pairs, *STEP*, parameter-like, visible)
For $S_1, S_2 \in \mathcal{B}_*^S$, the pair (S_1, S_2) is called a dynamic pair *if S_1 is not shorter than S_2 and if there exists a 0-1-word $\mathbf{w}$ which does not end with $\mathbf{v}$, such that $\{S_1, S_2\} = \{l_{\mathbf{w}0}^{\alpha,1-e}(S), l_{\mathbf{w}1}^{\alpha,1-e}(S)\}$.*

By the STEP of a dynamic pair (S_1, S_2) we understand the minimal number n with $h^n(S_1) = h^n(S_2) = S$. (It exceeds the length of the corresponding word $\mathbf{w}$ by one.) Similarly, for a boundary chord R of the $Gap\langle S\rangle$ let the STEP of R be the minimal $n \in \mathbb{N}_0$ with $h^n(R) = S$.

A dynamic pair (S_1, S_2) is said to be parameter-like *if S_1 lies behind S and none of the iterates of S_1 and S_2 separates S and S_1.*

It is called visible *(from S) if S_1 lies behind S and there is no dynamic pair (R_1, R_2) whose STEP is less than the STEP of (S_1, S_2) such that R_1 separates S_1 from S.*

Remarks. 1. The statements of Proposition 2.43(ii) concerning lengths of chords show that no dynamic pair of *STEP* not greater than another dynamic pair (S_1, S_2) lies between S_1 and S_2.

2. Visible dynamic pairs are parameter-like by definition, and at first glance the concept of visibility of dynamic pairs is a natural counterpart to visibility of chords in $\mathcal{B}_*$. However, we will see that it is a little too strong for our purposes.

If $\beta_1 \sim \beta_2$ and if $\beta_1\beta_2$ belongs to an invariant lamination, then $\beta_1\beta_2 \in \mathcal{B}$. This follows from the characterizations of $\sim$ in (3.5) and of $\mathcal{B}$ in (3.7). Therefore, Lemma 3.30(iii) and Corollary 3.2 yield the following statement, which says that those dynamic chords being interesting for our considerations are elements of $\mathcal{B}$.

Proposition 3.61. *All boundary chords of $Gap\langle S\rangle$ belong to $\mathcal{B}$, and a dynamic pair (S_1, S_2) behind S is parameter-like iff both S_1 and S_2 are elements of $\mathcal{B}$.* ■

For all chords $R = \gamma\delta \in \mathcal{B}$, in particular for all boundary chords R of $Gap\langle S\rangle$ and members R of parameter-like dynamic pairs we will use the notation $\hat{R} := \hat{\gamma} = \hat{\delta}$. This is justified by Theorem 3.1(iv).

Translation by dynamics. Parts of the subsequent proofs are based on the fact that h 'translates' much of the local structure of $\eth\mathcal{B}^S$, and this translation in dynamics can be considered as the root of translation-like symmetries of the (Abstract) Mandelbrot set.

By Propositions 2.43 we have the following statement:

Theorem 3.62. (Dynamical Translation Principle)
Let $s_1s_2\ldots s_l$ be a non-empty 0-1-word. Then h^m maps the whole interval behind $l^{\alpha,1-e}_{\mathbf{v}s_1\mathbf{v}s_2\ldots\mathbf{v}s_l\mathbf{v}e}(S)$ onto the whole interval behind $l^{\alpha,1-e}_{\mathbf{v}s_2\ldots\mathbf{v}s_l\mathbf{v}e}(S)$ in a homeomorphic way.

In particular, h^m transforms the set of all dynamic pairs behind the chord $l^{\alpha,1-e}_{\mathbf{v}s_1\mathbf{v}s_2\ldots\mathbf{v}s_l\mathbf{v}e}(S)$ into the set of all dynamic pairs behind $l^{\alpha,1-e}_{\mathbf{v}s_2\ldots\mathbf{v}s_l\mathbf{v}e}(S)$ decreasing the STEP by m. A dynamic pair (S_1,S_2) behind $l^{\alpha,1-e}_{\mathbf{v}s_1\mathbf{v}s_2\ldots\mathbf{v}s_l\mathbf{v}e}(S)$ is parameter-like if the dynamic pair $(h^m(S_1),h^m(S_2))$ is, and visible if and only if $(h^m(S_1),h^m(S_2))$ is. ■

Clearly, each chord $l^{\alpha,1-e}_{\mathbf{v}s_1\mathbf{v}s_2\ldots\mathbf{v}s_l\mathbf{v}e}(S)$ for $s_1s_2\ldots s_l; l \geq 1$ is mapped to $l^{\alpha,1-e}_{\mathbf{v}e}(S)$ after lm iterates. Thus the geometry behind $l^{\alpha,1-e}_{\mathbf{v}e}(S)$ is of special interest. Note that a dynamic pair (S_1,S_2) behind some $l^{\alpha,1-e}_{\mathbf{v}s_1\mathbf{v}s_2\ldots\mathbf{v}s_l\mathbf{v}e}(S)$ can be parameter-like even if $(h^m(S_1),h^m(S_2))$ is not. Later we will give an example for this.

Example 3.63. Consider $S = \alpha\overline{\alpha}$ with $\alpha = \frac{5}{31}$ and $\overline{\alpha} = \frac{6}{31}$, where $PER(\alpha) = 5, \mathbf{v}^\alpha = 0010$ and $e^\alpha = 1$. The corresponding invariant lamination $\eth\mathcal{B}^S$ is

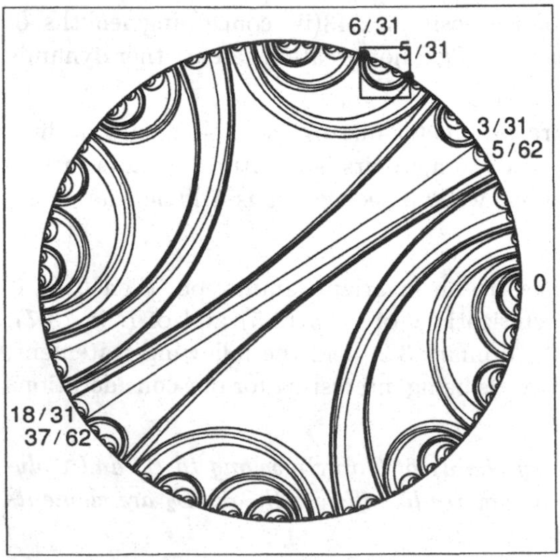

Fig. 3.7. The lamination $\eth\mathcal{B}^{\frac{5}{31}\frac{6}{31}}$ with window 1

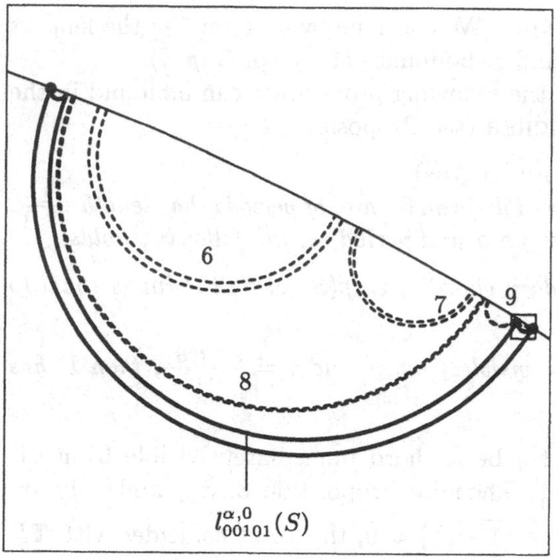

Fig. 3.8. View at $\eth\mathcal{B}_{\frac{5}{31}\frac{6}{31}}$ in window 1, with window 2

illustrated by Figure 3.7. Figures 3.8 and 3.9 show magnified parts of Figure 3.7. Boundary chords of the critical value gap are indicated by thick arcs. Further, the members of visible dynamic pairs behind boundary chords different from $S = \frac{5}{31}\frac{6}{31}$ are drawn as thick dashed arcs, and they are labelled by their *STEP*'s.

Figures 3.8 and 3.9 show that h^5 shifts visibility: the picture behind $l^{\alpha,0}_{0010100101}(S)$ in Figure 3.9 looks like the picture behind $l^{\alpha,0}_{00101}(S)$ in Figure 3.8.

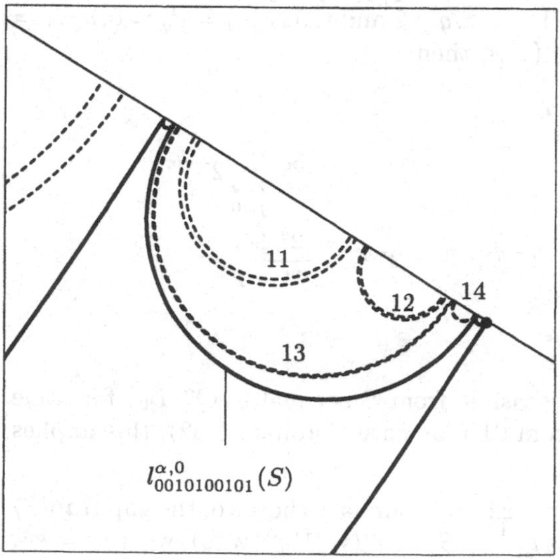

Fig. 3.9. View at $\eth\mathcal{B}_{\frac{5}{31}\frac{6}{31}}$ in window 2

The lengths of some special chords. We continue by determining the lengths of immediately visible chords and of boundary chords of $Gap\langle S\rangle$.

Note that statement (ii) of the following proposition can be found in the paper [92] by LAU and SCHLEICHER (see Proposition 2.7).

Proposition 3.64. (Some chord lengths)
Each chord in $\mathcal{B}_*$ immediately visible from 01 *and of period q has length $\frac{1}{2^q-1}$. Further, if $S = \alpha\overline{\alpha} \in \mathcal{B}_*$ has length a and period m, the following holds:*

- *(i) The length of each boundary chord of $Gap\langle S\rangle$ of STEP lm is equal to $\frac{2^m-2}{2^{lm}}a$.*
- *(ii) If $B \in \mathcal{B}_*$ is immediately visible from S and $q = \frac{PER(B)}{m}$, then B has length $\frac{(2^m-1)^2}{2^{qm}-1}a$.*

Proof: Let $B_0 = \gamma\overline{\gamma}$ of period q be a chord immediately visible from 01. We can assume that $B_0 \neq \frac{1}{3}\frac{2}{3}$. Then by Proposition 3.18 γ and $\overline{\gamma}$ lie on a common orbit. Since $\widehat{B_0}^{-} = \overline{\mathbf{v}^{\gamma}(1-e^{\gamma})} = \overline{0}$, this orbit coincides with $\mathbb{T}^{\gamma}_{\overline{0}}$, and the chords $B_0, h(B_0), \ldots, h^{q-1}(B_0)$ form the boundary chords of $\mathbb{D}^{\gamma}_{\overline{0}}$. The longest one is $\dot{B}_0 = h^{q-1}(B_0)$, and with $a_0 = d(B_0)$ it holds $d(\dot{B}_0) = \frac{1}{2} - \frac{a_0}{2}$.

For $i = 0,1,2,\ldots,q-2$ let I_i be the whole open interval in $\mathbb{T}$ lying behind $h^i(B_0)$. Further, let I_{q-1} be the longer open interval with the endpoints $\dot{\gamma}$ and $\dot{\overline{\gamma}}$. For each $i = 0,1,2,\ldots,q-2$, the map h transforms I_i into I_{i+1} in an orientation-preserving, hence homeomorphic way (compare Lemma 2.9). This shows $\frac{1}{2} + \frac{a_0}{2} = 1 - d(\dot{B}_0) = 2^{q-1}a_0$ implying $a_0 = \frac{1}{2^q-1}$.

(ii): Assume that $\gamma < \overline{\gamma}$ and that γ and $\overline{\gamma}$ have binary expansions $\overline{\mathbf{c}}$ and $\overline{\mathbf{d}}$ with $\mathbf{c} = c_1c_2\ldots c_q$ and $\mathbf{d} = d_1d_2\ldots d_q$, respectively. Then $\frac{1}{2^q-1} = \overline{\gamma} - \gamma = \sum_{j=1}^{q} 2^{-j}(d_j - c_j)\frac{2^q}{2^q-1}$, which implies $1 = \sum_{j=1}^{q}(d_j - c_j)2^{q-j}$. From this, one easily sees that $c_j = d_j$ for $j = 1,2,\ldots,q-2$ and that $c_{q-1} = d_q = 0, d_{q-1} = c_q = 1$. So if $\alpha < \overline{\alpha}$ and $B = Q^{\alpha}(B_0)$, then

$$\begin{aligned}
d(B) &= q^{\alpha}(\overline{\mathbf{d}}) - q^{\alpha}(\overline{\mathbf{c}}) \\
&= (2^m-1)(\overline{\alpha} - \alpha)(2^{-(q-1)m} - 2^{-qm})\sum_{j=0}^{\infty} 2^{-qmj} \\
&= (2^m-1)\ a\ (2^{-(q-1)m} - 2^{-qm})\frac{2^{qm}}{2^{qm}-1} \\
&= \frac{(2^m-1)^2}{2^{qm}-1}\ a.
\end{aligned}$$

Since each $B \in \mathcal{B}_*$ immediately visible from S is equal to $Q^{\alpha}(B_0)$ for some $B_0 \in \mathcal{B}_*$ immediately visible from 01 (compare Corollary 3.32), this implies statement (ii).

(i): According to Proposition 3.22, the boundary chords of the gap $Gap\langle S\rangle$ coincide exactly with the chords $l^{\alpha,1-e}_{\mathbf{w}0\langle\alpha\rangle}(S) = q^{\alpha}(\mathbf{w}0\overline{1})q^{\alpha}(\mathbf{w}1\overline{0})$, where $e = e^S$, and by definition of q^α we have

$$\begin{aligned} d(l_{0\langle\alpha\rangle}^{\alpha,1-e}(S)) &= q^\alpha(1\overline{0}) - q^\alpha(0\overline{1}) \\ &= (2^m-1)(\overline{\alpha}-\alpha)\left(2^{-m} - \sum_{i=2}^{\infty} 2^{-mi}\right) \\ &= (2^m-1)\, a \left(\frac{1}{2^m} - \frac{1}{2^m(2^m-1)}\right) = \frac{2^m-2}{2^m}\, a. \end{aligned}$$

Each preimage of $l_{0\langle\alpha\rangle}^{\alpha,1-e}(S)$ in $\mathcal{B}_*^S$ bounds the critical gap and must be shorter than $\frac{a}{2}$. The rest follows by successive application of Lemma 2.8. ■

Correspondence: b) Immediately visible parameter chords $B \mapsto$ special boundary chords R_B of Gap⟨S⟩. It is suggestive to think of the boundary chords of *Gap*⟨*S*⟩ different from S as immediately visible from S in the dynamical sense. This is supported by the following statement:

Proposition 3.65. *Let $B \in \mathcal{B}_*$ of period qm for some $q \geq 2$ be immediately visible.*

Then behind B there exists a unique boundary chord R_B of the gap Gap⟨S⟩ of STEP $(q-1)m$, and every other boundary chord of Gap⟨S⟩ behind B has STEP greater than $(q-1)m$. R_B satisfies the following properties:

(i) Each visible chord in $\mathcal{B}_*$ behind B lies behind R_B and has period between $(q-2)m+1$ and qm.

(ii) The endpoints of R_B have kneading sequence $(\mathbf{v}^S e^S)^{q-1}\overline{\mathbf{v}^S(1-e^S)}$.

Proof: We start as in the proof of Proposition 3.64, by looking at a chord B_0 immediately visible from 01, and we assume that the intervals I_i are defined as at the beginning of the corresponding proof.

The intervals I_i are mutually disjoint, and h^{q-1} maps the interval I_0 homeomorphically onto I_{q-1}. The latter contains 0 and so there exists a unique 0-1-word $\mathbf{w}$ of length $q-2$ such that the preperiodic point with binary expansions $\mathbf{w}0\overline{1}$ and $\mathbf{w}1\overline{0}$ is contained in I_0 and has kneading sequence $0^{q-1}\overline{1}$. Obviously, behind B_0 there is no further preperiodic point mapped to 0 after less than q iterates, and since $d(B_0) = \frac{1}{2^q-1}$, behind B_0 there is no space for a chord in $\mathcal{B}_*$ of period not greater than q.

As in the last proof, use that $B = Q^\alpha(B_0)$ for some $B_0 \in \mathcal{B}_*$ immediately visible from 01. Assume that B_0 is as above and let $R_B := q^\alpha(\mathbf{w}1\overline{0})q^\alpha(\mathbf{w}0\overline{1})$. Then (ii) follows immediately from Theorem 3.31.

If a chord $Q \in \mathcal{B}_*$ is visible and behind B, then clearly its period is less than qm, and by Lemma 3.10 and the already shown it cannot separate B and R_B. On the other hand, by Proposition 3.64 the difference between the lengths of B and R_B is

$$\left(\frac{(2^m-1)^2}{2^{qm}-1} - \frac{2^m-2}{2^{(q-1)m}}\right) d = \frac{2^{qm}+2^{2m}-2^{m+1}}{2^{qm}(2^{qm}-1)}\, d < \frac{2d}{2^{qm}-1} < \frac{1}{2^{qm}-1}$$

with $d = d(S)$. Since the chord Q is not shorter than $\frac{1}{2^{qm}-1}$, it must lie behind R_B. Moreover, since by Proposition 3.64(i) R_B has length

$$\frac{2^m-2}{2^{(q-1)m}}d < \frac{1-2^{-m}}{2^{(q-2)m+1}} < \frac{1}{2^{(q-2)m+1}-1},$$

the period of Q must be greater than $(q-2)m+1$. ■

Note that by Proposition 3.64 the lengths of a parameter chord B and the dynamic chord R_B are nearly equal, and that the correspondence $B \mapsto R_B$ includes a correspondence $PER(B) = qm \mapsto STEP\ (q-1)m$ of R_B.

Example 3.66. The only boundary chord of $Gap\langle S\rangle$ of period $2q$ is $Q^\alpha(\frac{1}{3}\frac{2}{3})$, and there are two boundary chords of period $3q$: $Q^\alpha(\frac{1}{7}\frac{2}{7})$ and $Q^\alpha(\frac{5}{7}\frac{6}{7})$. One easily sees that $B \mapsto R_B$ maps $Q^\alpha(\frac{1}{3}\frac{2}{3})$ to $l^{\alpha,1-e}_{\mathtt{ve}}(S)$ and $\{Q^\alpha(\frac{1}{7}\frac{2}{7}), Q^\alpha(\frac{5}{7}\frac{6}{7})\}$ to $\{l^{\alpha,1-e}_{\mathtt{v0ve}}(S), l^{\alpha,1-e}_{\mathtt{v1ve}}(S)\}$.

Let us add the interesting fact that visible chords in $\mathcal{B}_*$ behind some fixed immediately visible chord must have different periods.

Proposition 3.67. (Immediately visible chords as separators)
Let $B_1, B_2 \in \mathcal{B}_*$ be visible but not immediately visible of the same period. Then there exist different immediately visible chords $Q_1, Q_2 \in \mathcal{B}_*$ such that B_1 lies behind Q_1 and B_2 behind Q_2.

Proof: Assume that the above statement is false for some $B_1 = \alpha_1\overline{\alpha_1}$ and some $B_2 = \alpha_2\overline{\alpha_2}$ of common kneading sequence, and let $n = PER(B_1) = PER(B_2)$.

Since $\mathcal{B}$ is closed, there exists a unique shortest chord $R = \alpha\gamma \in \mathcal{B}$ separating S from B_1 and B_2, and clearly this chord is the longest boundary chord of a unique gap G in $\mathcal{B}$. So by Theorem 3.17 α is periodic and $\gamma = \overline{\alpha}$, or α is preperiodic.

In the first case one of the chords B_1 or B_2 lies behind some chord B in $\mathcal{B}_*$ immediately visible from R and of period not less than $3PER(R)$. (Clearly, there is only one chord of period $2PER(R)$ immediately visible from R.) This chord is visible from B, hence by Proposition 3.65 (i) of period greater than $PER(R)$, contradicting its visibility from S.

If α is preperiodic, let us consider the Abstract Julia set $J = \mathbb{T}/\approx^\alpha$ and the map $H = h/\approx^\alpha$ on it. It holds $\gamma \in [\alpha]_{\approx^\alpha} = [\alpha]_\sim$ (see Theorems 3.17 and 3.19) and J is a dendrite. Let $\mathbf{C}$ be the simple curve in J which connects $x_1 := [\alpha_1]_{\approx^\alpha}$ and $x_2 := [\alpha_2]_{\approx^\alpha}$. Since the points in the equivalence class $[\alpha]_{\approx^\alpha}$ have the same kneading sequence (see Theorem 3.1), each of its iterates must lie behind $\frac{\alpha}{2}\frac{\gamma+1}{2}$ or $\frac{\alpha+1}{2}\frac{\gamma}{2}$.

On the other hand, $h^{n-1}(\alpha_1)$ and $h^{n-1}(\alpha_2)$ are behind $\frac{\alpha}{2}\frac{\gamma}{2}$ or $\frac{\alpha+1}{2}\frac{\gamma+1}{2}$, implying the following: No point contained in an equivalence class on the unique simple curve from $H^{n-1}(x_1)$ to $H^{n-1}(x_2)$ lies behind $\frac{\alpha}{2}\frac{\gamma+1}{2}$ or $\frac{\alpha+1}{2}\frac{\gamma}{2}$.

Therefore, $H^{n-1}(\mathbf{C})$ cannot be a simple curve, and there exists a least $k < n-1$ such that H is not injective on $H^k(\mathbf{C})$. Since $H^k(\mathbf{C})$ is a simple curve with endpoints $H^k(x_1)$ and $H^k(x_2)$, one easily sees that $h^k(\alpha_1)$ and $h^k(\alpha_2)$ must be separated by $\frac{\alpha}{2}\frac{\alpha+1}{2}$ and by $\frac{\gamma}{2}\frac{\gamma+1}{2}$. This shows $\widehat{B_1}(k) \neq \widehat{B_2}(k)$, which is impossible by Corollary 3.12. ∎

Correspondence: c) Parameter-like dynamic pairs $(S_1, S_2) \mapsto$ *'periodic' parameter chords* $PChord((S_1, S_2)) \in \mathcal{B}_*$. Now we come to the most important correspondence. In particular, we justify the term 'parameter-like' dynamic pair.

Proposition 3.68. (Parameter-like dynamic pairs determine 'periodic' parameter chords)
Let (S_1, S_2) *be a parameter-like dynamic pair of some STEP* $n \in \mathbb{N}$. *Then* S_1 *and* S_2 *belong to* $\mathcal{B}$ *and there exists a unique chord* $Q =: PChord(S_1, S_2) \in \mathcal{B}_*$ *of period not greater than* n *which separates* S_1 *and* S_2. *The chord* Q *has period* n, *and it holds* $\widehat{S_1}|n-1 = \widehat{S_2}|n-1 = \widehat{Q}|n-1$.

Proof: Let $S_1 = \beta_1\gamma_1$ and $S_2 = \beta_2\gamma_2$ with $\beta_1 < \beta_2 < \gamma_2 < \gamma_1$ be given as above. First of all, $S_1, S_2 \in \mathcal{B}$ by Proposition 3.61. Beyond this, none of the iterates of S_1, S_2 is longer than $\dot{S}, \ddot{S}$, and each of them lies behind $\frac{\beta_2}{2}\frac{\gamma_2+1}{2}$ or behind $\frac{\beta_2+1}{2}\frac{\gamma_2}{2}$.

The latter can be obtained as follows: Obviously, $\frac{\beta_1}{2}\frac{\gamma_1}{2}$ and $\frac{\beta_2}{2}\frac{\gamma_2}{2}$ as well as $\frac{\beta_1+1}{2}\frac{\gamma_1+1}{2}$ and $\frac{\beta_2+1}{2}\frac{\gamma_2+1}{2}$ belong to $\mathcal{B}_*^S$. If some iterate of S_1 or S_2 crossed one of the chords $\frac{\beta_2}{2}\frac{\gamma_2+1}{2}, \frac{\beta_2+1}{2}\frac{\gamma_2}{2}$, it would cross both chords but not $\dot{S}$ or $\ddot{S}$. Then its image would separate S and the dynamic pair (S_1, S_2), contradicting that (S_1, S_2) is parameter-like.

Now one easily sees that $\widehat{S_1}|n-1 = \widehat{S_2}|n-1$ but $\widehat{S_1}(n) \neq \widehat{S_2}(n)$. Moreover, according to Proposition 2.43 (ii), the whole open intervals I_1, I_2 in $\mathbb{T}$ between S_1 and S_2 are shorter than $\frac{1}{2^n}$. Thus, for each $l \leq n$ both I_1 and I_2 contain at most one periodic point of period l (see Lemma 3.4). Precisely, again by Lemma 3.4 there is one in I_1 and one in I_2 for $l = n$ since $\widehat{S_1}(n) \neq \widehat{S_2}(n)$, but none for $l < n$ since $\widehat{S_1}(l) = \widehat{S_2}(l)$ for such l.

Define $PChord(S_1, S_2)$ to be the chord which connects the two periodic points obtained in I_1 and I_2. Since $S_1, S_2 \in \mathcal{B}$, the chord $PChord(S_1, S_2)$ belongs to $\mathcal{B}_*$. Clearly, it has period n. ∎

The following proposition provides some sufficient conditions for dynamic pairs being parameter-like.

Proposition 3.69. *Let* $R \neq S$ *be a boundary chord of STEP* $(q-1)m; q \geq 2$ *of the gap* $Gap(S)$, *and let* (S_1, S_2) *be a dynamic pair of STEP less than* qm *behind* R. *Then* (S_1, S_2) *is parameter-like in each of the following cases:*

(i) $q > 3$,

(ii) $q = 3$ *and the STEP of* (S_1, S_2) *is not greater than* $2m$,

(iii) $d(S) = \frac{1}{2^m-1}$.

Proof: Each chord behind R lies behind $l_{\mathbf{ve}}^{\alpha,1-e}(S)$ after $(q-2)m$ iterates. So if a dynamic pair (S_1, S_2) behind R fails to be parameter-like, its *STEP* must exceed $2(q-2)m$. This is impossible in the cases (i) and (ii).

In the case (iii) we have $d(S), d(h(S)), \ldots, d(h^{m-3}(S)) < \frac{1}{4}$. Since the iterates of S do not meet the open region between $\dot{S}$ and $\ddot{S}$, this implies the following: If a chord $\tilde{R}$ lies behind $l_{\mathbf{ve}}^{\alpha,1-e}(S)$, hence behind S, then $h^j(\tilde{R})$ is behind $h^j(S)$ and not between $\dot{S}$ and $\ddot{S}$ for $j < m-1$. Thus $h^j(S)$ fails to be behind S for $j = 1, 2, \ldots, n-1$.

So the number of iterates from S_1 and S_2 to S must be at least qm if (S_1, S_2) is not parameter-like. This contradicts our assumption. ■

The other direction: d) (Not immediately) visible 'periodic' parameter-chords $Q \mapsto$ dynamic pairs $DPair(Q)$. Now we want to assign to a given 'periodic' parameter a dynamic pair. In view of Theorem 1.24, we focus our attention to the visible 'periodic' parameter chords.

Fix some visible $Q \in \mathcal{B}_*$, and let l be minimal with the property that $h^l(Q)$ separates $\dot{S}$ and $\ddot{S}$. Clearly, $l \leq PER(Q) - 1$. Recall that the chords $S, \dot{S}$ and $\ddot{S}$ must belong to $\eth\mathcal{B}^Q$, implying that the iterates of Q do not cross these chords, and fix symbols $s_1, s_2, \ldots, s_l \in \{0,1\}$ such that $Q = l^{\alpha}_{s_1 s_2 \ldots s_l}(h^l(Q))$. Further let S_1 be the longer of the chords $l^{\alpha,1-e}_{s_r s_{r+1} \ldots s_l}(\dot{S}), l^{\alpha,1-e}_{s_r s_{r+1} \ldots s_l}(\ddot{S})$ and S_2 be the shorter one. We show that (S_1, S_2) forms a dynamic pair whose members are separated by Q.

For each r with $1 \leq r \leq l$ the chords $l^{\alpha,1-e}_{s_r s_{r+1} \ldots s_l}(\dot{S}), l^{\alpha,1-e}_{s_r s_{r+1} \ldots s_l}(\ddot{S})$ belonging to $\mathcal{B}_*^S$ are separated by $h^{r-1}(Q)$. This is an immediate consequence of the orientation-invariance of h (Lemma 2.9). Assume that (S_1, S_2) fails to be a dynamic pair. Then $s_{l-m+2} s_{l-m+3} \ldots s_l = \mathbf{v}$ and $\{h^{l-m+1}(S_1), h^{l-m+1}(S_2)\} = \{S, l_{\mathbf{ve}}^{\alpha,1-e}(S)\}$.

Therefore, $h^{l-m+1}(Q)$ separates the boundary chords S and $l_{\mathbf{ve}}^{\alpha,1-e}(S)$ of the critical value gap, implying that the initial subword of length $m-1$ of $\sigma^{l-m+1}(\overline{\mathbf{ve}})$ is equal to $\mathbf{v}$. (Since Q is visible, $\mathbf{v}^Q$ is an initial subword of $\overline{\mathbf{ve}}$). So by Lemma 3.11 m divides $l+1$.

There is some $q \geq 2$ such that Q lies behind an immediately visible chord of period qm. If $j \leq (q-2)m$, then $h^j(Q)$ is shorter than $l_{\mathbf{ve}}^{\alpha,1-e}(S)$. So we have $l - m + 1 > (q-2)m$ implying $l + 1 \geq qm$. This is impossible since $l \leq PER(Q) - 1$. Let us summarize:

Proposition 3.70. (Visible 'periodic' parameter chords determine dynamic pairs)
Let $Q \in \mathcal{B}_*$ behind S be visible but not immediately visible. Then there exists a unique dynamic pair $(S_1, S_2) =: DPair(Q)$ such that Q separates S_1 and S_2. Its STEP n is not greater than $PER(Q)$ and for $j < n$ the chord $h^j(Q)$ separates $h^j(S_1)$ and $h^j(S_2)$. ■

Domains with complete correspondence. The two maps $PChord : (S_1, S_2) \mapsto PChord((S_1, S_2))$ and $DPair : Q \mapsto DPair(Q)$ relate dynamical structure and parts of $\mathcal{B}_*$ in the two different directions. Our aim is to find maximal domains where one map is inverse to the other one. A first result is suggested by the above discussion.

Proposition 3.71. (Partial reflexivity)
The following statements are valid:

(i) If a dynamic pair (S_1, S_2) is parameter-like and $PChord((S_1, S_2))$ is visible, then $DPair(PChord((S_1, S_2))) = (S_1, S_2)$.

(ii) If $Q \in \mathcal{B}_*$ is visible but not immediately visible and $DPair(Q)$ is parameter-like, then $PChord(DPair(Q)) = Q$.

Proof: The statement (i) is obvious by Propositions 3.68 and 3.70. So we have to show (ii).

Let $(S_1, S_2) := DPair(Q)$ be of $STEP$ $n = jm + k$ with $j \geq 0$ and $0 \leq k < m$, and let $\tilde{Q} := PChord(DPair(Q))$. Since Q is visible, a chord in $\mathcal{B}_*$ of minimal possible period separating S_1 and S_2 is also visible. By Proposition 3.68 such a chord must coincide with the chord $\tilde{Q}$ of period n, and $k > 0$ by Corollary 3.12. Note that $PER(Q) \geq n$ by Proposition 3.70.

Assuming that the chords Q and $\tilde{Q}$ are different, Q separates S_1 and $\tilde{Q}$. Let r be the minimum of periods of all chords in $\mathcal{B}_*$ separating S_1 and $\tilde{Q}$. Then there is a unique chord of period r, which obviously must be visible, and clearly $r > n$.

Now by Lemma 3.10 and Corollary 3.12 we have $\widehat{\tilde{Q}}^{-}|r-1 = \widehat{S_1}|r-1 = \overline{\mathbf{ve}}|r-1$ and $\widehat{\tilde{Q}}^{-}(r) \neq \widehat{S_1}(r)$. Thus from $\widehat{\tilde{Q}}^{-} = \overline{(\mathbf{ve})^j v_1 v_2 \ldots v_{k-1} v_k}$ and $\widehat{S_1} = (\mathbf{ve})^j v_1 v_2 \ldots \overline{v_k \mathbf{v}(1-e)}$ one easily shows $r \geq (j+1)m + k$. So we have $(\mathbf{ve})^j v_1 v_2 \ldots v_k \mathbf{v} = (\mathbf{ve})^{j+1} v_1 v_2 \ldots v_{k-1}$, hence $\mathbf{v} = v_{k+1} \ldots v_{m-1} e v_1 \ldots v_{k-1}$, which contradicts Lemma 3.11(ii). ■

For which parameter-like dynamic pairs (S_1, S_2) is $PChord((S_1, S_2))$ visible? The answer to this question is provided by Proposition 3.73. It is already prepared by the last statement in Proposition 3.68 and by the characterization of visibility in Corollary 3.12, which give rise to the following notion:

Definition 3.72. (Semi-visible dynamic pairs)
Let $B \in \mathcal{B}_*$ be immediately visible. Then a dynamic pair (S_1, S_2) of $STEP$ n is said to be semi-visible *if it is parameter-like, and if the sequence $\overline{\mathbf{ve}}$ and the kneading sequence of S_1 have the same initial subwords of length n.*

Proposition 3.73. *For a parameter-like dynamic pair (S_1, S_2) the chord $PChord((S_1, S_2))$ is visible iff (S_1, S_2) is semi-visible.* ■

We want to justify the notion 'semi-visible'.

Proposition 3.74. *Let $B \in \mathcal{B}_*$ be immediately visible. Then each visible dynamic pair behind R_B is semi-visible, and PChord maps the set of all visible dynamic pairs behind R_B injectively into the set of visible chords in $\mathcal{B}_*$ behind B.*

Proof: Let (S_1, S_2) be a visible dynamic pair behind R_B of some *STEP* n, hence a parameter-like one. We show that $Q = PChord((S_1, S_2))$ is visible. The rest follows from Proposition 3.71(i) and Proposition 3.73.

Assuming the contrary, there would exist some visible chord $\tilde{Q} \in \mathcal{B}_*$ separating B and Q. By Proposition 3.68 and Lavaurs' lemma (Corollary 3.8) $\tilde{Q}$ would have period less than n and would not separate S_1 and S_2. The longer chord in the dynamic pair $DPair(\tilde{Q})$ would separate R_B and S_1, contradicting visibility of (S_1, S_2). ∎

It would have been nice to have a complete correspondence between semi-visible dynamic pairs and visible but not immediately visible 'parameter' chords. However, there are cases where complete correspondences fail. An example for this is provided below. In the general case we cannot show more than the following

Theorem 3.75. ('Dynamic' semi-visibility and 'parameter' visibility)
Let $B \in \mathcal{B}_*$ be immediately visible and of period $qm; q \geq 3$. Then PChord maps the set of all semi-visible dynamic pairs behind R_B bijectively onto the set of all visible chords in $\mathcal{B}_*$ behind B. The STEP of a semi-visible dynamic pair (S_1, S_2) behind R_B coincides with the period of $PChord((S_1, S_2))$, and at most one of the iterates of (S_1, S_2) lies behind $l^{\alpha,1-e}_{\mathbf{v0ve}}(S)$ or $l^{\alpha,1-e}_{\mathbf{v1ve}}(S)$.

Proof: By Proposition 3.73 it remains to be shown that the restriction of *PChord* to the semi-visible dynamic pairs is surjective. To get a preimage of some visible $Q \in \mathcal{B}_*$ behind B we show that $DPair(Q)$ is parameter-like. Then by Proposition 3.71(ii) $Q = PChord(DPair(Q))$.

Let $Q \in \mathcal{B}_*$ behind B be visible and let $(S_1, S_2) = DPair(Q)$. Then the *STEP* n of (S_1, S_2) is less than qm. Clearly, $(h^{(q-3)m}(S_1), h^{(q-3)m}(S_2))$ lies behind $l^{\alpha,1-e}_{\mathbf{v0ve}}(S)$ or $l^{\alpha,1-e}_{\mathbf{v1ve}}(S)$, and $(h^{(q-2)m}(S_1), h^{(q-2)m}(S_2))$ behind $l^{\alpha,1-e}_{\mathbf{ve}}(S)$.

If for some $j > 0$ the dynamic pair $(h^{(q-2)m+j}(S_1), h^{(q-2)m+j}(S_2))$ were behind $l^{\alpha,1-e}_{\mathbf{v0ve}}(S)$ or $l^{\alpha,1-e}_{\mathbf{v1ve}}(S)$, then also $h^{(q-2)m+j}(Q)$ would be. So by Proposition 2.43(ii) the initial subword of $\sigma^{(q-2)m+j}(\overline{\mathbf{ve}})$ of length $m-1$ would coincide with $\mathbf{v}$, and Lemma 3.11(ii) would imply $j = m$. Therefore $(h^{qm}(S_1), h^{qm}(S_2))$ would be behind $l^{\alpha,1-e}_{\mathbf{ve}}(S)$, what obviously is false. ∎

3.4.2 Consequences for the (Abstract) Mandelbrot set

This subsection is devoted to the Translation Principle for the (Abstract) Mandelbrot set mentioned in Chapter 1.2.2. We discuss it in the setting of

Chapters 2 and 3. As the main statement we get Theorem 3.78, the 'periodic chord' version of Theorem 1.24 (see discussion at the end of Chapter 1).

Definition 3.76. (Translation-invariance)
Let $S \in \mathcal{B}_* \cup \{01\}$ be of period m, and let $B_1, B_2 \in \mathcal{B}_*$ be immediately visible from S of periods $q_1 m$ and $q_2 m$, respectively. Then B_1 and B_2 are said to be translation-equivalent *if there exists an embedding-preserving bijection from the set of chords in $\mathcal{B}_*$ behind B_1 being visible from S onto the set of chords in $\mathcal{B}_*$ behind B_2 being visible from S, which increases periods by $(q_2 - q_1)m$.*

Theorem 3.75 turns semi-visible dynamic pairs into visible 'periodic' parameter chords. Therefore, in order to prove 'translation invariance' of visibility for the latter chords we need 'translation invariance' of semi-visibility for dynamic pairs. This is realized by the following statement.

Lemma 3.77. *Let $S \in \mathcal{B}_* \cup \{01\}$ be of period m, and let $B_1, B_2 \in \mathcal{B}_*$ be immediately visible from S of period $q_1 m$ and $q_2 m$, respectively, where $q_1, q_2 \geq 3$. Further, let $h^{(q_1-3)m}(R_{B_1}) = h^{(q_2-3)m}(R_{B_2})$, and let (S_1, S_3) and (S_2, S_4) be dynamic pairs of STEP's $k_1 < q_1 m$ and $k_2 < q_2 m$ behind R_{B_1} and R_{B_2}, respectively.*

If $h^{(q_1-3)m}(S_1) = h^{(q_2-3)m}(S_2)$ and $h^{(q_1-3)m}(S_3) = h^{(q_2-3)m}(S_4)$, then it holds $k_2 - k_1 = (q_2 - q_1)m$, and (S_1, S_3) is semi-visible iff (S_2, S_4) is.

Proof: $k_2 - k_1 = (q_2 - q_1)m$ is obvious. So let us assume that (S_1, S_3) is semi-visible. Then by the last statement of Proposition 3.75 only the $(q_1 - 3)m$-th iterate of (S_1, S_3) lies behind $l^{\alpha,1-e}_{\mathbf{v0ve}}(S)$ or $l^{\alpha,1-e}_{\mathbf{v1ve}}(S)$ and so only the $(q_2-3)m$-th iterate of (S_2, S_4) (see also Theorem 3.62). The first consequence is that also in the case $q_2 = 3$ the chords S_2, S_4 belong to $\mathcal{B}_*$. This follows from Corollary 3.2 since no iterate of S_2, S_4 separates S from S_2 or S_4.

Clearly, the chord $h^{(q_1-2)m-1}(S_1) = h^{(q_2-2)m-1}(S_3)$ lies behind $l^{\alpha,1-e}_{\mathbf{eve}}(S)$ or $l^{\alpha,1-e}_{(1-e)\mathbf{ve}}(S)$. If some of its iterates is between $\dot{S}$ and $\ddot{S}$, then again behind one of the chords $l^{\alpha,1-e}_{\mathbf{eve}}(S)$ or $l^{\alpha,1-e}_{(1-e)\mathbf{ve}}(S)$. Since by Proposition 3.65(ii) and Theorem 3.62 the kneading sequence of S_2 begins with $(\mathbf{ve})^{q_2-2}$, the initial subwords of length k_2 of $\overline{\mathbf{ve}}$ and the kneading sequence of S_2 coincide. ■

The combination of Theorem 3.75 and Lemma 3.77 provides the statement announced above. As already mentioned it covers Theorem 1.24.

Theorem 3.78. (Partial Translation Principle; abstract version)
Let $S \in \mathcal{B}_* \cup \{01\}$ be of period m. Then each chord $B \in \mathcal{B}_*$ of period not less than $3m$ being immediately visible from S is translation-equivalent to one of the two of period $3m$. More precisely, two chords $B_1, B_2 \in \mathcal{B}_*$ being immediately visible from S and of period $q_1 m, q_2 m$ with $q_1, q_2 \geq 3$ are translation-equivalent if $h^{(q_1-3)m}(R_{B_1}) = h^{(q_2-3)m}(R_{B_2})$ ■

As an immediate consequence of Theorem 3.78 we obtain

Corollary 3.79. (Partial Translation Principle for Internal Addresses) *For each admissible internal address $1 \to \ldots \to m$ the following holds: If for some positive $r < m$ one of the internal addresses $1 \to \ldots \to m \to qm + r$ with $q = 2, 3, 4, \ldots$ is admissible, then all these addresses are.* ■

A counter-example. The chords of periods $2m$ and $3m$ immediately visible from some chord in $\mathcal{B}_*$ of period m need not be translation-equivalent. In order to give a counter-example, let us discuss the case $S = \alpha\overline{\alpha}$ with $\alpha = \frac{13}{31}$ and $\overline{\alpha} = \frac{18}{31}$. The period of S is 5, and it holds $\mathbf{v} := \mathbf{v}^S = 0100$ and $e := e^S = 1$. For simplicity, we indicate *STEP*'s and periods of dynamic pairs and chords considered by a superscript. The two boundary chords of $Gap\langle S\rangle$ we are interested in are $R^5 = l^{\alpha,1-e}_{01001}(S) = \frac{421}{992}\frac{571}{992}$ and $R^{10} = l^{\alpha,1-e}_{0100101001}(S) = \frac{13317}{31744}\frac{13467}{31744}$.

The starting point of the counter-example is an observation made by BRUIN and SCHLEICHER: one finds a periodic point with kneading sequence $\overline{0100101001*}$ (behind S), but (as already mentioned in Example 3.16) no periodic point in $\mathbb{T}$ has kneading sequence $\overline{01001*}$.

A chord $Q_1 \in \mathcal{B}_*$ with kneading sequence $\overline{0100101001*}$ has internal address $1 \to 2 \to 4 \to 5 \to 11$, hence is visible from some $\tilde{S} \in \mathcal{B}_*$ of period 5. Consider two chords B_1, B_2 immediately visible from $\tilde{S}$ of periods $5(q+1); q > 1$ and $5q$ such that Q_1 is behind B_1. If B_1 and B_2 were translation-equivalent, we would have a chord in $\mathcal{B}_*$ behind B_2 of period 6 being visible from $\tilde{S}$, hence of kneading sequence $\overline{01001*}$.

In order to find the reason for the failure of the kneading sequence $\overline{01001*}$, we explored the geometry behind S by means of a computer and obtained the following: $Q^3 = \frac{3}{7}\frac{4}{7}$ is a visible chord in $\mathcal{B}_*$ behind R^5. The chords $Q^8 = \frac{107}{255}\frac{36}{85}$ and $Q^{11} = \frac{867}{2047}\frac{868}{2047}$ are the only visible ones in $\mathcal{B}_*$ of period not greater than 11 behind R^{10}.

Further, one has $DPair(Q^3) = (S_1^3, S_2^3)$ with $S_1^3 = \frac{53}{124}\frac{71}{124}$ and $S_2^3 = \frac{111}{248}\frac{137}{248}$, $DPair(Q^8) = (S_1^8, S_2^8)$ with $S_1^8 = \frac{1665}{3968}\frac{1683}{3968}$ and $S_2^8 = \frac{3335}{7936}\frac{3361}{7936}$, and $DPair(Q^{11}) = (S_1^{11}, S_2^{11})$ with $S_1^{11} = \frac{13445}{31744}\frac{13463}{31744}$ and $S_2^{11} = \frac{26895}{63488}\frac{26921}{63488}$.

The mutual position of the chords $S, R^5, R^{10}, Q^3, Q^8, Q^{11}, S_1^3, S_2^3, S_1^8, S_2^8, S_1^{11}$ and S_2^{11} is sketched in Figure 3.10.

All three dynamic pairs considered are parameter-like and so by Proposition 3.71 we have $PChord(DPair(Q^3)) = Q^3$, $PChord(DPair(Q^8)) = Q^8$ and $PChord(DPair(Q^{11})) = Q^{11}$. Moreover, one easily checks that (S_1^3, S_2^3) and (S_1^8, S_2^8) are visible.

The crucial point is that the dynamic pair $(S_1^6, S_2^6) := (h^5(S_1^{11}), h^5(S_2^{11})) = (\frac{549}{992}\frac{567}{992}, \frac{1103}{1984}\frac{1129}{1984})$ is not parameter-like (see Proposition 3.61): as Figure 3.10 shows, S_1^6 crosses the chord Q^3, hence it cannot belong to $\mathcal{B}_*$.

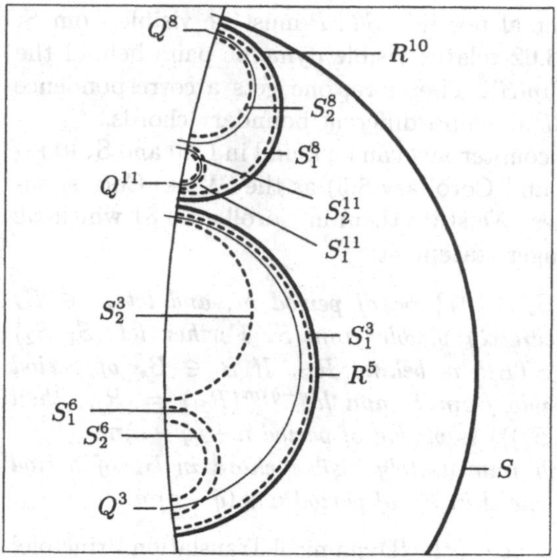

Fig. 3.10. Counter-example: $S = \frac{13}{31}\frac{18}{31}$

So (S_1^6, S_2^6) does not generate a (visible) chord $PChord(((S_1^6, S_2^6)) \in \mathcal{B}_*$ (of period 6). Also note that the dynamic pair (S_1^{11}, S_2^{11}) is semi-visible by Proposition 3.73, but fails to be visible since it lies between S_1^8 and S_2^8.

The graph in Figure 3.11 illustrates the structure of visibility from S. Its nodes represent the chord S, all immediately visible chords of periods 10, 15 and 20 and all visible chords behind them. All periods are indicated, and the incidence structure is induced by the 'behind'-relation in $\mathcal{B}_*$. The graph shows that the immediately visible chords of periods 10 and 15 are not translation-equivalent. Note that in the two cases of period 15 the obtained subtrees coincide, but their embeddings into the plane are different.

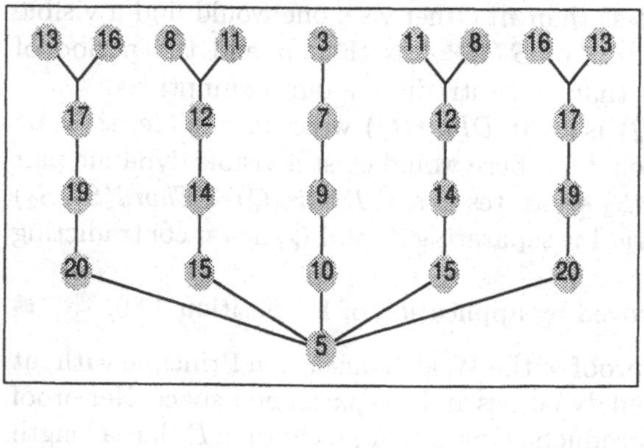

Fig. 3.11. Tree of visibility from $S = \frac{13}{31}\frac{18}{31}$

Weaker statements. Let $S \in \mathcal{B}_* \cup \{01\}$, and let R be a boundary chord of $Gap\langle S\rangle$ different from S. By the last statement in Proposition 2.43(ii) a dynamic pair behind R of minimal possible *STEP* must be visible from S. On the other hand, Theorem 3.62 relates visible dynamic pairs behind the various boundary chords of $Gap\langle S\rangle$. Therefore, one gets a correspondence between minimal possible *STEP*'s behind different boundary chords.

This fact and its parametric counterpart can be found in LAU and SCHLEICHER [92] (see Proposition 8.4 and Corollary 8.5) as the 'Weak' Correspondence and Translation Principles. We state them in Corollary 3.81 which we deduce from the following stronger statement:

Proposition 3.80. *Let $S \in \mathcal{B}_* \cup \{01\}$ be of period m, and let $B \in \mathcal{B}_*$ of period $qm; q \geq 2$ be immediately visible from S. Further let (S_1, S_2) be a visible dynamic pair of STEP n behind R_B. If $\tilde{B} \in \mathcal{B}_*$ of period $\tilde{q}m; \tilde{q} \geq 2$ is immediately visible from S and $h^{(q-\tilde{q})m}(R_B) = R_{\tilde{B}}$, then $PChord((h^{(q-\tilde{q})m}(S_1), h^{(q-\tilde{q})m}(S_2)))$ is visible of period $n + (\tilde{q} - q)m$.*

More generally, behind each immediately visible chord in $\mathcal{B}_*$ of period $\tilde{q}m; \tilde{q} \geq 2$ there exists a visible chord in $\mathcal{B}_*$ of period $n + (\tilde{q} - q)m$.

Proof: The first part is a consequence of the 'Dynamical Translation Principle' (see Theorem 3.62), and Proposition 3.74. From the first part and the fact that each chord $l^{\alpha,1-e}_{\mathbf{v}s_1\mathbf{v}s_2\ldots\mathbf{v}s_l\mathbf{v}e}(S)$ for $s_1 s_2 \ldots s_l; l \geq 1$ is mapped to $l^{\alpha,1-e}_{\mathbf{v}e}(S)$ after lm iterates one obtains the second part. ■

Corollary 3.81. (Weak Correspondence and Translation Principles) *Let $S \in \mathcal{B}_* \cup \{01\}$ be of period m, and let $B \in \mathcal{B}_*$ be immediately visible from S. Further, let $Q \in \mathcal{B}_*$ be behind B of minimal possible period n. Then $DPair(Q)$ is visible of minimal possible STEP behind B. In particular, $PChord(DPair(Q)) = Q$ and the STEP of $DPair(Q)$ is n.*

If B_1 and B_2 are immediately visible and have periods q_1m and q_2m, respectively, then the difference between the minimal periods of chords in $\mathcal{B}_*$ behind B_2 and B_1 is $(q_2 - q_1)m$.

Proof: We verify the first part of the corollary. First of all, behind B there is no dynamic pair of *STEP* less than n. Otherwise, one would find a visible dynamic pair (S_1, S_2) behind B of *STEP* less than n and the period of $PChord(S_1, S_2)$ would be less than n, contradicting our assumptions.

So the *STEP* of $DPair(Q)$ is n. If $DPair(Q)$ were not visible, then by the first remark after Definition 3.60 there would exist a visible dynamic pair (S_1, S_2) of *STEP* n such that S_2 separates B and $DPair(Q)$. $PChord(S_1, S_2)$ would be a chord in $\mathcal{B}_*$ of period n separating B and Q, again contradicting our assumptions.

Now the second part is proved by application of Proposition 3.80. ■

Remark. KAUKO [75] gives a proof of the Weak Translation Principle without using a correspondence between dynamics and the parameter space. Her proof provides some interesting by-products. One is that no chord in $\mathcal{B}_*$ has a length between $\frac{1}{2^n-1}$ and $\frac{1}{2^n+1}$ for some $n \in \mathbb{N}$.

The narrow case. If S has length $\frac{1}{2^m-1}$, then behind S there is no space for chords in $\mathcal{B}_*$ of period less than or equal to m. Otherwise, Theorem 3.7 provides at least one such chord. In the case $d(S) = \frac{1}{2^m-1}$ LAU and SCHLEICHER proved the statement that all chords immediately visible from S are translation-equivalent (see [92], Proposition 10.3). We want to give a proof on the base of the considerations above.

Definition 3.82. *A chord $S \in \mathcal{B}_* \cup \{01\}$ of period m is said to be* narrow *if its length is $\frac{1}{2^m-1}$, or equivalently, if all chords in $\mathcal{B}_*$ behind S have period greater than m.*

Proposition 3.83. *If $S \in \mathcal{B}_* \cup \{01\}$ is narrow, then the map $(S_1, S_2) \mapsto PChord(S_1, S_2)$ forms a bijection from the set of all dynamic pairs visible from S onto the set of all chords in $\mathcal{B}_*$ visible but not immediately visible from S, and all chords in $\mathcal{B}_*$ immediately visible from S are translation-equivalent.*

Proof: Let m be the period of S, and fix some immediately visible chord B of period $qm; q \geq 2$ and some visible but not immediately visible $Q \in \mathcal{B}_*$ of some period $n(< qm)$ behind B. We have statement (iii) of Proposition 3.69. So $(S_1, S_2) = DPair(Q)$ is parameter-like, and by Proposition 3.71 it holds $PChord(DPair(Q)) = Q$. In particular, the *STEP* of (S_1, S_2) is n.

It remains to be shown that (S_1, S_2) is visible. If there were a parameter-like dynamic pair (R_1, R_2) of *STEP* not greater than n for which R_2 would separate S and S_1, then the period of $PChord(R_1, R_2)$ would be less or equal to n, contradicting visibility of Q.

So assume that (S_1, S_2) lies between R_1, R_2 for a parameter-like dynamic pair (R_1, R_2). (Clearly, S_1, S_2 cannot separate R_1 and R_2.) Then by Proposition 3.65 (S_1, S_2) and (R_1, R_2) lie behind R_B of *STEP* $(q-1)m$, hence the dynamic pairs $(h^{(q-2)m}(S_1), h^{(q-2)m}(S_2))$ and $(h^{(q-2)m}(R_1), h^{(q-2)m}(R_2))$ are behind $l_{\mathbf{v}e}^{\alpha,1-e}(S)$.

$PChord(h^{(q-2)m}(S_1), h^{(q-2)m}(S_2))$ and $PChord(h^{(q-2)m}(R_1), h^{(q-2)m}(R_2))$ have periods greater than m. Therefore the *STEP*'s of both dynamic pairs (S_1, S_2) and (R_1, R_2) lie between $(q-1)m$ and qm. By Proposition 2.43(ii), the whole intervals on $\mathbb{T}$ between R_1 and R_2 are not longer than $\frac{1/(2^m-1)}{2^{(q-1)m+1}} = \frac{2^{m-2}}{(2^m-1)(2^{qm}-1)}$, but the two chords S_1, S_2 are not shorter than $\frac{1-1/(2^m-1)}{2^{qm}-1} = \frac{2^m-2}{(2^m-1)(2^{qm}-1)}$. This is impossible. ■

Figure 3.12 illustrates Proposition 3.83 for $S = \frac{5}{31}\frac{6}{31}$ (compare Example 3.63 and Figures 3.8 and 3.9). Here chords in $\mathcal{B}_*(S)$ are drawn in the usual way, but the elements of $\mathcal{B}_*$ are represented by arcs crossing the unit circle a little.

From the Weak Translation Principle (see second part of Corollary 3.81) we derive a supplementary statement, which was found by LAU and SCHLEICHER (see [92], Theorem 10.2) and underlines the simple structure of visibility in the narrow case.

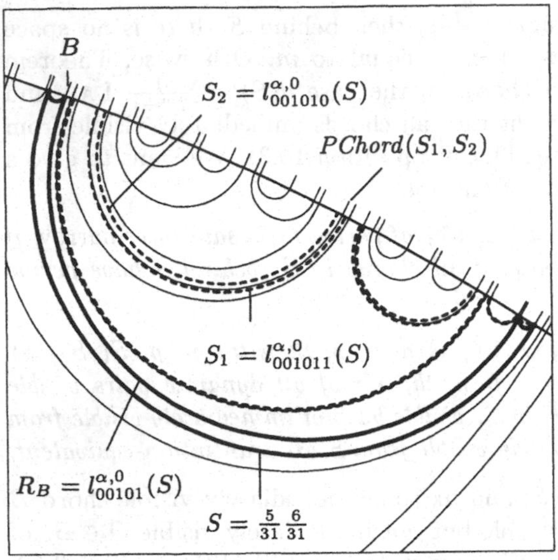

Fig. 3.12. Corresponding chords

Proposition 3.84. *If $S \in \mathcal{B}_*$ of period m is narrow, then for each $n > m$ there exists exactly one chord in $\mathcal{B}_*$ of period n being visible from S.*

Proof: First let $n = m + k$ with $k < m$. The chord S has length $\frac{1}{2^m-1}$, and it holds $\frac{2^k}{2^{m+k}-1} < \frac{1}{2^m-1} < \frac{2^k+1}{2^{m+k}-1}$. Since $\sim$ identifies points of period $m+k$ pairwisely, exactly 2^{k-1} chords in $\mathcal{B}_*$ of period $m+k$ are behind S. By Lemma 3.10 their kneading sequences start with $\mathbf{v}e$.

On the other hand, exactly 2^{k-1} 0-1-sequences start with $\mathbf{v}e$ and have length $m + k - 1$. Therefore, if behind S different chords in $\mathcal{B}_*$ of period $m + k$ have different kneading sequences, among them exactly one is visible.

So assume that there are two chords $B_1, B_2 \in \mathcal{B}_*$ behind S of period $m+k$ of the same kneading sequences and with minimal possible $k < m$. Let l be the next to the last entry of the common internal addresses of B_1 and B_2. By the considerations in Chapter 3.1.3 one finds a chord $B \in \mathcal{B}_*$ of period l separating both B_1 and B_2 from S or being equal to S. Obviously, $l \geq m$, and so B_1 and B_2 are not visible from B.

By Proposition 3.67 at most one of them can lie behind the unique chord R of period $2l$ immediately visible from B. On the other hand, by the Weak Translation Principle (see second part of Corollary 3.81) R is the only chord in $\mathcal{B}_*$ being immediately visible from B and allowing chords in $\mathcal{B}_*$ of periods less than $2m$ behind it. (Recall that no chord in $\mathcal{B}_*$ behind S has period less than m.) This is impossible.

Let us summarize. For each $n = m + 1, m + 2, \ldots, 2m - 1$ there exists exactly one chord in $\mathcal{B}_*$ of period n being visible from S. Again by the Weak Translation Principle (see second part of Corollary 3.81), such a chord lies behind the unique chord $\mathcal{B}_*$ of period $2m$ (being immediately visible from S). So the above statement follows from Proposition 3.83. ■

We finish with the following statement being an immediate consequence of Proposition 3.84.

Corollary 3.85. (Admissible internal addresses in the narrow case)
If an internal address $1 \to \ldots \to m$ *is admissible by some narrow chord* $S \in \mathcal{B}_*$, *then all internal addresses* $1 \to \ldots \to m \to n$ *for* $n > m$ *are admissible.* ■

4. Abstract and concrete theory

The theory of Abstract Julia sets and of the Abstract Mandelbrot set was developed in a self-contained manner. We now illustrate this theory in the complex plane. In particular, we complete verification of the statements given in Chapter 1. The whole discussion follows the motto 'Look at the abstract models, and ask for what remains true in the complex plane'.

4.1 Quadratic iteration

4.1.1 Julia equivalences and Julia sets

We continue the listing of well-known statements on quadratic polynomials started in Chapter 1.1.3. Much of the following is valid beyond the case of quadratic polynomials, but we do not need the general statements.

V. Classification of Fatou components. For a given $c \in \mathbb{C}$, the complement of the Julia set J_c in $\overline{\mathbb{C}}$ is said to be the *Fatou set* of p_c, and its connectedness components are called the *Fatou components.* A Fatou component either contains the fixed point ∞ or is a subset of the filled-in Julia set. In the latter case we call it *bounded.*

Each non-repelling periodic point $z \in \mathbb{C}$ of some period m belongs to the closure of a bounded Fatou component. Namely, for $g = p_c^m$ the following holds:

1. If the point z is attractive, then it lies in a bounded Fatou component K, which is g-invariant. For all points $x \in K$ it holds $\lim_{n\to\infty} g^n(x) = z$. Therefore, K is called the *immediate basin* of z.
2. If z is parabolic, there exist a $q \in \mathbb{N}$ and mutually different bounded Fatou components $K_1, K_2, \ldots, K_q = K_0$ satisfying the following properties: For all $j = 1, 2, \ldots, q$ it holds $g(K_{j-1}) = K_j, z \in \partial K_j \subseteq J_c$ and $\lim_{n\to\infty} g^n(x) = z$ if $x \in K_j$. (Note that all sets $p_c^n(K_0); n = 1, 2, 3, \ldots, mq$ are mutually disjoint.) Each $K_j; j = 1, 2, \ldots, q$ is called an *immediate basin* of z.

3. If z is irrationally indifferent, then either $z \in J_c$ or $z \in K_c \setminus J_c$. In the first case z is called a *Cremer point*. In the second case z belongs to a bounded Fatou component K on which g is conformally conjugate to an irrational rotation with rotation number ν (relative to 2π) given by $(p_c^m)'(z) = e^{2\pi\nu i}$: there exists a conformal bijection ϕ from K onto the open unit disk satisfying $g(x) = \phi^{-1}((p_c^m)'(z)\phi(x))$ for all $x \in K$. The set K is called a *Siegel disk*, and the point z a *Siegel point*.

Clearly, in each case K is periodic, but the fact that the orbit of each bounded Fatou component contains the immediate basin of some attractive or parabolic periodic point or some Siegel disk is highly non-trivial. It follows from SULLIVAN's famous theorem on non-wandering domains: each Fatou component of a rational map is preperiodic or periodic (see [163]).

VI. Local connectivity of some Julia sets. For simplicity, we call a parameter $c \in M$ *hyperbolic* (*superattractive, parabolic, irrationally indifferent*) of *period* m if p_c has an attractive (superattractive, parabolic, irrationally indifferent) orbit of period m. By the *multiplier* of such a parameter c we mean the multiplier of the corresponding periodic orbit. (Recall that there exists at most one non-repelling orbit for some p_c.)

If a parameter c is hyperbolic or parabolic, or if c is preperiodic with respect to p_c, then J_c is locally connected (see [43, 114, 25, 166]). This will be of high importance in the following discussion. Note that parameters c being preperiodic with respect to p_c are called *Misiurewicz points*.

VII. Periodic and preperiodic points depending on the parameter. In Chapter 1.2.2 we discussed the dependence of periodic points on the parameter. Let us generalize the discussion to the preperiodic case. For this, fix some pair (c_0, z_0), where z_0 is a preperiodic point of preperiod n and period m for some p_{c_0}; $c_0 \in \mathbb{C}$. ($n = 0$ provides the periodic case.)

Clearly, the map g defined by $g(c,z) = p_c^{m+n}(z) - p_c^n(z)$ satisfies $g(c_0, z_0) = 0$, and we have

$$\frac{\partial g}{\partial z}(c_0, z_0) = ((p_{c_0}^m)'(p_{c_0}^n(z_0)) - 1)2^n \prod_{i=0}^{n-1} p_{c_0}^i(z_0).$$

The term $(p_{c_0}^m)'(p_{c_0}^n(z_0))$ is no more than the multiplier of the n-th iterate of z_0, and so $\frac{\partial g}{\partial z}(c_0, z_0)$ is equal to 0 iff this iterate is parabolic of multiplier 1 or if 0 lies on the orbit of z_0. (Later it will turn out that both at the same time is impossible.)

Otherwise, the implicit function theorem defines a locally invertible holomorphic map $z = z(c) = z_{(c_0,z_0)}(c)$ in some neighborhood U of c_0 assigning to $c \in U$ a preperiodic point of preperiod n and period m and satisfying $z_{(c_0,z_0)}(c_0) = z_0$.

It is easy to see that $z_{(c_0,z_0)}$ is bounded on U if U is bounded. Furthermore, recall that in the periodic case where $n = 0$ the map assigning to each periodic point $z_{(c_0,z_0)}(c); c \in U$ its multiplier is holomorphic.

The change of periodic or preperiodic behavior is responsible for the complicated structure of quadratic iteration. So we will be especially interested in where the parabolic parameters of multiplier 1 and Misiurewicz points c_0 are located. At this place we note the well-known fact that there are only finitely many of given period m and preperiod n: algebraic geometry shows that the equation system $p_c^m(z) - z = 0, (p_c^m)'(z) - 1 = 0$ is solvable for only finitely many c, and the fact that $p_c^n(c) = p_c^{n+m}(c)$ is valid for only finitely many c is obvious.

Now we want to have a closer look at the relation between locally connected Julia sets and Julia equivalences. We say that $\approx^\alpha$ for $\alpha \in \mathbb{T}$ is the *landing pattern* for a parameter c if for all β_1, β_2 the following holds:

$$\beta_1 \approx^\alpha \beta_2 \Longleftrightarrow \mathbf{R}_c^{\beta_1}, \mathbf{R}_c^{\beta_2} \text{ land at the same point, i.e. } z_c^{\beta_1} = z_c^{\beta_2}.$$

Clearly, if $\approx^\alpha$ is the landing pattern for some c, then J_c is (connected and) locally connected, and according to the discussion in Chapter 1.1.4 the map $[\beta]_{\approx^\alpha} \mapsto c^\beta$ conjugates the topological dynamical systems $(J_c, p_c, -)$ and $(\mathbb{T}, h, ')/\approx^\alpha$.

Remark. If J_c were locally connected and contained a Cremer point z, then z would have an irrational external angle β. (As noted in Chapter 1.1.3, IV., dynamic rays $\mathbf{R}_c^\beta$ for rational β land at repelling or parabolic periodic or at preperiodic points.) So infinitely many iterates of β would form external angles of z, which is impossible since Julia equivalences have only finite equivalence classes.

This shows that locally connected quadratic Julia sets do not contain a Cremer point, which was first proved by SULLIVAN and DOUADY (see [159], also [97]). However, the existence of a Siegel point can imply local connectivity as well as local non-connectivity of the Julia set. We will say more about this in Chapter 4.1.4.

Obviously, simple closed curves in a locally connected Julia set J_c must be boundaries of bounded Fatou components, and a 'critical', i.e. $'$-symmetric, equivalence class of a Julia equivalence corresponds to the critical point 0 in the complex plane.

According to Proposition 2.66, an Abstract Julia set $(\mathbb{T}, h, ')/\approx^\alpha$ contains simple closed curves only if $\alpha \in \mathbb{P}_*$ or $\alpha \in \mathbb{P}_\infty$. In the first case each simple closed curve lands at $\mathbb{T}^\alpha_{*_m}/\approx^\alpha = [\mathbb{T}\langle\alpha\rangle]_{\approx^\alpha}/\approx^\alpha$ with $m = PER(\alpha)$ after finitely many iterates with respect to $h/\approx^\alpha$, and in the second case at $\mathbb{P}^\alpha_\infty/\approx^\alpha$.

Moreover, $c \in K_c \setminus J_c$ if the parameter c is hyperbolic or parabolic and $c \in J_c$ if c is irrationally indifferent, and $\approx^\alpha$ has a $'$-symmetric equivalence class iff $\alpha \in \mathbb{T} \setminus \mathbb{P}_*$.

So we have the following statements, which in particular relate the type of a quadratic polynomial with locally connected Julia set to the type of the corresponding landing pattern:

Theorem 4.1. (Realization types for Julia equivalences)
If for $\alpha \in \mathbb{T}, c \in \mathbb{C}$ the topological dynamical systems $(\mathbb{T}, h, ')/\approx^\alpha$ and $(J_c, p_c, -)$ are conjugate, then J_c is locally connected and just one of the following cases is valid:

a) Case 1. If $\alpha \in \mathbb{P}_*$, then the parameter c is hyperbolic or parabolic. The following table shows which objects in the Abstract Julia set correspond to which objects in the complex plane, and it defines an important point on the boundary of the critical value Fatou component:

Abstract Julia set	*complex plane*
$m = PER(\alpha)$	*minimum number of iterates mapping the critical value Fatou component to itself*
$\mathbb{T}^\alpha_{*m}/\approx^\alpha = [\mathbb{T}\langle\alpha\rangle]_{\approx^\alpha}/\approx^\alpha$	*boundary of the critical value Fatou component A*
$[\alpha]_{\approx^\alpha}$	*unique fixed point r of p_c^m on the boundary of A, called the* root *of A*

b) Case 2. If $\alpha \in \mathbb{P}_\infty$, then p_c possesses a Siegel disk. The following table shows which objects in the Abstract Julia set correspond to which objects in the complex plane:

Abstract Julia set	*complex plane*
$[\alpha]_{\approx^\alpha}$	*critical value c*
$\mathbb{P}^\alpha_\infty/\approx^\alpha$	*boundary ∂G of the Siegel disk G with $c \in \partial G$*
$PER(\hat{\alpha})$	*minimum number of iterates mapping this G to itself*

c) Case 3. If $\alpha \in \mathbb{T} \setminus \mathbb{P}$, then $J_c = K_c$, each periodic orbit is repelling, $[\alpha]_{\approx^\alpha}$ corresponds to c, and J_c does not contain closed curves. ■

Remark. As noted above, in the locally connected case simple closed curves in J_c are boundaries of bounded Fatou components. In this special case, SULLIVAN's celebrated result on the non-existence of wandering components follows immediately from Theorem 2.11 on non-wandering triangles. This theorem also implies that in quadratic Julia sets a point with more than two external angles, whose orbit does not contain the critical point, is 'non-wandering', i.e. has a finite orbit.

Theorem 4.1 indicates what the landing patterns for different types of parameters c are. We add to which points in J_c the corresponding main points are related.

Corollary 4.2. (Finding main points of the landing pattern)
For $c \in \mathbb{C}$ the following statements are valid:

(i) If the parameter c is hyperbolic or parabolic, then the landing pattern for c is given either by $\approx^0$, the trivial Julia equivalence, or by $\approx^\alpha = \approx^{\overline{\alpha}}$ for some periodic α. In the latter case α is determined by the root r of the critical value Fatou component A as follows: $\mathbf{R}_c^\alpha$ and $\mathbf{R}_c^{\overline{\alpha}}$ are the unique dynamic rays landing at r, whose union with r bounds an open part of $\mathbb{C}$ containing A and no dynamic ray with landing point r.

(ii) If the parameter c is not hyperbolic or parabolic and if J_c is locally connected, then the landing pattern for c is given by $\approx^\alpha$ for each external angle α of c. ■

From the viewpoint of topological dynamics the parabolic and the hyperbolic case need not be distinguishable. A Julia equivalence $\approx^\alpha$ for a periodic $\alpha \in \mathbb{T}$ can form the landing pattern for both a parabolic and a hyperbolic parameter. (Later we will see that this is valid for all periodic $\alpha \in \mathbb{T}$.)

However, a substantial difference of the dynamical behavior in the complex plane appears in the neighborhood of the periodic point z_c^α, the root of the critical value Fatou component. Whereas in the hyperbolic case the point z_c^α must be a repelling periodic point, in the parabolic case it is parabolic itself. This can be seen as follows: The critical orbit is attracted by the parabolic periodic orbit, which is contained in J_c. If α, hence the critical value Fatou component A, has period m, then $p_c^{im}(c) \in A$ for all $i \in \mathbb{N}$. Therefore, the boundary of A contains a parabolic point. As the only fixed point of p_c^m it must coincide with z_c^α.

The question whether two associated periodic points lie on a common orbit or not played an important role in the abstract theory (compare Theorem 2.44, Proposition 2.65 and Corollary 3.18).

Definition 4.3. *A periodic $\alpha \in \mathbb{T}$ is called* primitive *if α and $\overline{\alpha}$ lie on different orbits, and* non-primitive *otherwise.*

Now we obtain the following interpretation of Theorem 2.44 in the complex plane.

Corollary 4.4. *Let $\alpha \in \mathbb{T}$ be periodic and let $\approx^\alpha$ be the landing pattern for some parameter c. Then c is parabolic iff z_c^α is, and the following is valid:*

(i) If α is primitive, then the period of z_c^α with respect to p_c is equal to the period of α. In this case, z_c^α is contained in the boundary of exactly one bounded Fatou component, namely the critical value one.

(ii) If α is non-primitive, then the period of z_c^α is equal to $PER(\overline{\mathbf{v}^\alpha(1-e^\alpha)})$ and the number of bounded Fatou components whose boundaries contain z_c^α is equal to $\frac{PER(\alpha)}{PER(\mathbf{v}^\alpha(1-e^\alpha))} > 1$. ■

4.1.2 Julia equivalences and the Mandelbrot set (II)

1. Continuity of periodic dynamic ray landing, periodic points and multiplier maps. Now we come to the proof of the structure theorems for the Mandelbrot set stated in Chapter 1.2. DOUADY and HUBBARD [43] gave the proof of their results in an analytic way. We want to fill out the combinatorics developed: from the abstract theory we have an idea what could be happened; and we really find much of this in the complex plane.

The link between the abstract theory and quadratic dynamics in the complex plane is provided by the landing of certain dynamic rays. So in particular we have to investigate the landing behavior of dynamic rays $\mathbf{R}_c^\beta$ for fixed periodic $\beta \in \mathbb{T}$ in dependence on the parameter c. In the main parts of this subsection we want to combine the Julia equivalence approach with ideas to be found in GOLDBERG, MILNOR [58, 117] and SCHLEICHER [150].

Remark. Much of the following could be given in the very powerful languages of Hubbard trees developed by DOUADY and HUBBARD [43] and modified by other authors, or of orbit portraits introduced by GOLDBERG and MILNOR [58, 117]. Hubbard trees represent the embedding of finite critical orbits in the plane, and orbit portraits code the landing of dynamic rays related to the orbits of periodic points in $\mathbb{T}$.

Both concepts go beyond the quadratic case and reduce the dynamics to its substantial parts. However, as in the whole text we want to focus our attention to the whole combinatorial picture of the dynamics.

First of all, recall the notation z_c^β for the landing point of $\mathbf{R}_c^\beta$ for some $c \in \mathbb{C}$ and some $\beta \in \mathbb{T}$ (if there is one). Further, if two dynamic rays $\mathbf{R}_c^{\beta_1}$ and $\mathbf{R}_c^{\beta_2}$ land at the same point $z_c^{\beta_1} = z_c^{\beta_2}$, then we will consider the curve $\mathbf{C}_c^{\beta_1\beta_2}$ being the union of $\mathbf{R}_c^{\beta_1}, \mathbf{R}_c^{\beta_2}$ and $z_c^{\beta_1} = z_c^{\beta_2}$. In this way, we assign to special chords $S = \beta_1\beta_2$ simple curves $\mathbf{C}_c^S := \mathbf{C}_c^{\beta_1\beta_2}$ separating the complex plane.

Fix some parameter $c_0 \in \mathbb{C}$ and assume that the dynamic ray $\mathbf{R}_{c_0}^\beta$ for some (rational) $\beta \in \mathbb{T}$ lands at $z_0 = z_{c_0}^\beta \in J_{c_0}$. Then for $z \in \mathbf{R}_{c_0}^\beta$ the map $c \mapsto \Phi_c^{-1}(\Phi_{c_0}(z))$ is holomorphic if c is sufficiently near to c_0. On the other hand, if β is periodic and the multiplier of z_0 is different from 1, then in a some neighborhood of c_0 the holomorphic map $c \mapsto z_{(c_0,z_0)}(c)$ assigns to each parameter c a periodic point of p_c. One may ask whether $z_{(c_0,z_0)}(c)$ coincides with the landing point z_c^β of $\mathbf{R}_c^\beta$ in a some neighborhood of c_0.

Indeed, this was shown by GOLDBERG and MILNOR for the case that z_0 is repelling (see [58], Lemma B.1.) The statement (ii) of the following proposition noted by SCHLEICHER [150] can easily be obtained by taking backward iterates of repelling periodic points, and the last statement follows from continuity of $c \mapsto \Phi_c^{-1}(\Phi_{c_0}(z))$ near the c_0 considered.

Proposition 4.5. (Continuous landing of rational dynamic rays)
Let $c_0, z_0 \in \mathbb{C}$ and let $\beta \in \mathbb{T}$ be given such that $\mathbf{R}^\beta_{c_0}$ lands at z_0. Then the following holds:

(i) If z_0 is a repelling periodic point of p_{c_0} (and so β is periodic), then for c in some neighborhood of c_0 the dynamic ray $\mathbf{R}^\beta_c$ lands at the point $z_{(c_0,z_0)}(c)$.

(ii) If z_0 is preperiodic to some repelling periodic point (and so β is preperiodic) and $p^n_{c_0}(z_0) \neq c_0$ for all $n \in \mathbb{N}$, then for c in some neighborhood of c_0 the dynamic ray $\mathbf{R}^\beta_c$ lands at the point $z_{(c_0,z_0)}(c)$.

Moreover, let z_0 be given as in (i) or (ii), let β_1, β_2 be periodic or preperiodic such that $\mathbf{R}^{\beta_1}_c, \mathbf{R}^{\beta_2}_c$ land at $z_{(c_0,z_0)}(c)$ for c in a neighborhood of c_0.

Then a small closed disk lying on one side of $\mathbf{C}^{\beta_1\beta_2}_{c_0}$ does not change the side of $\mathbf{C}^{\beta_1\beta_2}_c$ if c is sufficiently near to c_0. ■

Two statements will play a key role in the following discussion: one is the above result of GOLDBERG and MILNOR and the other one is the following version of a result of SCHLEICHER (see [150], Lemma 3.7) given in our language.

Lemma 4.6. (Orbit separation lemma)
For periodic $\beta \in \mathbb{T}$ there exists a chord B in $\mathcal{B}_*$ separating each point in $[\beta]_{\approx^\beta}$ from each point in some iterate of $[\beta]_{\approx^\beta}$ different from $[\beta]_{\approx^\beta}$.

Proof: If β is simple, then by Corollary 3.18 there exists a sequence $(B_i)_{i\in\mathbb{N}}$ of chords in $\mathcal{B}_*$ converging to $\beta\overline{\beta}$ such that $\beta\overline{\beta}$ lies behind each $B_i; i \in \mathbb{N}$. By Theorem 3.1 the endpoints of each $B_i; i \in \mathbb{N}$ are $\approx^\beta$-equivalent. Since the iterates of β and $\overline{\beta}$ do not lie between β and $\overline{\beta}$, the above separation statement is obvious for some $B = B_i$ sufficiently near to $\beta\overline{\beta}$.

Now assume that β is non-simple and has period n. Then there exists a chord $B = \delta\overline{\delta} \in \mathcal{B}_* \cup \{01\}$ of some period m from which $Q := \beta\overline{\beta}$ is immediately visible. So β lies in $\mathbb{T}(\delta) \subset \overline{\mathbb{T}\langle\delta\rangle}$. Further, by Corollary 3.12 $m = PER(\hat{B}^+) = PER(\hat{Q}^-) = PER(\mathbf{v}^\beta(1-e^\beta))$ and m divides n.

According to Lemma 2.54 the equivalence class $[\beta]_{\approx^\beta}$ coincides with $\mathbb{T}^\beta_{\mathbf{v}^\beta(1-e^\beta)}$ and has period m. This implies $[\beta]_{\approx^\beta} = \{\beta, h^m(\beta), h^{2m}(\beta), \ldots, h^{n-m}(\beta)\}$. Now the above separation property is provided by Proposition 3.23(iii). ■

For the following compare Chapter 1.2.1 (II./IV.): By definition, a parameter c lies at the parameter ray $\mathbf{R}^\alpha$ for some given $\alpha \in \mathbb{T}$ iff c does not belong to the filled-in Julia set K_c and $\Phi(c) = \Phi_c(c)$ lies at the 'standard' dynamic ray $\mathbf{R}^\alpha_0$. Further, the dynamic ray $\mathbf{R}^\alpha_c = \Phi_c^{-1}(\mathbf{R}^\alpha_0)$ for $\alpha \in \mathbb{T}$ is not defined iff $\Phi(c) = \Phi_c(c)$ belongs to $\mathbf{R}_0^{h^n(\alpha)}$ for some $n \in \mathbb{N}$, in other words, iff c is at the parameter ray $\mathbf{R}^{h^n(\alpha)}$ for some $n \in \mathbb{N}$.

So for periodic $\alpha \in \mathbb{T}$ the set $\mathbb{C}^\alpha$ obtained from the complex plane by taking away all parameter rays $\mathbf{R}^\beta$ for some β in the orbit of α or $\overline{\alpha}$ is of particular interest. For all c contained in this set the landing point z_c^α is defined and depends continuously on c, as the following statement shows:

Corollary 4.7. *For periodic $\beta \in \mathbb{T}$, the maps $c \mapsto z_c^\beta$ and $c \mapsto z_c^{\overline{\beta}}$ defined on $\mathbb{C}^\beta$ are continuous and the set of all c with $z_c^\beta = z_c^{\overline{\beta}}$ is closed in $\mathbb{C}^\beta$.*

Proof: Let us show the continuity of the above maps. Then the second statement is a standard consequence.

Let β be periodic of period m, and let $(c_n)_{n\in\mathbb{N}}$ be a sequence in $\mathbb{C}^\beta$ converging to some $c \in \mathbb{C}^\beta$. By Proposition 4.5 we only have to consider the case where the periodic point z_c^β is not repelling, hence is parabolic. Then for some periodic γ the Julia equivalence $\approx^\gamma$ forms the landing pattern for c. Since z_c^γ and z_c^β must lie on the only parabolic orbit, in the following we can assume that that $\beta = \gamma$.

If $d, z \in \mathbb{C}$ and $|z| > 2|c|, 3$, then $|p_d(z)| = |z^2(1 - \frac{d}{z^2})| > 3|z|(1 - \frac{1}{|z|}) > 2|z|$, hence the orbit of z with respect to p_d is unbounded. Therefore, the sequence $(z_{c_n}^\beta)_{n\in\mathbb{N}}$ must be bounded, and obviously each accumulation point z of that sequence is periodic for p_c of some period dividing m. It suffices to show that such z must coincide with z_c^β.

Fix such z and some periodic $\eta \in \mathbb{T}$ with $z = z_c^\eta$ and assume that $z \neq z_c^\beta$, in other words that $\eta \notin [\beta]_{\approx^\beta}$. Then, again by Proposition 4.5, also the point z is parabolic, hence it lies in the orbit of z_c^β.

Now choose a chord $B = \delta\overline{\delta}$ as given in Lemma 4.6. Clearly, by Theorem 3.1 (see (vi)) $\mathbf{R}_c^\delta$ and $\mathbf{R}_c^{\overline{\delta}}$ land at the same repelling periodic point and $\mathbf{C}_c^{\delta\overline{\delta}}$ separates the dynamic ray $\mathbf{R}_c^\beta$ from z. This contradicts the last statement of Proposition 4.5. ■

Another immediate consequence of Proposition 4.5 is the following statement, which we will frequently apply.

Corollary 4.8. *Let c_0, c_1 be complex parameters and $\mathbf{C}$ a simple curve connecting them. Then the following statements are valid:*

(i) If $\alpha \in \mathbb{T}$ is periodic of period m and $\mathbf{C} \subset \mathbb{C}^\alpha$, and $\mathbf{C}$ does not contain a parabolic parameter of some period dividing m, then $z_{c_0}^\alpha = z_{c_0}^{\overline{\alpha}}$ iff $z_{c_1}^\alpha = z_{c_1}^{\overline{\alpha}}$.

(ii) If $\beta_1, \beta_2 \in \mathbb{T}$ are preperiodic points of common preperiods n and periods m, and $\mathbf{C}$ is disjoint to all parameter rays $\mathbf{R}^{h^i(\beta_1)}, \mathbf{R}^{h^i(\beta_2)}; i \in \mathbb{N}$, to all Misiurewitz points of preperiod less than n and period dividing m and to all parabolic parameters of period dividing m, then $z_{c_0}^{\beta_1} = z_{c_0}^{\beta_2}$ iff $z_{c_1}^{\beta_1} = z_{c_1}^{\beta_2}$. ■

Now consider a hyperbolic component W, i.e. a component of the set of hyperbolic parameters, of period m. For $c \in W$ let $Mult_W(c)$ be the multiplier of c. For each $c_0 \in W$ and each attractive periodic point z_0 of p_{c_0} the map $z_{(c_0,z_0)}$ given at the beginning of this subsection is well-defined as a holomorphic map on W, and we have $Mult_W(c) = 2^m \prod_{i=0}^{m-1} p_c^i(z_{(c_0,z_0)}(c))$.

If $(c_i)_{i\in\mathbb{N}}$ is a sequence in W which does not accumulate in W, then obviously $(Mult_W(c_i))_{i\in\mathbb{N}}$ has only accumulation points of absolute value 1, since otherwise on the boundary of W there would be a hyperbolic parameter. It is well-known that a map between domains U and V is proper iff every sequence in U with no accumulation point in U is mapped to a sequence with no accumulation point in V. Thus we have the following

Lemma 4.9. (Multiplier map is proper)
$Mult_W$ maps a hyperbolic component W in a proper holomophic way onto the open unit disk. ■

$Mult_W$ has a unique extension to a continuous map on the closure of W. To see this, fix some $c_0 \in W$ and some attractive periodic point z_0 of p_{c_0}. For $c \in \partial W$ and a sequence $(c_n)_{n\in\mathbb{N}}$ contained in W and converging to c let $Mult_W(c) := (p_c^m)'(z) = 2^m \prod_{i=0}^{m-1} p_c^i(z)$ where z is any accumulation point of $(z_n)_{n\in\mathbb{N}}$ with $z_n = z_{(c_0,z_0)}(c_n)$ for all $n \in \mathbb{N}$.

z is periodic for p_c of a period r dividing m. Either $r = m$ and $Mult_W(c)$ is the multiplier of z, or $r < m$ and z is also an accumulation point of $(p_{c_n}^r(z_n))_{n\in\mathbb{N}}$. In the latter case $Mult_W(c)$ is equal to 1 and by the chain rule the multiplier of z a root of 1. In each case z lies in the only non-repelling periodic orbit of p_c.

2. Landing of 'periodic' parameter rays. On the analogy of the dynamic case we want to introduce two notations for objects related to landing parameter rays: c^α for the landing point of a parameter ray $\mathbf{R}^\alpha; \alpha \in \mathbb{T}$, and $\mathbf{C}^S = \mathbf{C}^{\alpha_1\alpha_2}$ for the curve corresponding to a chord $S = \alpha_1\alpha_2$ with $c^{\alpha_1} = c^{\alpha_2}$ and being the union of the parameter rays $\mathbf{R}^{\alpha_1}, \mathbf{R}^{\alpha_2}$ with their landing point $c^{\alpha_1} = c^{\alpha_2}$.

The first step to realizing parts of the abstract Mandelbrot set in the complex plane is given by the following statement. The proof is due to GOLDBERG, MILNOR, DOUADY and HUBBARD (see [58], Theorem C.7).

Lemma 4.10. ('Periodic' parameter rays land)
For periodic $\alpha \in \mathbb{T}$ the parameter ray $\mathbf{R}^\alpha$ lands at a parabolic parameter c, and z_c^α is a parabolic periodic point for p_c.

Proof: Fix some c_0 in $\mathrm{cl}\mathbf{R}^\alpha \setminus \mathbf{R}^\alpha$. Then the landing point of $\mathbf{R}_{c_0}^\alpha$ cannot be repelling. Otherwise, $\mathbf{R}_c^\alpha$ would land at a repelling orbit for $c \in \mathbf{R}^\alpha$ sufficiently near to c_0, which is impossible since $\mathbf{R}_c^\alpha$ is not defined for $c \in \mathbf{R}^\alpha$. There are only countably many parabolic parameters but $\mathrm{cl}\mathbf{R}^\alpha \setminus \mathbf{R}^\alpha$ is connected, hence uncountable or a one-point set. So we are finished. ■

Lemma 4.10 says that a given periodic $\alpha \in \mathbb{T}$ provides a well-defined parabolic parameter c^α. By the above considerations, there exists a periodic $\beta \in \mathbb{T}$ such that $\approx^\beta$ is the landing pattern for c^α. We show that $\alpha \in [\beta]_{\approx^\beta}$.

Let $B = \delta\overline{\delta}$ be given as in Lemma 4.6. Clearly, $z^\delta_{c^\alpha} = z^{\overline{\delta}}_{c^\alpha}$ is repelling, hence $z^\delta_c = z^{\overline{\delta}}_c$ for each parameter c sufficiently near to c^α. Moreover, the critical value c^α of p_{c^α} lies in the open part of the plane which is bounded by $\mathbf{C}^{\delta\overline{\delta}}_{c^\alpha}$ and does not contain 0.

For $c \in \mathbf{R}^\alpha$ the argument of $\Phi(c) = \Phi_c(c)$ is $2\pi\alpha$. If such c is sufficiently near to c^α, then $z^\delta_c = z^{\overline{\delta}}_c$. Thus by Proposition 4.5 α lies behind B, and Lemma 4.6 shows that α is an element of $[\beta]_{\approx^\beta}$. So we have

Corollary 4.11. *Let $\alpha, \beta \in \mathbb{T}$ be periodic. If $\approx^\beta$ is the landing pattern for c^α, then $\alpha \approx^\beta \beta$.* ■

We show now that for each periodic $\alpha \in \mathbb{T}$ an identification of α and $\overline{\alpha}$ can be realized in the complex plane.

Lemma 4.12. (Existence of sufficiently many identifications)
For each periodic $\alpha \in \mathbb{T}$ there exists a periodic $\gamma \in \mathbb{T}$ between α and $\overline{\alpha}$ such that for the (parabolic) parameter $c = c^\gamma$ the point z^α_c is repelling, and $z^\alpha_c = z^{\overline{\alpha}}_c$.

Proof: We can assume that $\alpha \neq 0$. It is no problem to fix a prime number $k > 1$ and a periodic point γ of period k between α and $\overline{\alpha}$. Choose a periodic β such that $\approx^\beta$ is the landing pattern for $c = c^\gamma$.

We show that β is primitive. By Corollary 4.11, this implies $\beta \in \{\gamma, \overline{\gamma}\}$, hence $\approx^\gamma = \approx^\beta$. Since γ lies between α and $\overline{\alpha}$, by Theorem 3.1 one obtains $\alpha \approx^\gamma \overline{\alpha}$, in other words $z^\alpha_c = z^{\overline{\alpha}}_c$.

Assume that β fails to be primitive. Then $[\beta]_{\approx^\beta}$ is contained in the orbit of β and $CARD([\beta]_{\approx^\beta})$ divides k. Therefore, since $1 < CARD([\beta]_{\approx^\beta})$ and k is a prime number, $[\beta]_{\approx^\beta}$ must coincide with the orbit of γ.

According to Lemma 2.54 this means $[\beta]_{\approx^\beta} = \mathbb{T}^\beta_{\mathrm{v}^\beta(1-e^\beta)} = \mathbb{T}^\beta_{\overline{0}}$, implying $\hat{\beta} = \overline{0^{k-1}*}$. So by Theorem 3.17 γ lies in the intersection of the 'big' gap of $\mathcal{B}$ with $\mathbb{T}$. This contradicts the assumption that γ is between α and $\overline{\alpha}$. ■

Remark. It is a crucial point in the proof of the following proposition to guarantee that certain combinatorial patterns can be realized in the complex plane, which is provided by Lemma 4.12. MILNOR gave a similar Realization Theorem for orbit portraits (see [117], Theorem 2.4).

However, there is a more general approach to the realization of combinatorial patterns in complex dynamics, namely THURSTON's famous topological characterization of postcritically finite rational maps as branched coverings of the Riemann sphere. (Postcritically finite means that the set of iterates of critical points is finite.) The proof of THURSTON's statement is very deep (see DOUADY and HUBBARD [42]), but in the case of a polynomial there were given some simplifications:

Here combinatorics can be described by the systems of external rays landing at the critical points of the polynomial or some certain substitutes (as seen in the quadratic case). The abstract objects modelled from this were 'critical portraits', and it was shown that they can be realized. For the case that all critical points are preperiodic, see BIELEFELD, FISHER and HUBBARD [13], and for the general case, see POIRIER [134, 135]. Note that HUBBARD and SCHLEICHER [67] gave a special proof of THURSTON's theorem for quadratic polynomials leading to the iterative 'Spider algorithm', which allows to locate those parameters in the Mandelbrot set where the combinatorial patterns are realized.

Now we are in a position to prove that the landing of 'periodic' parameter rays is exactly that we expect from the abstract theory.

Proposition 4.13. *For periodic $\alpha \in \mathbb{T}$ the landing pattern for c^α is $\approx^\alpha$.*

Proof: Let $\approx^\beta$ for some periodic $\beta \in \mathbb{T}$ be the landing pattern for c^α. If $\alpha = 0$, it holds $0 \approx^\beta \beta$ according to Corollary 4.11. Besides 0 there is no fixed point (with respect to h) in $\mathbb{T}$. So Proposition 2.65 implies $\beta = 0$.

For $\alpha \neq 0$ of period m choose some γ between α and $\overline{\alpha}$ satisfying $z^\alpha_{c^\gamma} = z^{\overline{\alpha}}_{c^\gamma}$. This is possible by Lemma 4.12. There exists a simple curve in $\mathbb{C}^\alpha$ connecting c^γ with c^α and containing no parabolic point of period less or equal to m different from c^α. By Corollaries 4.8(i) and 4.7 it holds $z^\alpha_{c^\alpha} = z^{\overline{\alpha}}_{c^\alpha}$. So Corollary 4.11 yields $\alpha, \overline{\alpha} \in [\beta]_{\approx^\beta}$ implying $\approx^\beta = \approx^\alpha$ (see Proposition 2.65). ■

As an immediate consequence of Proposition 4.13 we get

Corollary 4.14. *If for periodic $\alpha, \gamma \in \mathbb{T}$ the parameter rays $\mathbf{R}^\alpha$ and $\mathbf{R}^\gamma$ land at the same point, then $\alpha \sim \gamma$, in other words $\gamma = \overline{\alpha}$.* ■

Fix some periodic $\alpha \in \mathbb{T} \setminus \{0\}$ of some period m and a γ between α and $\overline{\alpha}$ with $z^\alpha_{c^\gamma} = z^{\overline{\alpha}}_{c^\gamma}$ (compare Lemma 4.12).

An iterate of δ or $\overline{\delta}$ for some periodic $\delta \in \mathbb{T}$ cannot lie behind $\delta\overline{\delta}$ (see Lemma 2.60, condition (iii) and Proposition 2.62 (ii)). In particular, if each of the points $\delta, \overline{\delta}$ lies on the orbit of α or $\overline{\alpha}$, then $\delta\overline{\delta}$ does not separate α or $\overline{\alpha}$ from 0.

Therefore, Corollary 4.13 allows the following argumentation: In case that $\mathbf{R}^\alpha$ and $\mathbf{R}^{\overline{\alpha}}$ landed at different points, there would exist a curve in $\mathbb{C}^\alpha$ which connects c^γ with c^0 and contains no parabolic parameter different from c^0 of period dividing m. By Corollary 4.8(i) and Corollary 4.7 it would hold $z^\alpha_{c^0} = z^{\overline{\alpha}}_{c^0}$, which is false according to Proposition 4.13. This shows

Corollary 4.15. *For periodic $\alpha \in \mathbb{T}$ the parameter rays $\mathbf{R}^\alpha$ and $\mathbf{R}^{\overline{\alpha}}$ land at the same point.* ■

The following statement characterizes $\mathbf{Wake}^{\alpha\overline{\alpha}}$ by the landing behaviour of the dynamic rays $\mathbf{R}_c^\alpha$ and $\mathbf{R}_c^{\overline{\alpha}}$.

Proposition 4.16. *Let $\alpha \in \mathbb{T}$ be periodic. Then $z_c^\alpha = z_c^{\overline{\alpha}}$ for $c \in \mathbf{Wake}^{\alpha\overline{\alpha}}$ and $z_c^\alpha \neq z_c^{\overline{\alpha}}$ for $c \in \mathbb{C}^\alpha \setminus \mathbf{Wake}^{\alpha\overline{\alpha}}$.*

Proof: By Lemma 2.60 and Theorem 3.1 no iterate of α or $\overline{\alpha}$ lies between α and $\overline{\alpha}$. So by Lemma 4.12 each point $c \in \mathbf{Wake}^{\alpha\overline{\alpha}} \setminus \{c^\alpha\}$ can be connected by a simple curve $\mathbf{C} \subset \mathbf{Wake}^{\alpha\overline{\alpha}} \setminus \{c^\alpha\}$ with some parabolic parameter d, such that z_d^α is repelling, $z_d^\alpha = z_d^{\overline{\alpha}}$, and $\mathbf{C}$ does not contain a parabolic parameter of period less than or equal to $PER(\alpha)$ different from c and d. Corollaries 4.8(i), 4.7 and Proposition 4.5(i) show $z_c^\alpha = z_c^{\overline{\alpha}}$.

Let now $c \in \mathbb{C}^\alpha \setminus \mathbf{Wake}^{\alpha\overline{\alpha}}$. We can assume that c is not parabolic. Otherwise, we would have $c = c^\gamma$ for some periodic γ not in the small interval with endpoints α and $\overline{\alpha}$, hence it would hold $z_c^\alpha = z_{c^\gamma}^\alpha \neq z_{c^\gamma}^{\overline{\alpha}} = z_c^{\overline{\alpha}}$.

There is a simple curve in $\mathbb{C}^\alpha \setminus \mathbf{Wake}^{\alpha\overline{\alpha}}$ whose endpoints are c and c^δ for $\delta = 0$, or δ in the orbit of α or $\overline{\alpha}$, and whose inner points are different from parabolic parameters of period less than or equal to $PER(\alpha)$. Then $\approx^\delta$ does not identify α and $\overline{\alpha}$, and we are done by Corollaries 4.8(i) and 4.7. ■

Proposition 4.16 allows a simple dynamic description of combinatorial classes.

Corollary 4.17. *For each $\alpha \in \mathbb{T}$, the combinatorial class $Comb^\alpha$ consists of all $c \in M$ satisfying $z_c^\gamma = z_c^{\overline{\gamma}}$ for exactly those periodic $\gamma \in \mathbb{T}$ with α behind $\gamma\overline{\gamma}$ or $\alpha \in \{\gamma, \overline{\gamma}\}$.* ■

We finish by completing the correspondence between periodic points in $\mathbb{T}$ and parabolic parameters in the complex plane.

Proposition 4.18. *Each parabolic parameter c is the landing point of $\mathbf{R}^\alpha$ for some periodic $\alpha \in \mathbb{T}$.*

Proof: Let $\approx^\alpha$ for some periodic $\alpha \in \mathbb{T}$ be the landing pattern for c, and assume that $c \neq c^\alpha$. Then z_c^α is parabolic and c lies in $\mathbf{Wake}^{\alpha\overline{\alpha}}$.

There exists an open disk U around c which is completely contained in $\mathbf{Wake}^{\alpha\overline{\alpha}}$, such that for each $\tilde{c} \in U \setminus \{c\}$ the point $z_{\tilde{c}}^\alpha$ is repelling. By Proposition 4.5 the period of $z_{\tilde{c}}^\alpha$ does not change on $U \setminus \{c\}$ and if m denotes that period, then the map $\tilde{c} \mapsto (2^m \prod_{i=0}^{m-1} p_{\tilde{c}}^i(z_{\tilde{c}}^\alpha))^{-1}$ must be holomorphic on U with maximum absolute value 1 assumed in c. This contradicts the maximum principle. ■

In connection with Theorem 1.13 which is proved by Corollaries 4.14 and 4.15 now, let us highlight that associated periodic points α and $\overline{\alpha}$ are related in a purely combinatorial manner, which is provided by formula (2.7):

$$\overline{\alpha} = \alpha + \frac{2^{PER(\alpha)}}{2^{PER(\alpha)} - 1}(l_{\mathbf{v}^\alpha}^\alpha(\ddot{\alpha}) - \alpha).$$

Remarks. 1. The notation $\mathbf{Wake}^{\alpha\overline{\alpha}}$ is justified by the fact that the set described is usually called the wake of the parabolic parameter $c = c^{\alpha}$.

2. MILNOR [117] and SCHLEICHER [150] use counting arguments to show that each parabolic parameter different from 0 is the landing point of two 'periodic' parameter rays.

3. Hyperbolic components and proof of the Theorems 1.17, 1.18 and 1.20. For a given periodic $\alpha \in \mathbb{T}$ of period m fix some small disk U around c^{α} such that no parabolic parameter in $U \setminus \{c^{\alpha}\}$ has period dividing m and that z_c^{α} and $z_c^{\overline{\alpha}}$ have period m for $c \in U \setminus \mathrm{cl}\,\mathbf{Wake}^{\alpha\overline{\alpha}}$.

Further, fix some $c_0 \in U \setminus \mathrm{cl}\,\mathbf{Wake}^{\alpha\overline{\alpha}}$ and let $z_0 := z_{c_0}^{\alpha}$. Starting from the pair (c_0, z_0), by analytic continuation along curves in $U \setminus \{c_0\}$ one gets a multivalued map, which assigns to each $c \in U \setminus \{c^{\alpha}\}$ periodic points of p_c of period m. It can be continuously extended to c^{α} with unique value $z_{c^{\alpha}}^{\alpha}$.

The image V of the multivalued map obtained now must be a simply connected domain, and the inversion of that map a branched covering with branching point $z_{c^{\alpha}}^{\alpha}$. Call this inversion $c(z)$.

The map $z \mapsto 2^m \prod_{i=0}^{m-1} p_{c(z)}^i(z)$ assigns to $z \in V \setminus \{z_{c^{\alpha}}^{\alpha}\}$ the multiplier with respect to $p_{c(z)}$ and to $z_{c^{\alpha}}^{\alpha}$ a point of absolute value 1. It is holomorphic, and by the maximum value principle one easily obtains that c^{α} lies on the boundary of a hyperbolic component. So we have the following

Lemma 4.19. *For each periodic $\alpha \in \mathbb{T}$ there exists a hyperbolic component W with the following properties:*

$$c^{\alpha} \in \partial W, \text{ and the periods of } \alpha \text{ and } W \text{ coincide.} \quad \blacksquare \tag{4.1}$$

Remark. Lemma 4.19 implies that c^0 lies at the boundary of the main hyperbolic component, the only hyperbolic component of period 1, and according to Example 1.16 it holds $c^0 = \frac{1}{4}$.

Fix a hyperbolic component W satisfying (4.1). Since on the one hand $z_c^{\alpha} \neq z_c^{\overline{\alpha}}$ for $c \in \mathbb{C}^{\alpha} \setminus \mathbf{Wake}^{\alpha\overline{\alpha}}$ and on the other hand $z_{c^{\alpha}}^{\alpha} = z_{c^{\alpha}}^{\overline{\alpha}}$, by the inverse mapping theorem the derivative of $g(c,z) = p_c^m(z) - z$ by z must vanish at $(c^{\alpha}, z_{c^{\alpha}}^{\alpha})$. In other words, $Mult_W(c^{\alpha}) = 1$.

If α is primitive, then Corollary 3.18 provides $W \subset \mathbf{Wake}^{\alpha\overline{\alpha}}$. In the non-primitive case let k be the period of the parabolic periodic point $z_{c^{\alpha}}^{\alpha}$. According to Corollary 4.4 we have $k < m$.

The number of roots of the equation $p_c^k(z) - z = 0$ counted with their multiplicities is 2^k, independent of c. Hence for $c \in U \cap W$ the map p_c must have at least one repelling periodic orbit of period k more than $p_{c^{\alpha}}$. By Proposition 4.5(i) a 'new' repelling periodic point must have an external angle on the orbit of α. Since $k < n$, this is only possible if $z_c^{\alpha} = z_c^{\overline{\alpha}}$. So we see that $W \subset \mathbf{Wake}^{\alpha\overline{\alpha}}$. Let us summarize:

Lemma 4.20. *For a periodic $\alpha \in \mathbb{T}$ and a hyperbolic component W with (4.1) it holds $Mult_W(c^\alpha) = 1$ and $W \subset \mathbf{Wake}^{\alpha\overline{\alpha}}$.* ∎

Now let Γ be the set of all periodic $\gamma \in \mathbb{T}$ such that $c^\gamma (= c^{\overline{\gamma}})$ is a parabolic point in ∂W with $Mult_W(c^\gamma) \neq 1$. By our abstract theory (see Theorem 3.17) and the consideration on landing of 'periodic' parameter rays it holds $\Gamma \subseteq \mathbb{T}(\alpha)$. (Recall that $\mathbb{T}(\alpha)$ was the intersection of the infinite gap $Gap(\alpha)$ of the lamination $\mathcal{B}$ with $\mathbb{T}$.)

Let $\gamma \in \Gamma$ and fix a hyperbolic component U of the same period as γ with $c^\gamma \in \partial U$ and $Mult_U(c^\gamma) = 1$. There exist $p, q \in \mathbb{N}$ (being relatively prime) with $Mult_W(c^\gamma) = e^{2\pi\frac{p}{q}i}$, and by Proposition 3.38(ii) $PER(\gamma) = \tilde{q}m$ for some $\tilde{q} \in \mathbb{N}$. We deduce that $q = \tilde{q}$ by showing the following: For all r with $1 \leq r < \tilde{q}$, it holds $(p_{c^\gamma}^{rm})'(z_{c^\gamma}^\gamma) \neq 1$.

If this were false for minimal r, then $(p_{c^\gamma}^{rm})'(z_{c^\gamma}^\gamma) - 1 = 0$, and for $p_c; c \in U$ the equality $p_c^{rm}(z) - z = 0$ would provide at least one repelling period orbit of period rm more than for p_{c^γ}. Arguing as above, one could see that there would not be enough periodic external angles for the repelling periodic points of period rm.

According to Proposition 3.38, ν^α parameterizes $\mathbb{T}(\alpha)$ and, by the shown and again by Proposition 3.38, the denominator of $\nu^\alpha(\gamma) = \nu^\alpha(\overline{\gamma})$ is q when $Mult_W(c^\gamma) = e^{2\pi\frac{p}{q}i}$. So for a fixed q the number of all c^γ with $Mult_W(c^\gamma)$ of the form $e^{2\pi\frac{p}{q}i}$ is the same as the number of all fractions $\frac{p}{q}$, and one gets bijectivity of $Mult_W$ on the parabolic points in ∂W and

Proposition 4.21. *If W is a hyperbolic component with (4.1) for some periodic $\alpha \in \mathbb{T}$, then W is unique and $\mathbb{T}(\alpha) \cap \mathbb{Q} = \{\gamma \in \mathbb{T} \mid \gamma$ is an external angle of some parabolic parameter in $\partial W\}$.* ∎

Obviously, bijectivity of $Mult_W$ on the parabolic points in ∂W extends to bijectivity on the whole boundary of W. Since $Mult_W$ maps W in a proper holomorphic way onto the open unit disk and $\mathrm{cl}\,W$ continuously onto the closed unit disk (see Lemma 4.9 and text below it), $Mult_W$ must be a homeomorphism on $\mathrm{cl}\,W$. This shows Theorem 1.17.

Finally, since the periodic points $\gamma \in \mathbb{T}(\alpha)$ with $\nu^\alpha(\gamma) = \frac{p}{q}$ for fixed q and for p relatively prime to q lie in the same circular order as the parabolic parameters of multiplier $e^{2\pi\frac{p}{q}i}$ in ∂W, the following is valid:

Proposition 4.22. *For periodic $\alpha \in \mathbb{T}$ there is a unique hyperbolic component W with (4.1). For this W it holds $Mult_W(c^\gamma) = e^{2\pi\nu^\alpha(\gamma)i}$ for all $\gamma \in \mathbb{T}(\alpha) \cap \mathbb{Q}$, in particular, $r = c^\alpha$ is the only point on the boundary of W with $Mult_W(r) = 1$.* ∎

Clearly, for periodic $\alpha \in \mathbb{T}$ the hyperbolic components in Proposition 4.22 corresponding to α and $\overline{\alpha}$ coincide. Recall that the hyperbolic component W defined by Proposition 4.22 was denoted by $W^{\alpha\overline{\alpha}}$ (see Chapter 1.2.2).

We see that each infinite gap of $\mathcal{B}$ corresponds to a hyperbolic component and vice versa. If the longest chord of the gap is $\alpha\overline{\alpha} \in \mathcal{B}_* \cup \{01\}$, then the corresponding hyperbolic component is $W^{\alpha\overline{\alpha}}$. Theorem 1.18 and 1.20 are consequences of the classification of gaps of $\mathcal{B}$ in Theorem 3.17 and of Lemma 4.20 and Propositions 4.21 and 4.22.

Landing at the boundary of hyperbolic components. YOCCOZ showed that M is locally connected at each irrationally indifferent parameter (see [66]). His proof bases on the Pommerenke-Levin-Yoccoz inequality (see [129]), which in particular guarantees that each hyperbolic component has only finitely many sublimbs of diameter greater than a given $\epsilon > 0$.

Local connectivity at irrationally indifferent parameters is related to the fact that each parameter ray $\mathbf{R}^\alpha$ for non-periodic α with periodic kneading sequence lands at the boundary of a hyperbolic component, which we need now. We want to prove local connectivity at irrationally indifferent parameters by use of the combinatorics developed above.

In preparation for this we need the concept of an impression, coming from prime end theory. We give that concept only as far as we will use it. For a general discussion the reader is referred to [28].

Definition 4.23. *For $c \in M$ and $\beta \in \mathbb{T}$, by the* impression $Imp(\mathbf{R}_c^\beta)$ *of the dynamic ray $\mathbf{R}_c^\beta$ one understands the set of all $z \in \mathbb{C}$ satisfying the following property:*

$$\textit{There exists a sequence } (z_n)_{n\in\mathbb{N}_0} \textit{ with } \lim_{n\to\infty} z_n = z \textit{ and } \lim_{n\to\infty} \Phi_c(z_n) = e^{2\pi\beta i}. \tag{4.2}$$

Analogously, the impression $Imp(\mathbf{R}^\beta)$ *is defined for a parameter ray $\mathbf{R}^\beta$ substituting Φ_c in (4.2) by Φ.*

The impression of a dynamic and a parameter ray is a connected subset of J_c and ∂M, respectively, and one easily sees the following: If it consists of only one point z, then z is the landing point of the ray, and each neighborhood of z contains a simple curve in the complement of K_c (M) connecting two points in $J_c \setminus \{z\}$ ($\partial M \setminus \{z\}$) and cutting off a small piece of $\mathbb{C} \setminus K_c$ ($\mathbb{C} \setminus M$) whose boundary contains z. (Note that such curves are called cross-cuts in prime end theory.)

Now we can continue the above discussion. Fix some periodic α and some irrationally indifferent parameter c_0 on the boundary of $W^{\alpha\overline{\alpha}}$ of multiplier $e^{2\pi\nu i}$. By the above shown the set $\{c^\gamma | \gamma \in \mathbb{T}(\alpha)\}$ is dense in the boundary of $W^{\alpha\overline{\alpha}}$. Hence there exists a unique irrational $\delta \in \mathbb{T}(\alpha)$, such that c_0 lies in the impression $Imp(\mathbf{R}^\delta)$ of $\mathbf{R}^\delta$. Note that by Proposition 3.38 it holds $\nu = \nu^\alpha(\delta)$.

We show that $Imp(\mathbf{R}^\delta) = \{c_0\}$. Then obviously $\mathbf{R}^\delta$ lands at c_0, and in order to show local connectivity of M at c_0 beyond this, one can argue as follows:

For each open disk U around c_0 there exists some rational $\delta_1, \delta_2 \in \mathbb{T}(\alpha)$ (with $\delta_1 < \delta < \delta_2$), and two curves $\mathbf{C}_1 \subset U \cap W^{\alpha\overline{\alpha}}$ and $\mathbf{C}_2 \subset U \setminus M$ both connecting c^{δ_1} with c^{δ_2} such that the curve $\mathbf{C}_1 \cup \mathbf{C}_2 \cup \{c^{\delta_1}, c^{\delta_2}\}$ bounds a small open neighborhood of c. (The curve $\mathbf{C}_2$ can easily be constructed on the base of the remark below Definition 4.23).

To show that $Imp(\mathbf{R}^\delta) = \{c_0\}$, let z_0 be an irrationally indifferent periodic point of p_{c_0}. Then for a small neighborhood U of c_0 the periodic point $z_{(c_0,z_0)}(c)$ is repelling if $c \in U \setminus \mathrm{cl}\, W^{\alpha\overline{\alpha}}$ and attractive if $c \in U \cap W^{\alpha\overline{\alpha}}$. If the connected set $Imp(\mathbf{R}^\delta)$ were different from $\{c_0\}$, one could fix some $c_1 \in (U \setminus \mathrm{cl}\, W^{\alpha\overline{\alpha}}) \cap Imp(\mathbf{R}^\delta)$.

Clearly, c_1 would belong to M and the repelling periodic point $z_{(c_0,z_0)}(c_1)$ would have a periodic external angle β. It would be possible to connect c_1 with a hyperbolic parameter $c_2 \in U \cap W^{\alpha\overline{\alpha}}$ by a simple curve $\mathbf{C}$ being disjoint to all sets $\mathrm{cl}\,\mathbf{Wake}^{\eta\overline{\eta}}$ for η in the orbit of β. Clearly, z_c^β would be repelling for all $c \in \mathbf{C}$. Thus $z_{(c_0,z_0)}(c_2)$ would be repelling, which is false.

Theorem 4.24. (The boundary of hyperbolic components; internal and external angles)
Each irrationally indifferent parameter is the landing point c^γ of the parameter ray $\mathbf{R}^\gamma$ for a unique $\gamma \in \mathbb{T}$. That γ is non-periodic of a periodic kneading sequence. Conversely, $\mathbf{R}^\gamma$ lands at an irrationally indifferent parameter for each γ of that type.

For the hyperbolic component $W^{\alpha\overline{\alpha}}$ associated with a periodic $\alpha \in \mathbb{T}$ the following statements are valid:

(i) $\partial W^{\alpha\overline{\alpha}} = \{c^\gamma | \gamma \in \mathbb{T}(\alpha)\}$ and $[\gamma]_\sim \mapsto c^\gamma$ defines a homeomorphism from $\mathbb{T}(\alpha)/\sim$ onto $\partial W^{\alpha\overline{\alpha}}$.

(ii) $c \in \partial W^{\alpha\overline{\alpha}}$ has external angle γ if its internal angle with respect to $W^{\alpha\overline{\alpha}}$ is $\nu^\alpha(\gamma)$. ■

Remark. Up to the proof of $Imp(\mathbf{R}^\gamma) = \{c^\gamma\}$ for all $\gamma \in \mathbb{P}_\infty$ the statement that for $c \in M$ each repelling periodic point of p_c is the landing point of some periodic dynamic ray was not used.

Consider the hyperbolic component $W^{\alpha\overline{\alpha}}$ with root $c^\alpha = c^{\overline{\alpha}}$ for some given periodic $\alpha \in \mathbb{T}$, and let $c^\gamma = c^{\overline{\gamma}}$ for some periodic $\gamma \in \mathbb{T}$ be a bifurcation point on the boundary of $W^{\alpha\overline{\alpha}}$ having internal angle ν. In supplement to Theorem 4.24 we want to give algorithms for determining $\gamma, \overline{\gamma} \in \mathbb{P}_*$ from $\alpha, \overline{\alpha}$ and ν, and vice versa. The algorithms following immediately from the discussion in Chapter 3.2.4 are presented by a) and b).

a) If $\alpha, \overline{\alpha}$ and ν are given, then we can assume that $\alpha < \overline{\alpha}$ and that $l = PER(\alpha)$. One calculates one of the two continued fraction expansions $[; k_1, k_2, \ldots, k_n]$ of ν. If $k_1 > 1$, one determines the 0-1-word $b_1 b_2 \ldots b_r = \mathbf{a}((k_1 - 1)k_2 \ldots k_n)$, and if $k_1 = 1$, the 0-1-word $b_1 b_2 \ldots b_r = \mathbf{a}(k_2 k_3 \ldots k_n)^-$ (see 4. in Chapter 3.2.4, and compare formula (2.2)). With

$$\gamma = \alpha + (\overline{\alpha} - \alpha)\frac{2^{lr}(2^l - 1)}{2^{lr} - 1}\sum_{i=1}^{r} b_i 2^{-li} \tag{4.3}$$

one obtains one of the angles $\gamma, \overline{\gamma}$ in quest. (Formula (4.3) follows as the sum of a geometric series from $\gamma = q^\alpha(\overline{b_1 b_2 \dots b_r})$ and formula (3.8).)

b) If γ is given, one determines e^γ, best by use of Proposition 3.14(iii), and one sets $m = PER(\gamma)$ and $l = PER(\mathbf{v}^\gamma(1 - e^\gamma))$. Then one decomposes the initial subword of length m of the binary expansion of α into $\frac{m}{l}$ words of length l. The obtained decomposition contains two different types of words, here denoted by $c_1 c_2 \dots c_l$ and $d_1 d_2 \dots d_l$. We assume that $d_1 d_2 \dots d_l$ is greater than $c_1 c_2 \dots c_l$ in the lexicographic order.

Then we have $\alpha = \frac{2^l}{2^l - 1}\sum_{i=1}^{l}\frac{c_i}{2^i}$ and $\overline{\alpha} = \frac{2^l}{2^l-1}\sum_{i=1}^{l}\frac{d_i}{2^i}$. (One uses Proposition 3.27, and again one sums up a geometric series). If k denotes the number of subwords in the decomposition which coincide with $d_1 d_2 \dots d_l$, then $\nu = \frac{kl}{m}$.

4. Proof of Theorem 1.15. Before verifying Theorem 1.15, let us say how the combinatorial classes of M look like.

Proposition 4.25. (Classification of combinatorial classes)
There are the following two types of combinatorial classes:

a) One type corresponds to the hyperbolic component $W^{\alpha\overline{\alpha}}$ for some periodic α: it consists of this component, its root and all non-parabolic parameters on its boundary, and is equal to $Comb^\alpha = Comb^{\overline{\alpha}}$ and to $Comb^\gamma$ for all non-periodic $\gamma \in \mathbb{T}(\alpha)$.

b) The other type corresponds to some equivalence class E of $\mathbb{T}/\sim$ consisting of points with non-periodic kneading sequence: it is closed and equal to $Comb^\alpha$ for exactly those α which lie in E. (Recall that by Theorem 3.19(ii) such an equivalence class is equal to $[\alpha]_{\approx^\alpha}$.)

Proof: The existence of combinatorial classes as described in a) is easy to see by use of Theorem 4.24 and the above argumentation showing that M is locally connected at irrationally indifferent parameters (see discussion before Theorem 4.24).

If α has a non-periodic kneading sequence, then by Theorem 3.17 a point $\beta \notin [\alpha]_\sim$ is separated from α by a chord in $\mathcal{B}_*$ and a point $\beta \in [\alpha]_\sim$ is not. This provides combinatorial classes in b). ∎

Besides the curves $\mathbf{C}_c^{\beta_1\beta_2}$ for $c \in M$ and special chords $\beta_1\beta_2$, we need the equipotential curves $\mathbf{E}_c^r$ for given $r > 0$, which consist of all points in the complement of K_c with $\log|\Phi_c(z)| = r$. The map p_c transforms $\mathbf{E}_c^r$ into $\mathbf{E}_c^{2r}$, and the domain containing K_c and being bounded by $\mathbf{E}_c^r$ into that which is bounded by $\mathbf{E}_c^{2r}$.

Now we start to prove Theorem 1.15. First of all, note that by Corollary 4.17 different combinatorial classes provide different periodic landing patterns. For simplicity, introduce an equivalence relation $\approx$ on the rationals of $\mathbb{T}$: If $\beta_1, \beta_2 \in \mathbb{T} \cap \mathbb{Q}$, let $\beta_1 \approx \beta_2 :\iff z_c^{\beta_1} = z_c^{\beta_2}$. The parameter c will be specified below.

I. First let $\alpha \in \mathbb{T} \setminus \mathbb{P}$ and $c \in Comb^\alpha$. Further, let Γ be the set of all periodic $\gamma \in \mathbb{T}$ satisfying the following properties:

(i) $\gamma < \overline{\gamma}$,

(ii) α lies behind $\gamma\overline{\gamma}$,

(iii) $\gamma, \overline{\gamma}$ do not lie on the orbit of a point in $[\alpha]_\sim = [\alpha]_{\approx^\alpha}$ (which is only relevant if α is preperiodic).

By Theorems 3.19(ii) and 3.17, we can assume that α is the maximum of Γ, and we denote the minimum of $\{\overline{\gamma} | \gamma \in \Gamma\}$ by δ. Consider the set $\mathcal{B}^\Gamma := \{l_\mathbf{w}^\alpha(\gamma\overline{\gamma}) | \mathbf{w} \in \{0,1\}^*, \gamma \in \Gamma\}$. Recall that by Proposition 3.20 $\eth\mathcal{B}^\alpha$ is equal to the closure of $\mathcal{B}^\Gamma$. Clearly, by Corollary 4.17, $\gamma \approx \overline{\gamma}$ for all $\gamma \in \Gamma$, hence $\dot{\gamma} \approx \dot{\overline{\gamma}}$ for all $\gamma \in \Gamma$.

STEP 1. We show that the endpoints of each chord in $\mathcal{B}^\Gamma$ are $\approx$-equivalent:

In the case $\alpha = \delta$, this is simple. Here it exists a sequence of chords of the form $\dot{\gamma}\dot{\overline{\gamma}}; \gamma \in \Gamma$ converging to $\frac{\alpha}{2}\frac{\alpha+1}{2}$.

Two $\approx$-equivalent points in $\mathbb{T}$ lie both in $]\frac{\alpha}{2}, \frac{\alpha+1}{2}[$ or both in $]\frac{\alpha+1}{2}, \frac{\alpha}{2}[$. Otherwise, one would find points $\gamma_1, \gamma_2 \in \Gamma$ having different orbits such that $\dot{\gamma_1} \approx \dot{\gamma_2}$, hence $\gamma_1 \approx \gamma_2$. We can assume that $\gamma_1\overline{\gamma_1}$ is behind $\gamma_2\overline{\gamma_2}$. For $c \in \mathbf{Wake}^{\gamma_1\overline{\gamma_1}}$, the dynamic ray $\mathbf{R}_c^{\gamma_1}$ and $\mathbf{R}_c^{\gamma_2}$ land at repelling periodic points. This follows from Lemma 4.10, Proposition 4.18, and the fact that no iterate of γ_1 or γ_2 is between γ_1 and $\overline{\gamma_1}$. Therefore, by Proposition 4.5(i) and Corollary 4.7 the dynamic rays $\mathbf{R}_{\tilde{c}}^{\gamma_1}$ and $\mathbf{R}_{\tilde{c}}^{\gamma_2}$ with $\tilde{c} = c^{\gamma_1}$ would land at the same point. This is impossible, since the landing pattern for $\tilde{c}$ is $\approx^{\gamma_1}$ not identifying γ_1 and γ_2.

Further, p_c is two-to-one. Thus $l_s^\alpha(\beta_1) \approx l_s^\alpha(\beta_2)$ if $\beta_1, \beta_2 \in \mathbb{T}, \beta_1 \approx \beta_2$ and $s \in \{0,1\}$, and we are done by induction.

If $\alpha \neq \delta$ then the chords $\frac{\alpha}{2}\frac{\delta+1}{2}$ and $\frac{\alpha+1}{2}\frac{\delta}{2}$ are the longest in $\eth\mathcal{B}^\alpha$. Both are limits of sequences of chords $\dot{\gamma}\dot{\overline{\gamma}}, \ddot{\gamma}\ddot{\overline{\gamma}}$ with $\gamma \in \Gamma$. Here the proof that the endpoints of a chord in $\mathcal{B}^\Gamma$ are $\approx$-equivalent is more complicated.

Assume that this is false. Then for some $\gamma \in \Gamma$ one finds a backward iterate $\beta_1\beta_2$ of $\gamma\overline{\gamma}$ satisfying the following properties: $\beta_1\beta_2 \notin \mathcal{B}^\Gamma$ and $\beta_1 \approx \beta_2$, but $h^n(\beta_1\beta_2) \in \mathcal{B}^\Gamma$ for all $n \in \mathbb{N}$. Clearly, $\beta_1\beta_2$ must be longer than $\frac{\alpha+1}{2}\frac{\delta}{2}$ and β_1, β_2 are preperiodic of common preperiods n and periods m. (They cannot be periodic since otherwise β_2' would be preperiodic, but obviously $\beta_1 \approx^\alpha \beta_2'$.)

There exist some $\gamma \in \Gamma$ and some simple curve connecting c with the parabolic parameter c^γ and being disjoint to all $\mathbf{R}^{h^i(\beta_1)}, \mathbf{R}^{h^i(\beta_2)}; i \in \mathbb{N}$ and to all Misiurewitz points of preperiod less than n and period dividing m and all parabolic parameters of period dividing m. So by Corollary 4.8(ii) it holds $z_{c^\gamma}^{\beta_1} = z_{c^\gamma}^{\beta_2}$, but this contradicts $\beta_1 \not\approx^\gamma \beta_2$. Recall that $\approx^\gamma$ does not identify points of distance greater than $d(\dot{\gamma}, \dot{\overline{\gamma}})$.

STEP 2. We show that (pre)periodic $\approx^\alpha$-equivalence classes correspond to (pre)periodic points in J_c:

Consider a periodic $\approx^\alpha$-equivalence class of some period n, which by Theorem 2.48 (and Proposition 2.6) must be equal to $\mathbb{T}_{\mathbf{s}}^\alpha$ for some periodic 0-1-sequence $\mathbf{s}$. By Proposition 2.25 all boundary chords of $\mathbb{D}_{\mathbf{s}}^\alpha$ belong to $\eth\mathcal{B}^\alpha$. Let us define a special connected domain bounded by an equipotential curve and some curves of type $\mathbf{C}_c^{\beta_1\beta_2}$.

If $\mathbb{T}_{\mathbf{s}}^\alpha = \{\beta_1\}$ for some $\beta \in \mathbb{T}$, let $\gamma_1, \delta_1 \in \mathbb{T}$ be given such that $\gamma_1 < \beta_1 < \delta_1, \gamma_1 \approx \delta_1$ and the interval $]\gamma_1, \delta_1[$ and its first n iterates are disjoint to $\frac{\alpha}{2}\frac{\alpha+1}{2}$. If $\mathbb{T}_{\mathbf{s}}^\alpha = \{\beta_1, \beta_2, \ldots, \beta_k\}$ with $k > 1$ and $\beta_1 \curvearrowright \beta_2 \curvearrowright \ldots \curvearrowright \beta_k$, then let $\gamma_i, \delta_i; i = 1, 2, \ldots, k$ with the following properties be given: $\gamma_i \approx \delta_i$ for all $i = 1, 2, \ldots, k$ and $\delta_k \curvearrowright \beta_1 \curvearrowright \gamma_1 \curvearrowright \delta_1 \curvearrowright \beta_2 \curvearrowright \gamma_2 \curvearrowright \delta_2 \curvearrowright \ldots \curvearrowright \beta_k \curvearrowright \gamma_k$, and the connectedness component of $\mathbb{D} \setminus \bigcup_{i=1}^k \gamma_i\delta_i$ bounded by all $\gamma_i\delta_i$ and the first n iterates of that component are disjoint to $\frac{\alpha}{2}\frac{\alpha+1}{2}$.

Fix an equipotential curve $\mathbf{E} = \mathbf{E}_c^r$ for some $r > 0$ and let G be the domain bounded by $\mathbf{E}$ and the curves $\mathbf{C}_c^{\gamma_i\delta_i}; i = 1, 2, \ldots, k$. In the case $k = 1$ take the smaller domain.

By the construction of G, it holds $\mathrm{cl}\, G \subset p_c^n(G)$, and the restriction of p_c^n to G is a bijection. An application of the Schwarzian lemma to the inversion of that restriction shows that G contains a fixed point whose external angles are given precisely by $\beta_1, \beta_2, \ldots, \beta_k$.

Using this, it is not a problem to show that each preperiodic equivalence class of $\approx^\alpha$ corresponds with a preperiodic point in J_c, in particular, $[\frac{\alpha}{2}]_{\approx^\alpha}$ with the critical point 0 if α is preperiodic.

II. J_c is the union of all impressions $Imp(\mathbf{R}_c^\beta); \beta \in \mathbb{T}$. By the above discussion, we have the following

Proposition 4.26. *If $\alpha \notin \mathbb{P}$, then c is contained in the impression of the dynamic ray $\mathbf{R}_c^\gamma$ for some $\gamma \in \mathbb{T}$ with $\gamma \sim \alpha$.* ■

The Julia set J_c for a Misiurewicz point c is locally connected, and since c is preperiodic, the landing pattern for c is equal to $\approx^\alpha$ for a preperiodic $\alpha \in \mathbb{T}$. By the last statement in Proposition 3.20 and the above shown it follows $c \in Comb^\alpha$.

III. It remains to be considered the case that α is periodic (see Proposition 4.25 a)). For such α, the only parabolic parameter in $Comb^\alpha$ is c^α, and by the just shown there is no Misiurewicz point in $Comb^\alpha$. So by Corollary 4.7 the rational landing pattern cannot change in $Comb^\alpha \setminus \{c^\alpha\}$.

The landing pattern for c^α is $\approx^\alpha$, and for $c \in W^{\alpha\overline{\alpha}}$ one has a landing pattern $\approx^\gamma$ with periodic γ (see Corollary 4.2(i)) and with $\alpha \approx^\gamma \overline{\alpha}$. Since

γ is a main point of $\approx^\gamma$, the chord $\gamma\overline{\gamma}$ must lie behind $\alpha\overline{\alpha}$ or be equal to $\alpha\overline{\alpha}$. The first is impossible, since c lies in $\mathbb{C} \setminus \mathrm{cl}\,\mathbf{Wake}^{\gamma\overline{\gamma}}$, where $z_c^\gamma \neq z_c^{\overline{\gamma}}$. This completes the proof of Theorem 1.15, and as a by-product we get the following statement:

Proposition 4.27. *For all c in the hyperbolic component $W^{\alpha\overline{\alpha}}$ the landing pattern is $\approx^\alpha$.* ■

As an immediate consequence of Theorem 1.15 we get DOUADY and HUBBARD's landing result for 'preperiodic' parameter rays (see [43]).

Corollary 4.28. ('Preperiodic' parameter rays land nicely)
Each parameter ray $\mathbf{R}^\alpha$ for preperiodic $\alpha \in \mathbb{T}$ lands at a Misiurewicz point c^α with landing pattern $\approx^\alpha$ and with $Comb^\alpha = \{c^\alpha\}$. Conversely, each Misiurewicz point coincides with c^α for some preperiodic $\alpha \in \mathbb{T}$.

Proof: Let α be preperiodic with preperiod n and period m, and let $c \in Comb^\alpha$. By Theorem 1.15 the rational landing pattern for c is equal to $\approx^\alpha \cap(\mathbb{Q} \times \mathbb{Q})$. In particular, the dynamic rays $\mathbf{R}_c^{\frac{\alpha}{2}}$ and $\mathbf{R}_c^{\frac{\alpha+1}{2}}$ land at the same point. Clearly, the landing point must be 0, hence c a Misiurewicz point of preperiod n and period dividing m. Since there are only finitely many such Misiurewicz points and $Comb^\alpha$ must be connected, it follows that $Comb^\alpha = \{c^\alpha\}$ and that the parameter ray $\mathbf{R}^\alpha$ lands at c. The rest follows from Theorem 1.15. ■

Remark. SCHLEICHER takes the 'rational' landing patterns to define combinatorial classes (see [151]). He relates the combinatorial classes given in this manner to 'fibers' of M. These parts of M are separated by pairs of rational parameter rays connected by a common landing point or a simple curve within an interior component of M, and only differ from combinatorial classes at hyperbolic components. Theorem 1.15 is slightly stronger than the related statements in [151] because we use only 'periodic' parameter rays and give the rational landing patterns for the combinatorial classes explicitly.

5. Hyperbolicity and MLC Conjectures, and proof of Theorem 1.12. We have seen that 'rational' parameter rays land, and Corollaries 4.14 and 4.15 say that for periodic points $\alpha_1, \alpha_2 \in \mathbb{T}$ the rays $\mathbf{R}^{\alpha_1}, \mathbf{R}^{\alpha_2}$ land at the same parabolic parameter iff $\alpha_1 \sim \alpha_2$. Now Proposition 4.25 and Corollary 4.28 provide the first part of Theorem 1.12. Let us complete the proof of that theorem.

By Caratheodory's Theorem, M and ∂M are locally connected iff Φ extends continuously to the unit circle, implying that the parameter rays land in a continuous way. If two parameter rays $\mathbf{R}^{\gamma_1}$ and $\mathbf{R}^{\gamma_2}$ for $\gamma_1, \gamma_2 \in \mathbb{T}(\alpha)$ with periodic $\alpha \in \mathbb{T}$ land at the same point, then γ_1 and γ_2 are associated periodic points. Further, the set $\mathbb{P}$ of points with periodic kneading sequence is equal to $\bigcup_{\alpha \in \mathbb{P}_*} \mathbb{T}(\alpha)$ (see Theorem 3.34).

According to Theorem 3.19(i) $\sim$ is the only closed planar equivalence relation on $\mathbb{T}$ which identifies associated periodic points but no other points in $\mathbb{P}$. This gives the second part of Theorem 1.12.

For a moment assume that M is locally connected. Then ∂M is homeomorphic to $\mathbb{T}/\sim$ and the boundary of a component of the interior of M is a simple closed curve. By Theorem 4.24 and the classification of infinite gaps of $\mathcal{B}$ in Theorem 3.17(ii), each such simple closed curve corresponds to a hyperbolic component. This yields the following implication, which was first proved by DOUADY and HUBBARD [43].

Proposition 4.29. *M is locally connected $\Longrightarrow$ Each component of the interior of M is hyperbolic.* ■

One obtains a continuous map from the Mandelbrot set less all hyperbolic components onto the Abstract Mandelbrot set $\mathbb{T}/\sim$ by the following assignments (compare SCHLEICHER [151]):

$$\begin{cases} c^\alpha \mapsto [\alpha]_\sim & \text{for all } \alpha \in \mathbb{P}, \\ c \in Comb^\alpha \mapsto [\alpha]_\sim & \text{for all } \alpha \notin \mathbb{P}, \end{cases}$$

and local connectivity of M is related to the combinatorial classes as follows (compare [151]):

Proposition 4.30. *M is locally connected $\Longleftrightarrow$ Each combinatorial class $Comb^\alpha$ for non-preperiodic $\alpha \in \mathbb{T} \setminus \mathbb{P}$ is a one-point set.* ■

Remarks. 1. There are various languages for describing combinatorics of the Mandelbrot set and of Julia sets, having different explicitness. We have already mentioned the Hubbard trees of DOUADY and HUBBARD, the critical portraits discussed in BIELEFELD, FISHER and HUBBARD [13], and the orbit portraits of GOLDBERG and MILNOR (compare also POIRIER [132, 134, 135]).

LAU and SCHLEICHER's internal addresses describing the Mandelbrot set were considered in Chapter 3. In the complex plane an internal address of a hyperbolic component W corresponds to sequences $W_0, W_1, W_2, \ldots, W_k$ of hyperbolic components with the following properties: W_1 is the main hyperbolic component, W_k coincides with W, and W_{i-1} is behind W_i for $i = 1, 2, \ldots, n$. Fixing for each i the angle $\frac{p}{q}$ with W_i in the $\frac{p}{q}$-sublimb of W_{i-1}, one obtains the *angled internal address* defined by LAU and SCHLEICHER [92] (Definition 6.1). Note that angled internal addresses are injective on the set of hyperbolic components (see [92], Theorem 9.2). In our work, this follows from Proposition 3.67.

DOUADY and HUBBARD used also pinched disk models (compare [43, 41]). THURSTON's invariant laminations, in particular our Julia equivalences, give rather explicit descriptions, but for some purposes they contain too much (hidden) information. However, the different models provide similar structure results. So THURSTON's QML, LAVAUR's combinatorial description of landing of periodic parameter rays (compare Chapter 3.1.2) and the equivalence relation $\sim$ look completely different at first glance, but we have seen that they all provide the Mandelbrot set under the conjecture of local connectivity.

2. A rational map acting on $\overline{\mathbb{C}}$ is called *hyperbolic* if there exists a conformal metric on $\overline{\mathbb{C}}$ for which the derivative at each point of the Julia set is greater than 1. This is equivalent to the fact that each critical point is attracted by an attractive orbit (see [114, 107]). For a quadratic map p_c the point ∞ is the only critical point besides 0. Since ∞ is a superattractive fixed point, p_c is hyperbolic iff $c \notin M$ or c is attracted by a periodic orbit in $\mathbb{C}$.

This justifies the concept 'hyperbolic components', and the Hyperbolicity Conjecture can be expressed as follows: The set of hyperbolic quadratic polynomials is dense in the set of all quadratic polynomials. (This is a special case of the general Hyperbolicity Conjecture that hyperbolic maps are dense in the set of all rational maps.)

3. The Hyperbolicity Conjecture is supported by the result of MCMULLEN [107] that there does not exist a non-hyperbolic component of the interior of M containing a real number and ŚWIĄTEK's [164] slightly stronger result that hyperbolicity is dense in the real (see also GRACZYK and ŚWIĄTEK [59], and LYUBICH [101]).

4. The MLC Conjecture is still open but there has been substantial progress during the last years: YOCCOZ (see [115, 66]) proved local connectivity of M at all parameters c where p_c is not infinitely renormalizable, and LYUBICH [101, 99, 98] obtained local connectivity in special infinitely renormalizable cases. A presentation of their results would require a closer look at the geometry of quadratic dynamics. This would exceed the framework of the present book.

4.1.3 Renormalization of quadratic polynomials (II)

The aim of the following discussion is to relate renormalization of quadratic polynomials p_c to our abstract setting. First of all recall that if p_c is n-renormalizable with a renormalization (U, p_c^n), then from the topological viewpoint p_c^n acts like a quadratic polynomial in the bounded neighborhood U of 0. One easily sees that in the factorization

$$U \xrightarrow{p_c} p_c(U) \xrightarrow{p_c} \dots \xrightarrow{p_c} p_c^{n-2}(U) \xrightarrow{p_c} p_c^{n-1}(U) \supset \mathrm{cl}U \tag{4.4}$$

of p_c^n the first p_c is proper of degree two from U onto $p_c(U)$ and the other p_c are injective. In particular, $U = -U$. (Compare also the abstract counterpart (3.12) of (4.4).)

The dynamically interesting behavior of p_c^n restricted to U is concentrated on the small Julia set $J(p_c^n)$ - the boundary of the small filled-in Julia set

$K^{(0)} = K(p_c^n)$. Clearly, all n small filled-in Julia sets $K^{(i)} = p_c^i(K(p_c^n))$ and the corresponding small Julia sets $p_c^i(J(p_c^n))$ for $i = 0, 1, 2, \ldots, n-1$ are p_c^n-invariant.

By use of the Schwarzian lemma it is not hard to show that two different small filled-in Julia sets intersect in at most one point, which must be a repelling fixed point for p_c^n (see [108], Theorem 7.3).

In order to see more, let us look at the fixed points of a quadratic polynomial $p_{\tilde{c}}$ with $\tilde{c} \in M$. Clearly, there are two, which coincide iff $\tilde{c} = c^0 = \frac{1}{4}$ (compare Example 1.16). Then the double fixed point is given by $z_{\tilde{c}}^0 = \frac{1}{2}$. Usually the fixed point $z_{\tilde{c}}^0$ is called *BETA fixed point* and the other one *ALPHA fixed point.* (We use '*ALPHA*' and '*BETA*' instead of 'α' and 'β' since the Latin symbols are reserved for angles.)

If $\tilde{c} \neq c^0$ lies in the main hyperbolic component, then the *BETA* fixed point is repelling and the *ALPHA* fixed point is not. So the two fixed points can be distinguished purely from topological dynamics on $K_{\tilde{c}}$ (see discussion at the beginning of this chapter).

If $\tilde{c} \in Comb^\alpha \neq Comb^0$ for some $\alpha \in \mathbb{T}$, then besides the repelling fixed point $z_{\tilde{c}}^0$ - recall that $I^\alpha(0) = \overline{1}$ -, there is the repelling or parabolic fixed point z whose external angles are those $\beta \in \mathbb{T}$ with $I^\alpha(\beta) = \overline{0}$ or $\widetilde{I^\alpha}(\beta) = \overline{0}$. All those β lie on a periodic orbit of period greater than 1. So $K_{\tilde{c}} \setminus \{z\}$ splits into at least two different components. On the other hand, $K_{\tilde{c}} \setminus \{z_{\tilde{c}}^0\}$ is connected. This can be obtained by arguing as in the proof of Theorem 1.15:

There exists a point γ with $0 \neq \gamma \leq \alpha \leq \overline{\gamma}$ and $\gamma \approx^\alpha \overline{\gamma}$. Let $\beta_0 = \gamma, \delta_0 = \overline{\gamma}$, and if β_i, δ_i are already defined for $i \in \mathbb{N}_0$, then let $\beta_{i+1} = \frac{\beta_i}{2}$ and $\delta_{i+1} = \frac{\delta_i + 1}{2}$. Clearly, 0 lies behind $\beta_i\delta_i = l_{0^i}^\alpha(\gamma\overline{\gamma}) \in \eth\mathcal{B}^\alpha$ for all $i \in \mathbb{N}$, and $(\beta_i\delta_i)_{i\in\mathbb{N}}$ converges to 0.

Let G_i be the smaller domain bounded by the equipotential line $\mathbf{E}_{\tilde{c}}^{2^{-i}}$ and by $\mathbf{R}_{\tilde{c}}^{\beta_i\delta_i}$. For each $i \in \mathbb{N}$ the restriction of p_c to G_{i+1} maps G_{i+1} bijectively to G_i and it holds $z \in \mathrm{cl}G_{i+1} \subset G_i$. By application of the Schwarzian lemma to the inversion of that restriction one obtains that the diameter of G_i shrinks to 0 if i approaches to ∞, showing that $K_{\tilde{c}} \setminus \{z_{\tilde{c}}^0\}$ is connected.

By virtue of the Straightening theorem *ALPHA* and *BETA* fixed points can be given also in the small filled-in Julia sets for a given renormalization. The straightening provides a conjugate quadratic Julia set, and the above considerations justify that the following definition is correct (compare [108]):

Definition 4.31. *Let p_c be n-renormalizable and let z be a fixed point of p_c^n in one of the small filled-in Julia sets K. Then z is called the BETA* fixed point *if a topologically conjugacy from (K, p_c^m) onto $(K_{\tilde{c}}, p_{\tilde{c}})$ for any $\tilde{c} \in M$ map z to $z_{\tilde{c}}^0$, and ALPHA* fixed point *otherwise.*

One easily sees that either all intersection points of small filled-in Julia sets are *ALPHA* fixed points or not. Furthermore, we will need the following interesting fact:

Lemma 4.32. *If a quadratic polynomial p_c is n-renormalizable, then a repelling ALPHA fixed point in some small filled-in Julia sets has at least two external angles.*

Proof: Let z be the repelling $ALPHA$ fixed point in the small filled-in Julia set $K^{(0)} = K(p_c^n)$. By the Straightening theorem and the above considerations there exist a small neighborhood V of z, a simple curve $\mathbf{C} = \{z(t)|t \in [0,1]\}$ with $z(0) = z$ and some $q > 1$ satisfying the following properties: p_c^n is injective - hence orientation-preserving - on V, $\mathbf{C} \subset p_c^{qn}(\mathbf{C})$, and the q 'semi-open' curves $p_c^{in}(\mathbf{C} \setminus \{z\}); i = 0, 1, \ldots, q-1$ lie in q different components of $V \setminus K(p_c^n)$.

Fixing some external angle β of z, one easily sees that $p_c^n(\mathbf{R}^\beta) \neq \mathbf{R}^\beta$, hence that $h^n(\beta)$ is a further external angle of z. ■

As announced, we show that renormalization is actually a combinatorial phenomenon. Namely - with a small exception - it can completely be described within our abstract setting. Recall that a renormalization (U, p_c^n) is called crossed if there exist two small filled-in Julia sets which cross each other, and simple otherwise.

The following statement does not mention the case $\gamma \in \mathbb{P}_\infty$. This is not substantial because $\mathbb{P}_\infty \subset \bigcup_{\alpha \in \mathbb{P}_*} \mathbb{T}(\alpha)$: for each non-periodic $\gamma \in \mathbb{T}(\alpha)$, it holds $Comb^\gamma = Comb^\alpha$, and γ is n-renormalizable iff α is.

Theorem 4.33. (Renormalization and abstract renormalization)
Let $\gamma \in \mathbb{T} \setminus \mathbb{P}_\infty$ and $c \in Comb^\gamma$. Then for $n \in \mathbb{N} \setminus \{1\}$ the following statements are equivalent:

(i) p_c is n-renormalizable,

(ii) γ is n-renormalizable, and if the point c is a parabolic parameter, then its period is not less than n.

If (i) is valid, $J(p_c^n)$ is the union of the impressions of all $\mathbf{R}_c^\beta; \beta \in h^{-1}(\mathbb{T}_{*_n}^\gamma)$. Then, moreover, p_c is simply n-renormalizable iff γ is, and crossed otherwise.

Proof: '(ii) $\Longrightarrow$ (i)': On the assumption (ii), from $h^{-1}(\mathbb{T}_{*_n}^\gamma) = h^{n-1}(\mathbb{T}_{*_n}^\gamma)$ and its convex hull $h^{-1}(\mathbb{D}_{*_n}^\gamma) = h^{n-1}(\mathbb{D}_{*_n}^\gamma)$ we construct an open domain U such that (U, p_c^n) forms a renormalization of p_c.

By Corollary 3.57 all boundary chords of $h^{-1}(\mathbb{D}_{*_n}^\gamma)$ have rational endpoints and by Theorem 3.48 these endpoints are pairwisely $\approx^\gamma$-equivalent. This and Theorem 1.15 show that to each boundary chord $\beta\delta$ of $h^{-1}(\mathbb{D}_{*_n}^\gamma)$ there corresponds the simple closed curve $\mathbf{C}_c^{\beta\delta}$ dividing the complex plane into two parts.

Now let η be the length of a longest boundary chord of the sets $h^j(\mathbb{D}_{*_n}^\gamma)$; $j = 0, 1, \ldots, n-1$. Clearly, $\eta \geq \frac{1}{3}$. Further, let $\mu = 2^{1-n}(1 - 2\eta)$.

If S is a boundary chord of one of the sets $h^j(\mathbb{D}^{\gamma}_{*_n}); j = 0, 1, \ldots, n-1$ with $d(h(S)) < \mu$, then also $d(S) < \mu$. Otherwise, one would have $1 - 2\eta > \mu > d(h(S)) = 1 - 2d(S)$, which would contradict the definition of η. Therefore, the set of boundary chords of $h^{n-1}(\mathbb{D}^{\gamma}_{*_n})$ splits into two subsets: one contains all chords shorter than μ, and the other consists of finitely many chords not shorter than μ and is h^n-invariant.

Denote the chords in the second subset by $S_l = \beta_l\delta_l; l = 0, 1, 2, \ldots, k$, and assume that the points $\beta_1, \delta_1, \beta_2, \delta_2, \ldots, \beta_{k-1}, \delta_{k-1}, \beta_k = \beta_0, \delta_k = \delta_0$ are given in counter-clockwise cyclic order where $\beta_l = \delta_{l-1}$ for some $l = 1, 2, \ldots, k$ is allowed. If S is a boundary chord of $h^{-1}(\mathbb{D}^{\gamma}_{*_n})$ being shorter than μ, then its first $n-2$ iterates are shorter than $1 - 2\eta$.

This and the fact that each of the sets $h^j(\mathbb{T}^{\gamma}_{*_n}); j = 0, 1, \ldots, n-2$ lies in one of the intervals $]\frac{\gamma}{2}, \frac{\gamma+1}{2}[$ and $]\frac{\gamma+1}{2}, \frac{\gamma}{2}[$ show that also each of the first $n-2$ iterates of $\mathbb{T} \setminus \bigcup_{l=0}^{k-1}]\beta_l, \delta_l[$ is contained in one of the intervals.

Therefore, h^n maps $\mathbb{T} \setminus \bigcup_{l=0}^{k-1}]\beta_l, \delta_l[$ two-to-one onto a set containing $\mathbb{T} \setminus \bigcup_{l=0}^{k-1}]\beta_l, \delta_l[$. (Recall that $h^{n-1}(\mathbb{T}^{\gamma}_{*_n}) = h^{-1}(\mathbb{T}^{\gamma}_{*_n})$.) Moreover, for sufficiently small $\epsilon > 0$ the map h^n remains two-to-one on $\mathbb{T} \setminus \bigcup_{l=0}^{k-1} [\beta_l + \epsilon, \delta_l - \epsilon]$, and the following inclusion holds: $\mathbb{T} \setminus \bigcup_{l=0}^{k-1}]\beta_l + \epsilon, \delta_l - \epsilon[\subseteq h^n(\mathbb{T} \setminus \bigcup_{l=0}^{k-1} [\beta_l + \epsilon, \delta_l - \epsilon])$.

Let $z_l; l = 0, 1, 2, \ldots, k-1$ be the landing points of the dynamic rays $\mathbf{R}_c^{\beta_l}$. (Note that they need not be different on the whole.) By Corollary 3.57 among the chords $S_l = \beta_l\delta_l; l = 0, 1, 2, \ldots, k-1$ there is one whose orbit contains the other ones such that the corresponding z_l is a fixed point for p_c^n. We assume that this chord is S_0. All z_l different from z_0 lie in the backward orbit of z_0 with respect to p_c^n and are preperiodic.

Now for each $l = 0, 1, 2, \ldots, k-1$ let A_l be that closed part of the complex plane which is bounded by the curve $\mathbf{C}_c^{\beta_l\delta_l}$ and does not contain 0. Furthermore, let G be the simply connected domain which is bounded by a fixed equipotential line.

Assume that for each $l = 0, 1, 2, \ldots, k-1$ there exists a simply connected domain $U_l \subset G$ satisfying the following properties for all $l, l_1, l_2 \in \{0, 1, 2, \ldots, k-1\}$:

1. $z_l \in U_l$,
2. $U_{l_1} = U_{l_2}$ if $z_{l_1} = z_{l_2}$,
3. $U_{l_1} = -U_{l_2}$ if $z_{l_1} = -z_{l_2}$,
4. $z_{l_1} = z_{l_2}$ if $U_{l_1} \cap A_{l_2} \neq \emptyset$ or $U_{l_1} \cap U_{l_2} \neq \emptyset$,
5. $\mathrm{cl}(U_{l_1} \cap A_{l_1}) \subset p_c^n(U_{l_2})$, if $z_{l_1} = p_c^n(z_{l_2})$,
6. p_c^n is injective on U_l.

Then p_c^n would act on $\tilde{U} := (G \setminus \bigcup_{l=0}^{k-1} A_l) \cup \bigcup_{l=0}^{k-1} U_l$ as a map of degree two with $\tilde{U} \subseteq p_c^n(\tilde{U})$. One would obtain a renormalization (U, p_c^n) by construction of a 'fattened' domain U from $\tilde{U}$ as follows:

For sufficiently small $\epsilon > 0$, enlarge $\tilde{U}$ by all connectedness components of $G \setminus (\tilde{U} \cup \bigcup_{l=0}^{k-1} \mathbf{R}^{\beta_l+\epsilon} \cup \bigcup_{l=0}^{k-1} \mathbf{R}^{\delta_l-\epsilon})$ which contain a part of one of the dynamic rays $\mathbf{R}^{\beta_l}, \mathbf{R}^{\delta_l}; l = 0, 1, \ldots, q-1$.

In order to show the existence of the sets $U_l; l = 0, 1, 2, \ldots, k-1$ with properties 1.-6. one essentially needs that each neighborhood V_0 of z_0 contains a simply connected domain U_0 satisfying the property $\mathrm{cl}(U_0 \cap A_l) \subset p_c^n(U_0)$ for all l with $z_l = z_0$. The other U_l can be successively constructed from a sufficiently small U_0 which does not contain the points $p_c^j(0); j = 1, 2, \ldots, (k-1)n$ by slightly enlarging suitable connectedness components in the backward iterates of U. (This is possible since 0 cannot be periodic and be contained in J_c at the same time.)

So let V_0 be a neighborhood of z_0. If the fixed point z_0 of p_c^n is repelling, the existence of the required U_0 is obvious. If z_0 is parabolic, then by Proposition 4.25 γ is periodic, and it holds $c = c^\gamma$. Clearly, z_0 lies in the orbit of z_c^γ (see Lemma 4.10) and by (ii) n is the period of z_0 with respect to p_c.

Look at Corollary 4.4 and Theorem 3.50. If γ is non-primitive, then $n = PER(\mathbf{v}^\gamma(1-e^\gamma)) = SUBPER_n(\gamma) < PER(\gamma)$. The point γ is simply n-renormalizable and there exists some periodic α of period n such that $h^{-1}(\mathbb{D}^\gamma_{*_n}) = h^{-1}(\mathbb{D}^\alpha_{*_n})$. This is impossible, since the endpoints of each non-degenerate boundary chord of $\mathbb{D}^\alpha_{*_n}$ are backward iterates of α, hence not of γ. Therefore, γ must be primitive of period n. So with exception of the critical Fatou component K there is no bounded Fatou component whose boundary contains z_0.

It is well-known that in this case there exists a special set V with $\mathrm{cl}V \subseteq p_c^n(V) \cup \{z_0\}$ - a so-called repelling petal- in V_0, such that V together with the critical Fatou component K forms a neighborhood of z_0 (e.g., see [114], §7). Clearly, K is p_c^n-invariant and disjoint to all $A_l; l = 0, 1, 2, \ldots, k-1$, and the required U_0 can be obtained as the union of V and the intersection of K with a sufficiently small neighborhood of z_0.

Before we come to the implication '(i) $\Longrightarrow$ (ii)', let us show that on the assumption (ii), $J(p_c^n)$ is equal to the union of impressions of all $\mathbf{R}_c^\beta; \beta \in h^{-1}(\mathbb{T}^\gamma_{*_n})$. For simplicity we denote that union by J.

First of all, J is p_c^n-invariant, and if U is constructed as above, then J is contained in U. This shows $J \subseteq J(p_c^n)$. The $U_l; l = 0, 1, \ldots, k-1$ can be chosen arbitrarily small, and in each case one obtains a renormalization (U, p_c^n). From this one easily sees $J(p_c^n) = J$.

As a simple consequence of this and Lemma 3.54(iv),(v) one gets the following: On the assumption of (ii), p_c is simply n-renomalizable if γ is, and crossed otherwise.

'(i) $\Longrightarrow$ (ii)': Assume that p_c is n-renormalizable for some $n \in \mathbb{N} \setminus \{1\}$. First consider the case $c \in Comb^\gamma$ for some periodic $\gamma \in \mathbb{T}$ of period m, where the rational landing pattern for J_c is equal to the rational part of $\approx^\gamma$.

For each boundary chord $S = \beta_1\beta_2$ of the critical gap $h^{-1}(Gap\langle\gamma\rangle) = h^{m-1}(Gap\langle\gamma\rangle)$ of $\mathfrak{d}\mathcal{B}^\gamma$ we consider the curve $\mathbf{C}^S$ consisting of the two dynamic rays $\mathbf{R}_c^{\beta_1}, \mathbf{R}_c^{\beta_1}$ and their common landing point $z_c^S = z_c^{\beta_1} = z_c^{\beta_2}$. (For the structure of $h^{-1}(Gap\langle\gamma\rangle)$ see Proposition 2.43.)

All obtained curves $\mathbf{C}^S$ bound a closed connected part of the complex plane containing the point 0. Let us denote the intersection of K_c with this part by A. We know that $z_0 := z_c^{\dot{\gamma}\bar{\gamma}}$ is a fixed point of p_c^m and that all z_c^S for $S \neq \dot{\gamma}\bar{\gamma}$ are in the backward orbit of z_0 with respect to p_c^m.

If the connected small filled-in Julia set $K^{(0)} = K(p_c^n)$ contains a point outside from A, then it must contain a point z_c^S for some boundary chord S of $h^{-1}(Gap\langle\gamma\rangle)$. Further, if such a point z_c^S is in $K^{(0)}$, then also the point z_0 as some iterate of z_c^S with respect to p_c^{mn}.

γ is periodic of period m, hence m-renormalizable. In the parabolic case $c = c^\gamma$ the point z_0 - which lies on the orbit of z_c^γ - is parabolic. In this case A is the closure of the critical Fatou component (compare Theorem 4.1, Case 1). The critical Fatou component itself, as an immediate basin of a parabolic periodic point, does not contain a periodic point. All points on its boundary which are different from z_0 have a period greater than m. This can easily be seen from the abstract theory.

$K^{(0)}$ must contain a fixed point of p_c^n. If $n \leq m$, this point is equal to z_0 or it does not belong to A. In the latter $z_0 \in K^{(0)}$ by the argument above. Now let $n > m$. In order to show that $z_0 \in K^{(0)}$ in this case, assume that $K^{(0)} \subseteq A \setminus \{z_0\}$.

Then the $ALPHA$ fixed point in $K^{(0)}$ is repelling and by Lemma 4.32 it has at least two periodic external angles. The structure of $\approx^\gamma$ on $\mathbb{T}\langle\gamma\rangle = Gap\langle\gamma\rangle \cap \mathbb{T}$ and Theorem 1.15 allow no periodic point in A different from z_0 to have more than one periodic external angle, which contradicts our assumption.

In order to show that $z_0 \in K^{(0)}$ if $c \neq c^\gamma$ and $n \neq m$, one argues similarly: Here p_c has an attractive or irrationally indifferent periodic orbit of period m. Therefore, the $ALPHA$-fixed point is repelling if $n < m$, or if $n > m$ and the small filled-in Julia sets are disjoint. An intersection point of two different small filled-in Julia sets cannot be attractive or irrationally indifferent, implying that the $ALPHA$-fixed point is repelling for $n > m$ in the general. Now we can use that the $ALPHA$-fixed point has at least two external angles.

We show that $n > m$ is impossible. For this, note that A is p_c^m-invariant and so contains all $p_c^{im}(0)$ for $i \in \mathbb{N}_0$. Assume that $n > m$. Then by Proposition 2.43(i) it holds $n = km$ for some $k > 1$. (Recall that $Gap_*^\gamma = h^{-1}(Gap\langle\gamma\rangle)$.)

Consider the small filled-in Julia sets $K^{(im)} = p_c^{im}(K(p_c^n))$ for $i = 0, 1, \ldots, k-1$, which all contain z_0. Further, let B be that component of $\mathbb{C} \setminus \mathbf{C}^{\dot{\gamma}\bar{\gamma}}$ which contains A. Obviously, the map p_c^m permutes the mutually disjoint sets $K^{(im)} \cap B$ and fixes the dynamic rays $\mathbf{R}_c^{\dot{\gamma}}$ and $\mathbf{R}_c^{\dot{\overline{\gamma}}}$. Since p_c^m is an orientation-preserving homeomorphism in some neighborhood of z_0, it follows $p_c^m(K^{(0)}) = K^{(0)}$, which contradicts $n > m$.

$K^{(0)}$ is the only small filled-in Julia set containing both z_0 and $-z_0$, so the only one which can intersect each of the three connectedness components of $\mathbb{C} \setminus (\mathbf{C}^{\dot{\gamma}\dot{\overline{\gamma}}} \cup \mathbf{C}^{\ddot{\gamma}\ddot{\overline{\gamma}}})$. Each iterate $h^i(\gamma); i = 0, 1, \ldots, m-2$ lies behind $\dot{\alpha}\dot{\overline{\alpha}}$ or $\ddot{\alpha}\ddot{\overline{\alpha}}$. So one easily sees that γ is n-renormalizable.

It must be excluded that for parabolic c the period of the parabolic orbit is less than n since then z_0 is on that orbit and belongs to more than one small filled-in Julia set, which is impossible.

Let us come to the case $\gamma \notin \mathbb{P}$. According to Proposition 4.25 (see b)) and Proposition 4.26 we can assume that $c \in Imp(\mathbf{R}^{\gamma})$. So the union of the two dynamic rays $\mathbf{R}_c^{\frac{\gamma}{2}}, \mathbf{R}_c^{\frac{\gamma+1}{2}}$ and their impressions form a closed connected set $\widetilde{\mathbf{C}_c^{\frac{\gamma}{2}\frac{\gamma+1}{2}}}$, which contains c but by Proposition 3.20 no periodic point.

$\widetilde{\mathbf{C}_c^{\frac{\gamma}{2}\frac{\gamma+1}{2}}}$ divides the plane into two parts. Moreover, if p_c is n-renormalizable for $n > 1$, then each of the small filled-in Julia sets $p_c^i(K(p_c^n))$ for $i = 1, 2, \ldots, n-1$ lies in the same part as $\mathbf{R}_c^{h^{i-1}(\gamma)}$. So we are finished. ■

We want to direct our attention to the small difference between the abstract renormalization concept for periodic γ and that for the complex plane: The abstract theory cannot realize differences of the dynamics in a neighborhood of J_{c^γ} and of those in a neighborhood of J_c for $c \in W^{\gamma\overline{\gamma}}$. If γ is n-renormalizable, and the period of the parabolic parameter $c = c^\gamma$ is less than n, then a 'small' filled-in Julia set for p_c^n is recognized (by the abstract theory), but a renormalization fails. Clearly, in this case γ is not primitive.

Recall from Chapter 1.2.2 that DOUADY and HUBBARD's tuning provides for each hyperbolic component W of period m a homeomorphism $d \mapsto W * d$ from the Mandelbrot set onto a small copy of it.

In the abstract setting tuning is reflected by the map Q^α: If $W = W^{\alpha\overline{\alpha}}$ for some periodic α, then Q^α provides a homeomorphism from the Abstract Mandelbrot set $\mathbb{T}/\sim$ onto its 'small' copy $Q^\alpha(\mathbb{T})/\sim$ (see Theorem 3.31). In particular, for the points $c^\gamma; \gamma \in \mathbb{P}$ abstract and concrete tuning are the same; it holds $c^{Q^\alpha(\gamma)} = W * c^\gamma$.

The above discussion shows that p_c for $c = W * \frac{1}{4}$ need not be m-renormalizable. The renormalization fails iff the parabolic parameter c has period less than m, in other words iff α is not primitive.

For some parts of ∂M not belonging to the boundary of some hyperbolic component the abstract tuning can only be interpreted as a relation between combinatorial classes: $Comb^\gamma \mapsto Comb^{Q^\alpha(\gamma)}$. (For a further discussion see SCHLEICHER [152].)

DOUADY's [40] gave this relation as one between the binary expansions of external angles. In our abstract setting his result appears as Proposition 3.27. However, if M were locally connected, we could see tuning and its abstract setting as one and the same.

For a long time it was assumed that each $c \in M$ for which p_c is n-renormalizable is the result of tuning defined by some hyperbolic component of period m, but MCMULLEN found counter-examples (see [116, 66, 108]). There exist parameters c and numbers $n \in \mathbb{N} \setminus \{1\}$ such that p_c is n-renormalizable but $c \neq W * d$ for all hyperbolic components W and all $d \in M$. Exactly in this case is p_c n-renormalizable with crossed renormalization. This is a consequence of Theorems 4.33 and 3.50.

We emphasize that all information on the combinatorics of renormalization of p_c is hidden in each point $\alpha \in \mathbb{T}$ with $c \in Comb^\alpha$. For this information we refer to Chapter 3.2.

4.1.4 Non-realizable Julia equivalences

We know from Theorem 1.15 that the rational parts of Julia equivalences are realized in the complex plane, namely that $\approx^\alpha \cap (\mathbb{T} \times \mathbb{T})$ is the rational landing pattern for c if $c \in Comb^\alpha$. So the following question arises: For which α is the whole Julia equivalence $\approx^\alpha$ realizable?

If α is rational, the situation is obvious: $\approx^\alpha$ for periodic α is exactly the landing pattern for all parameters $c \in Comb^\alpha$ which are not irrationally indifferent (see Theorems 1.15, 4.1 and Proposition 4.25), hence hyperbolic or parabolic, and $\approx^\alpha$ for preperiodic α is exactly the landing pattern for the Misiurewicz point c^α (see Corollary 4.28). Note that the Spider algorithm by HUBBARD and SCHLEICHER [67] allows for a rational $\alpha \in \mathbb{T}$ to determine the Misiurewicz point c^α and the center of the hyperbolic component $W^{\alpha\overline{\alpha}}$, respectively. The *center* of a hyperbolic component W is the unique parameter c with $Mult_W(c) = 0$.

The really interesting case is that α is irrational. If a Julia equivalence $\approx^\alpha$ is non-realizable, then only two cases are possible: Either $\alpha \in \mathbb{T} \setminus \mathbb{P}$ and α is n-renormalizable for infinitely many $n \in \mathbb{N}$, which follows from Theorem 4.33, Theorem 1.15 and YOCCOZ's Theorem, or $\alpha \in \mathbb{P}_\infty$. We want to describe a class of points $\gamma \in \mathbb{P}_\infty$ for which $\approx^\gamma$ is the landing pattern for no $c \in M$.

First of all, recall the fact that the points of $\mathbb{P}_\infty$ correspond to the irrational indifferent parameters. Assume that $\approx^\gamma$ is the landing pattern for such a parameter c. Then there is a unique periodic $\alpha \in \mathbb{T}$ with $\alpha < \overline{\alpha}$, and with $\gamma \in \mathbb{T}(\alpha)$ or equivalently $c \in Comb^\alpha$ (compare Proposition 4.25).

On the one hand, c^γ is the unique parameter in $\partial W^{\alpha\overline{\alpha}}$ with internal angle $\nu^\alpha(\gamma)$, in other words the only parameter in $\partial W^{\alpha\overline{\alpha}}$ providing a periodic point of multiplier $e^{2\pi\nu^\alpha(\gamma)i}$.

On the other hand, the simple closed curve $\mathbb{P}^\gamma_\infty/\approx^\gamma$ in $\mathbb{T}/\approx^\gamma$ corresponds to the boundary of a Siegel disk of p_c, and $(\mathbb{P}^\gamma_\infty/\approx^\gamma, (h/\approx^\gamma)^m)$ is conjugate to an irrational rotation on the unit circle with rotation number $\nu^\alpha(\gamma)$ (relative to 2π). For this, see Proposition 2.66 and Theorem 3.36. Therefore, one easily obtains the following

Proposition 4.34. *Let $\gamma \in \mathbb{P}_\infty$. Then $\approx^\gamma$ is the landing pattern for c iff $c = c^\gamma$ and the Julia set J_{c^γ} is locally connected.* ■

Recall that besides the above periodic α there is a unique sequence $\mathbf{k} = k_1k_2k_3\ldots \in \mathbb{N}^\mathbb{N}$ with $\gamma = q^\alpha(\mathbf{a}(\mathbf{k}))$ or $\gamma = q^\alpha(\mathbf{a}(\mathbf{k})^-) = q^{\overline{\alpha}}(\mathbf{a}(\mathbf{k}))$, and that conversely each point $q^\delta(\mathbf{a}(\mathbf{k}))$ for some $\mathbf{k} \in \mathbb{N}^\mathbb{N}$ and some periodic $\delta \in \mathbb{T}$ lies in $\mathbb{P}_\infty$. (For the definition of $q^\delta(\mathbf{a}(\mathbf{k}))$ see (2.2) and (3.8).)

If $\gamma = q^\alpha(\mathbf{a}(\mathbf{k}))$, then $\nu^\alpha(\gamma)$ is equal to $\nu(\mathbf{a}(\mathbf{k}))$ and has continued fraction expansion $k_1 + 1k_2k_3\ldots$. If $\gamma = q^\alpha(\mathbf{a}(\mathbf{k})^-)$, then $\nu^\alpha(\gamma)$ is equal to $1-\nu(\mathbf{a}(\mathbf{k}))$ and has continued fraction expansion $1k_1k_2k_3\ldots$ (see Chapter 3.2.4, in particular 4., and the discussion preceding Lemma 3.37).

A striking result of PETERSEN [130, 131] says the following: if an irrationally indifferent parameter c has multiplier $e^{2\pi\nu i}$ and the continued fraction expansion of ν is bounded, then J_c is locally connected. So by the above discussion for all periodic $\alpha \in \mathbb{T}$ and all sequences $\mathbf{k} = k_1k_2k_3\ldots \in \mathbb{N}^\mathbb{N}$ with k_i bounded the Julia equivalence $\approx^\gamma$ for $\gamma = q^\alpha(\mathbf{a}(\mathbf{k}))$ is the landing pattern for c^γ.

On the other hand, we know that a Julia set J_c cannot be locally connected if p_c has a Cremer point. Thus $\approx^\gamma$ for $\gamma \in \mathbb{P}_\infty$ fails to be a landing pattern for some c if p_{c^γ} has a Cremer point.

Let us discuss the fixed point case. To each irrational $\nu \in \mathbb{T}$ there exists a unique quadratic map p_c with a unique irrationally indifferent fixed point z of multiplier $e^{2\pi\nu i}$. This is immediately clear by the considerations above. (In canonical linear coordinates, p_c has the form $x^2 + e^{2\pi\nu i}x$ with the unique fixed point 0.)

It was a long-standing problem to give a number-theoretic condition on ν deciding whether z is a Cremer point, i.e. $z \in J_c$, or a Siegel point, i.e. $z \in K_c \setminus J_c$, until YOCCOZ [172] gave the complete answer (compare also [125]): $z \in K_c \setminus J_c$ iff ν is a *Brjuno number*, i.e. iff it holds

$$\sum_{n=1}^{\infty} \frac{\log q_{n+1}}{q_n} < \infty \tag{4.5}$$

for the denominators q_n of the convergents of ν.

The continued fraction expansions of ν and $1-\nu$ show that ν is a Brjuno number iff $1-\nu$ is. So assume that $\nu \in]0, \frac{1}{2}]$ has continued fraction expansion $[;\kappa_1, \kappa_2, \ldots, \kappa_l(,\ldots)]$ and set $k_1k_2k_3k_4\ldots = \kappa_1 - 1\kappa_2\kappa_3\kappa_4\ldots$. Then by the discussion preceding Proposition 3.38 it holds $q_n = |\mathbf{a}(k_1k_2\ldots k_n)|$. Hence the following is valid:

Example 4.35. (A class of non-realizable Julia equivalences)
Let $k_1k_2k_3\ldots \in \mathbb{N}^\mathbb{N}$ and $\sum_{n=1}^{\infty} \log|\mathbf{a}(k_1k_2\ldots k_{n+1})|/|\mathbf{a}(k_1, k_2, \ldots, k_n)| = \infty$.

If α is the point in $\mathbb{T}$ with binary expansion $\mathbf{a}(k_1, k_2, k_3, \ldots)$, then the Julia equivalence $\approx^\alpha$ cannot be the landing pattern for some parameter $c \in M$.

Remarks. 1. Condition (4.5) was formulated and shown to be necessary for a Siegel fixed point by the Australian mathematician T. M. CHERRY in 1964. His result remained unpublished, and one year later BRJUNO [21] proved it beyond the quadratic case. Finally, the implication that in the quadratic case an irrationally indifferent fixed point whose multiplier $e^{2\pi\nu i}$ satisfies (4.5) must be a Siegel point was given by YOCCOZ [172] in 1988.

2. The understanding of quadratic polynomials p_c for irrationally indifferent parameters is far from complete. For lists of open problems, see [105, 143] and the references therein. Also we refer to the papers [60] by GRISPOLAKIS, MAYER and OVERSTEEGEN, [84] by KIWI and [126, 128, 127] by PEREZ-MARCO, which discuss the topological structure of Julia sets for (quadratic) polynomials with irrationally indifferent periodic points beyond the locally connected case.

3. The topological structure of the Julia set of a quadratic map with irrationally indifferent fixed point z (or periodic point) of multiplier $e^{2\pi\nu i}$ depends on the number-theoretic properties of ν in a sensitive way. Roughly speaking, the better ν can be approximated by rationals in the Diophantine sense, the more complicated is the structure of the Julia set.

For example, we have seen that a bounded continued fraction expansion of ν implies local connectivity of the Julia set and that in the fixed point case local connectivity of the Julia set implies the Brjuno condition for ν.

Moreover, HERMAN has shown that there are cases where the boundary of a Siegel disk around z does not contain the critical point. It is well-known that then J_c fails to be locally connected, which is also a consequence of Theorem 4.1 b).

On the other hand, generalizing a result of HERMAN [62], ROGERS [145] recently proved the following Theorem: If a polynomial has a Siegel disk whose rotation number satisfies a Diophantine condition, i.e. if for some real numbers $r > 0$ and $s \geq 2$ and for all fractions $\frac{p}{q}$ it holds $|\nu - \frac{p}{q}| > \frac{r}{q^s}$, then the boundary of the Siegel disk contains a critical point.

LEVIN and VAN STRIEN (see [96]) have shown local connectivity for real $c \in M$. For such c the Julia set is invariant under complex conjugation. So $z \mapsto \overline{\Phi_c(\overline{z})}; z \in \overline{\mathbb{C}} \setminus K_c$ maps $\overline{\mathbb{C}} \setminus K_c$ conformally onto the unit disk, and from the uniqueness of the Riemann map it follows that complex conjugation transforms each dynamic ray $\mathbf{R}_c^\beta; \beta \in \mathbb{T}$ into the dynamic ray $\mathbf{R}_c^{1-\beta}$.

Therefore, if some Julia equivalence $\approx^\alpha$ forms the landing pattern for some real $c \in M$, then $\approx^\alpha$ is also completely invariant with respect to the map $\beta \mapsto 1 - \beta; \beta \in \mathbb{T}$.

Proposition 4.36. *Each Julia equivalence $\approx^\alpha$ being invariant with respect to $\beta \mapsto 1 - \beta; \beta \in \mathbb{T}$ forms the landing pattern for some real c.*

Proof: $\alpha \in \mathbb{P}_\infty$ is impossible since $\frac{\alpha}{2} \approx^\alpha \frac{\alpha+1}{2}$ would imply that the four points $\frac{\alpha}{2}, \frac{\alpha+1}{2}, 1-\frac{\alpha}{2}, 1-\frac{\alpha+1}{2}$ are in the same equivalence class. This contradicts Theorem 2.50.

If $\gamma \approx^\alpha \overline{\gamma}$ for some periodic $\gamma \in \mathbb{T}$, then $1-\gamma \approx^\alpha 1-\overline{\gamma} = \overline{1-\gamma}$. So the point α must be in both small intervals with endpoints $\gamma, \overline{\gamma}$ and $1-\gamma, 1-\overline{\gamma}$, respectively, hence all four endpoints are $\approx^\alpha$-equivalent. Therefore, by Proposition 2.65 it follows $\overline{\gamma} = 1-\gamma$. (More than one pair of associated points in one equivalence class is impossible.)

Let c be the minimum of all parameters c^γ for periodic γ with $\gamma \approx^\alpha \overline{\gamma}$, hence $\overline{\gamma} = 1-\gamma$ (which are obviously real). Then $c \in Comb^\alpha$, such that $\approx^\alpha$ is the landing pattern for c. ∎

Let us finish with a remark on the common landing of rational dynamic rays. If α is rational, then J_c for $c = c^\alpha$ is locally connected and the landing pattern for c is completely described by Theorems 2.47 and 2.53. For $c = c^\alpha$ with $\alpha \in \mathbb{P}_\infty$ or $c \in Comb^\alpha$ with $\alpha \in \mathbb{T} \setminus \mathbb{P}$, where J_c need not be locally connected, Theorems 2.50 and 2.47 provide the following simple description of rational dynamic ray landing: For rational β_1, β_2 it holds

$$z_c^{\beta_1} = z_c^{\beta_2} \iff \text{The iterates of } \beta_1\beta_2 \text{ do not cross } \tfrac{\alpha}{2}\tfrac{\alpha+1}{2}.$$

4.2 Miscellaneous

4.2.1 Abstract Julia sets as shift-invariant factors

Theorems 2.48, 2.51 and 2.57 show that the elements of (Abstract) Julia sets can often be considered as classes of points whose itineraries are (substantially) equal to some 0-1-sequence.

This and the fact that with a few exceptions each 0-1-sequence appears as an itinerary (see Proposition 2.56) suggests a description of Abstract Julia sets as shift-invariant factors of $\{0,1\}^{\mathbb{N}}$, where the identification is induced by the 'wild card' symbol $*$. Besides the forward-shift σ, we consider the 'backward shifts' τ_0 and τ_1 defined by $\tau_i(s_1s_2s_3\ldots) = is_1s_2\ldots$ for $s_1s_2s_3\ldots \in \{0,1\}^{\mathbb{N}}$ and $i = 0,1$ (see Appendix A.2).

Actually, there are natural descriptions of $\mathbb{T}/\approx^\alpha$ for $\alpha \notin \mathbb{P}_\infty$, which only depend on the kneading sequence of α. The case $\alpha \in \mathbb{P}_\infty$ need not be considered since it allows that points of the same itinerary are not identified by $\approx^\alpha$.

Let us start with the case $\alpha \notin \mathbb{P}$, where the kneading sequence $\mathbf{s} = \hat{\alpha}$ does not contain the symbol $*$. In this case we have the following obvious assignment of $\approx^\alpha$-equivalence classes to 0-1-sequences induced by Theorem 2.48:

$$\begin{cases} \widetilde{\mathbb{T}^\alpha_{\mathbf{s}}} \mapsto \{\mathbf{s}\} & \text{for } \mathbf{s} \text{ not ending with } 0\hat{\alpha} \text{ or } 1\hat{\alpha}, \\ \widetilde{\mathbb{T}^\alpha_{\mathbf{w}0\hat{\alpha}}} \cup \widetilde{\mathbb{T}^\alpha_{\mathbf{w}1\hat{\alpha}}} \mapsto \{\mathbf{w}0\hat{\alpha}, \mathbf{w}1\hat{\alpha}\} & \text{for all } \mathbf{w} \in \{0,1\}^*. \end{cases} \tag{4.6}$$

So let $\equiv_{\mathbf{s}}$ be the equivalence relation on $\{0,1\}^{\mathbb{N}}$ whose non-trivial equivalence classes are of the form $\{\mathbf{w}0\mathbf{s}, \mathbf{w}1\mathbf{s}\}$ for some 0-1-word $\mathbf{w}$. Actually, $\equiv_{\mathbf{s}}$ is invariant on $(\{0,1\}^{\mathbb{N}}, \tau_0, \tau_1, \sigma)$. This is not hard to see, but we refer to Appendix A.2, in particular to Proposition A.9.

The map (4.6) forms a conjugacy of the topological dynamical systems $(\mathbb{T}, h)/\approx^{\alpha}$ and $(\{0,1\}^{\mathbb{N}}, \sigma)/\equiv_{\hat{\alpha}}$. Indeed, it is continuous since $\overset{\frown}{I^{\alpha}}(\beta) \neq \overset{\frown}{I^{\alpha}}(\beta)$ iff the orbit of β contains $\frac{\alpha}{2}$ or $\frac{\alpha+1}{2}$ and since in this case $\{\overset{\frown}{I^{\alpha}}(\beta), \overset{\frown}{I^{\alpha}}(\beta)\} = \{\mathbf{w}0\hat{\alpha}, \mathbf{w}1\hat{\alpha}\}$ for some $\mathbf{w} \in \{0,1\}^*$. As a bijection on a compact Hausdorff space the map forms a homeomorphism.

Example 4.37. The kneading sequence of $\frac{1}{2}$ is $0\overline{1}$. Thus the corresponding abstract Julia set is obtained from $\{0,1\}^{\mathbb{N}}$ by identifying $\mathbf{w}00\overline{1}$ and $\mathbf{w}10\overline{1}$ for each 0-1-word $\mathbf{w}$ (compare considerations below Definition 1.3). Note that the same description can be given for the Tent map (see Example A.7).

Let us consider the case $\alpha \in \mathbb{P}_*$. For this, let $e = e^{\alpha}, \mathbf{v} = \mathbf{v}^{\alpha}$. We deduce an equivalence relation on $\{0,1\}^{\mathbb{N}}$ by looking at the discontinuities of $I^{\alpha}(\cdot)$. $\overset{\frown}{I^{\alpha}}(\beta) \neq \overset{\frown}{I^{\alpha}}(\beta)$ holds exactly for those $\beta \in \mathbb{T}$ whose orbits contain $\dot{\alpha}$ or $\ddot{\alpha}$. For $\beta = l^{\alpha}_{\mathbf{w}}(\dot{\alpha})$ with α-regular $\mathbf{w} \in \{0,1\}^*$ it holds $\{\overset{\frown}{I^{\alpha}}(\beta), \overset{\frown}{I^{\alpha}}(\beta)\} = \{\mathbf{w}(1-e)\mathbf{v}, \mathbf{w}e\overline{\mathbf{v}}\}$, and for $\beta = l^{\alpha}_{\mathbf{w}}(\ddot{\alpha})$ with $\mathbf{w} \in \{0,1\}^*$ one has the equality $\{\overset{\frown}{I^{\alpha}}(\beta), \overset{\frown}{I^{\alpha}}(\beta)\} = \{\mathbf{w}e\overline{\mathbf{v}(1-e)}, \mathbf{w}(1-e)\overline{\mathbf{v}e}\}$.

This provides the equivalence relation $\equiv_{\overline{\mathbf{v}*}}$ on $\{0,1\}^{\mathbb{N}}$ determined by the following identifications:

$$\mathbf{w}(1-e)\overline{\mathbf{v}e} \equiv_{\overline{\mathbf{v}*}} \mathbf{w}e\overline{\mathbf{v}(1-e)} \text{ for all 0-1-words } \mathbf{w}, \text{ and} \tag{4.7}$$

$$\mathbf{w}e\overline{\mathbf{v}} \equiv_{\overline{\mathbf{v}*}} \mathbf{w}\overline{(1-e)\mathbf{v}} \text{ for all } \alpha\text{-regular 0-1-words } \mathbf{w}. \tag{4.8}$$

Note that in the case $\alpha = 0$ the equivalence relation $\equiv_{\overline{\mathbf{v}*}}$ is no more than $\equiv_0$, the equivalence relation defined by the binary expansion.

One easily sees that the two sets $\{(\mathbf{w}(1-e)\overline{\mathbf{v}e}, \mathbf{w}e\overline{\mathbf{v}(1-e)}) \mid \mathbf{w} \in \{0,1\}^*\}$, $\{(\mathbf{w}e\overline{\mathbf{v}}, \mathbf{w}(1-e)\mathbf{v}) \mid \mathbf{w} \in \{0,1\}^*, \mathbf{w}\ \alpha\text{-regular}\}$ can only accumulate at the diagonal of the product space $\{0,1\}^{\mathbb{N}} \times \{0,1\}^{\mathbb{N}}$, which implies that $\equiv_{\overline{\mathbf{v}*}}$ is closed as a subset of $\{0,1\}^{\mathbb{N}} \times \{0,1\}^{\mathbb{N}}$.

By Theorem 2.57, the assignments

$$\begin{cases} \mathbb{T}^{\alpha}_{\mathbf{s}} \mapsto \{\mathbf{s}\} & \text{for } \mathbf{s} \in \{0,1\}^{\mathbb{N}} \text{ not ending} \\ & \text{with } \overline{\mathbf{v}e} \text{ or } \overline{\mathbf{v}(1-e)}, \\ \mathbb{T}^{\alpha}_{\mathbf{w}e\overline{\mathbf{v}(1-e)}} \mapsto \{\mathbf{w}e\overline{\mathbf{v}(1-e)}, \mathbf{w}(1-e)\overline{\mathbf{v}e}\} & \text{for all } \mathbf{w} \in \{0,1\}^*, \\ \mathbb{T}^{\alpha}_{\mathbf{w}\overline{(1-e)\mathbf{v}}} \mapsto \{\mathbf{w}\overline{(1-e)\mathbf{v}}, \mathbf{w}e\overline{\mathbf{v}}\} & \text{for all } \alpha\text{-regular } \mathbf{w} \in \{0,1\}^* \end{cases}$$

yield a bijective map from $\mathbb{T}/\approx^{\alpha}$ onto $\{0,1\}^{\mathbb{N}}/\equiv_{\hat{\alpha}}$. One obtains continuity of that map in a similar way as for $\alpha \in \mathbb{T} \setminus \mathbb{P}$. Namely, $\equiv_{\overline{\mathbf{v}*}}$ was just defined as in (4.7) and (4.8) to remove discontinuities. So let us summarize:

Theorem 4.38. ($\mathbb{T}/\approx^\alpha; \alpha \in \mathbb{T} \setminus \mathbb{P}_\infty$ as shift-invariant factor)
For each $\alpha \in \mathbb{T} \setminus \mathbb{P}_\infty$, the two topological dynamical systems $(\mathbb{T}, h)/\approx^\alpha$ and $(\{0,1\}^{\mathbb{N}}, \sigma)/\equiv_{\hat{\alpha}}$ are conjugate via the map $[\beta]_{\approx^\alpha} \mapsto [\widetilde{I^\alpha}(\beta)]_{\equiv_{\hat{\alpha}}}; \beta \in \mathbb{T}$. ■

Remarks. 1. C. PENROSE [123, 124] has worked out an interesting theory of 'glueing spaces', which contains the invariant factor $(\{0,1\}^{\mathbb{N}}, \sigma)/\equiv_{\mathbf{s}}$ for non-periodic 0-1-sequences as the main case and as a tool for describing quadratic Julia sets. In particular, the statement of Theorem 4.38 for $\alpha \in \mathbb{T} \setminus \mathbb{P}$ and the fact that $(\{0,1\}^{\mathbb{N}}, \sigma)/\equiv_{\mathbf{s}}$ is a dendrite for non-periodic $\mathbf{s}$ can be found in [123, 124]. (For the latter, compare also [5]).

PENROSE has shown that many of the statements which we gave for Abstract Julia sets $\mathbb{T}/\approx^\alpha; \alpha \in \mathbb{T} \setminus \mathbb{P}$ are valid for the more general spaces $(\{0,1\}^{\mathbb{N}}, \sigma)/\equiv_{\mathbf{s}}$ with non-periodic $\mathbf{s}$. For example, a remarkable statement (see [124], Theorem 2.4) says roughly that points of ramification order greater than two must be periodic or preperiodic, or must have some iterate equal to $\mathbf{s}$ (compare Corollary 2.20).

Moreover, PENROSE considered a space of 0-1-sequences which is more general than the Abstract Mandelbrot set. He investigated bifurcation and renormalization there, and as already mentioned his 'principal non-periodic function' is a generalized version of an internal address.

2. KAMEYAMA [74] studied 'invertible postcritically finite self-similar sets' and used THURSTONS's characterizations of rational maps to relate them to connected Julia sets of polynomials. In the terminology of Appendix A.2 those 'sets' are invariant factors of $(\{1,2,\ldots,k\}^\infty, \sigma, \tau_1, \tau_2, \ldots, \tau_k)$ with respect to some equivalence relation $\equiv$. (Here $\tau_i; i = 1,2,\ldots,k$ denotes the map defined by $\tau_i(s_1 s_2 s_3 \ldots) = i s_1 s_2 \ldots$ for $s_1 s_2 s_3 \ldots \in \{1,2,\ldots,k\}^{\mathbb{N}}$.)

$\equiv$ satisfies the following properties: Two sets $\phi_\equiv(\tau_i(\{1,2,\ldots,k\}^\infty))$ and $\phi_\equiv(\tau_j(\{1,2,\ldots,k\}^\infty))$ for $i \neq j$ overlap each other in at most finitely many points, which must be preperiodic with respect to $\sigma/\equiv$.

KAMEYAMA has shown that each simple connected invertible postcritically finite self-similar set is homeomorphic to the Julia set of a polynomial. His work could be interesting in view of the question of which kneading sequences can be realized by quadratic dynamics.

4.2.2 Conjugate (Abstract) Julia sets

It is possible that different Julia equivalences $\approx^\alpha$ and $\approx^\gamma$ provide Abstract Julia sets $(\mathbb{T}, h, ')/\approx^\alpha$ and $(\mathbb{T}, h, ')/\approx^\gamma$ being conjugate.

For example, a simple conjugacy is induced by the symmetry $\beta \mapsto 1 - \beta \bmod 1$ on $(\mathbb{T}, h)$: for all $\alpha, \beta_1, \beta_2 \in \mathbb{T}$ it holds $\beta_1 \approx^\alpha \beta_2$ iff $1-\beta_1 \approx^{1-\alpha} 1-\beta_2$, and the conjugacy obtained is no more than the complex conjugacy on a concrete Julia set. Other examples can be constructed by use of Theorem 4.38. Our aim is to give a classification of conjugate Abstract Julia sets. In preparation for this, let us give two simple facts.

At first, for an equivalence class E of a Julia equivalence it holds $E = E'$ or $E \cap E' = \emptyset$, and $h(E) = h(E')$. This reduces conjugacy of $(\mathbb{T}, h, ')/\approx^\alpha$ and $(\mathbb{T}, h, ')/\approx^\gamma$ for some $\alpha, \gamma \in \mathbb{T}$ to conjugacy of $(\mathbb{T}, h)/\approx^\alpha$ and $(\mathbb{T}, h)/\approx^\gamma$.

Furthermore, for periodic $\delta, \eta \in \mathbb{T}$ with conjugate Abstract Julia sets $\mathbb{T}/\approx^\delta$ and $\mathbb{T}/\approx^\eta$ we establish a relation between Julia equivalences containing $\approx^\delta$ and $\approx^\eta$, respectively, which provides conjugacy.

We want to use the following notations for special equivalence relations constructed from two other ones (see Appendix A.1, V.): If $\equiv_1$ is an equivalence relation on a set X and $\equiv_2$ is one on $X/\equiv_1$, then $\equiv_2 \circ \equiv_1$ means that equivalence relation $\equiv$ on X defined by $x_1 \equiv x_2 :\Longleftrightarrow [x_1]_{\equiv_1} \equiv_2 [x_2]_{\equiv_1}$. As in Chapter 3.3.3, for two equivalence relations $\equiv_1$ and $\equiv_2$ on some X with $\equiv_1 \subseteq \equiv_2$, by $\equiv_2 / \equiv_1$ is meant the equivalence relation on $X/\equiv_1$ identifying $[x_1]_{\equiv_1}$ and $[x_2]_{\equiv_1}$ if $x_1 \equiv_2 x_2$.

Lemma 4.39. (Lifting of a conjugacy)
Let $\delta, \eta \in \mathbb{P}_*$, and let $\approx$ be a Julia equivalence with $\approx^\delta \subset \approx$. Further, let θ be a conjugacy from $(\mathbb{T}, h, ')/\approx^\delta$ onto $(\mathbb{T}, h, ')/\approx^\eta$. Then $\widetilde{\approx}$ defined by

$$\beta_1 \,\widetilde{\approx}\, \beta_2 \Longleftrightarrow \theta^{-1}([\beta_1]_{\approx^\eta}) \approx / \approx^\delta \theta^{-1}([\beta_2]_{\approx^\eta})$$

for all $\beta_1, \beta_2 \in \mathbb{T}$, is a Julia equivalence. Moreover, $\approx^\eta \subset \widetilde{\approx}$, and $[E]_\approx \mapsto [\theta(E)]_{\widetilde{\approx}}; E \in \mathbb{T}/\approx^\delta$ defines a conjugacy from $(\mathbb{T}, h, ')/\approx$ onto $(\mathbb{T}, h, ')/\widetilde{\approx}$.

Proof: On the assumptions of the lemma we can identify $(\mathbb{T}, h, ')/\approx^\delta$ with $(\mathbb{T}, h, ')/\approx^\eta$ and $\widetilde{\approx}$ with $(\approx / \approx^\delta) \circ \approx^\eta$. So by the Propositions A.4 and A.5 $\widetilde{\approx}$ is a completely invariant equivalence relation on $(\mathbb{T}, h, ')$ whose equivalence classes are finite. It remains to be shown that $\widetilde{\approx}$ is planar.

Let $\beta_1 \widetilde{\approx} \beta_2$, and let $\gamma_1 \approx \gamma_2$ be given such that $\theta([\gamma_1]_\approx) = [\beta_1]_{\widetilde{\approx}}$ and $\theta([\gamma_2]_\approx) = [\beta_2]_{\widetilde{\approx}}$. Then there exists a shortest sequence $\alpha_0 = \gamma_1, \alpha_1, \alpha_2, \ldots, \alpha_n = \gamma_2$ such that $\alpha_i \approx \alpha_{i-1}$ for all $i = 1, 2, \ldots, n$ and in each case either $\alpha_i \approx \alpha_{i-1}$, or $\alpha_i \not\approx \alpha_{i-1}$ and the chord $\alpha_i\alpha_{i-1}$ is completely contained in an infinite gap of $\eth\mathcal{B}^\delta$. (Otherwise, $\approx$ would have infinite equivalence classes.)

Therefore, since θ transforms each simple closed curve of $(\mathbb{T}, h, ')/\approx^\delta$ into a simple closed curve in $(\mathbb{T}, h, ')/\approx^\eta$, for each $i = 0, 1, 2, \ldots, n$ we find a point $\widetilde{\alpha_i} \in \theta([\alpha_i]_\approx)$ such that the following is satisfied: It does not hold $\widetilde{\alpha_i} \,\widetilde{\approx}\, \widetilde{\alpha_{i-1}}$, and $\widetilde{\alpha_i}\widetilde{\alpha_{i-1}}$ is contained in an infinite gap of $\eth\mathcal{B}^\eta$ iff $\alpha_i \not\approx \alpha_{i-1}$ and $\alpha_i\alpha_{i-1}$ lies in an infinite gap of $\eth\mathcal{B}^\delta$.

Using this and again the fact that δ preserves simple closed curves, the planarity of $\widetilde{\approx}$ is easily shown. ■

Now let us answer the question when two Abstract Julia sets are conjugate (compare (3.10) and Theorem 3.34).

Theorem 4.40. (Conjugacy of Abstract Julia sets)
For $\alpha, \gamma \in \mathbb{T}$ the Abstract Julia sets $(\mathbb{T}, h, ')/\approx^\alpha$ and $(\mathbb{T}, h, ')/\approx^\gamma$ are conjugate iff one of the following conditions is satisfied:

(i) It holds $\alpha, \gamma \notin \mathbb{P}_\infty$ *and* $\hat{\alpha} = \hat{\gamma}$,

(ii) it holds $\alpha, \gamma \in \mathbb{P}_\infty, \hat{\alpha} = \hat{\gamma}$ *and* $\mathbf{a}(\alpha) = \mathbf{a}(\gamma)$.

Proof: First of all, note that conjugacy of $(\mathbb{T}, h, ')/\approx^\alpha$ and $(\mathbb{T}, h, ')/\approx^\gamma$ implies either $\alpha, \gamma \in \mathbb{T} \setminus \mathbb{P}$ or $\alpha, \gamma \in \mathbb{P}_*$ or $\alpha, \gamma \in \mathbb{P}_\infty$: only for $\delta \in \mathbb{T} \setminus \mathbb{P}$ does the Abstract Julia set $\mathbb{T}/\approx^\delta$ contain a simple closed curve and only for $\delta \in \mathbb{P}_*$ is there a periodic point on some simple closed curve (see Proposition 2.66).

Now let θ be a conjugacy from $(\mathbb{T}, h, ')/\approx^\alpha$ onto $(\mathbb{T}, h, ')/\approx^\gamma$ and let ϕ_α and ϕ_γ be the projections from $\mathbb{T}$ onto $\mathbb{T}/\approx^\alpha$ and $\mathbb{T}/\approx^\gamma$, respectively.

For $\delta \in \mathbb{P}_*$, the set $[\mathbb{T}\langle\delta\rangle]_{\approx^\delta}/\approx^\delta$ is the image of the only $'/\approx^\delta$-invariant simple closed curve and $[\delta]_{\approx^\delta}$ the only periodic point on this curve. Therefore, θ transforms $[\mathbb{T}\langle\alpha\rangle]_{\approx^\alpha}/\approx^\alpha$ into $[\mathbb{T}\langle\gamma\rangle]_{\approx^\gamma}/\approx^\gamma$ and $[\alpha]_{\approx^\alpha}$ into $[\gamma]_{\approx^\gamma}$ if $\alpha(,\gamma) \in \mathbb{P}_*$. Clearly, then $\theta([\dot{\alpha}]_{\approx^\alpha}) = [\dot{\gamma}]_{\approx^\gamma}$ and $\theta([\ddot{\alpha}]_{\approx^\alpha}) = [\ddot{\gamma}]_{\approx^\gamma}$.

Since for $\delta \notin \mathbb{P}_*$ the point $[\delta]_{\approx^\delta}$ is the only one with exactly one preimage, it follows $\theta([\alpha]_{\approx^\alpha}) = [\gamma]_{\approx^\gamma}$ for $\alpha \notin \mathbb{P}_*$.

Finally, for $\delta \in \mathbb{P}_\infty$ the equivalence class $[\delta]_{\approx^\delta}$ consists of only one point, hence $\mathbb{P}^\delta_\infty/\approx^\delta$ is the only simple closed curve containing $[\delta]_{\approx^\delta}$. Consequently, if $\alpha(,\gamma) \in \mathbb{P}_\infty$, then θ transforms the simple closed curve $\mathbb{P}^\alpha_\infty/\approx^\alpha$ into the simple closed curve $\mathbb{P}^\gamma_\infty/\approx^\gamma$.

If $\delta \in \mathbb{T} \setminus \mathbb{P}_*$, then $I^\eta(\delta) = \widehat{\eta} = \widehat{\delta}$ for each element η of $[\delta]_{\approx^\delta}$ (see Theorem 3.1 and (3.2)). This is only possible if no iterate of δ lies in $]\frac{\delta_{min}}{2}, \frac{\delta_{max}}{2}[$ or $]\frac{\delta_{min}+1}{2}, \frac{\delta_{max}+1}{2}[$, where δ_{min} and δ_{max} denote the least and the greatest element of $[\delta]_{\approx^\delta}$, respectively. For $\delta \in \mathbb{P}_* \setminus \{0\}$ with $\delta < \overline{\delta}$ the orbit of δ is disjoint to the intervals $]\frac{\delta}{2}, \frac{\overline{\delta}}{2}[$ and $]\frac{\delta+1}{2}, \frac{\overline{\delta}+1}{2}[$ (see Proposition 2.62 and Lemma 2.60).

Now one easily shows the following: it holds $\theta(\phi_\alpha([\frac{\alpha_{max}}{2}, \frac{\alpha_{min}+1}{2}])) = \phi_\gamma([\frac{\gamma_{max}}{2}, \frac{\gamma_{min}+1}{2}])$ and $\theta(\phi_\alpha([\frac{\alpha_{max}+1}{2}, \frac{\alpha_{min}}{2}])) = \phi_\gamma([\frac{\gamma_{max}+1}{2}, \frac{\gamma_{min}}{2}])$ for $\alpha \in \mathbb{T}\setminus\mathbb{P}_*$, and $\theta(\phi_\alpha([\frac{\overline{\alpha}}{2}, \frac{\alpha+1}{2}])) = \phi_\gamma([\frac{\overline{\gamma}}{2}, \frac{\gamma+1}{2}])$ and $\theta(\phi_\alpha([\frac{\overline{\alpha}+1}{2}, \frac{\alpha}{2}])) = \phi_\gamma([\frac{\overline{\gamma}+1}{2}, \frac{\gamma}{2}])$ for $\alpha \in \mathbb{P}_*$, where we assume that $\alpha < \overline{\alpha}, \gamma < \overline{\gamma}$. This implies $\hat{\alpha} \cong \hat{\gamma}$.

For $\alpha \in \mathbb{T} \setminus \mathbb{P}_*$ this means no more than $\hat{\alpha} = \hat{\gamma}$. If $\alpha \in \mathbb{P}_*$, the ramification order of the points $[\alpha]_{\approx^\alpha}$ and $[\gamma]_{\approx^\gamma}$ in the Abstract Julia sets must be equal, hence it holds $CARD([\alpha]_{\approx^\alpha}) = CARD([\gamma]_{\approx^\gamma})$.

Further, by Theorem 2.44, Corollary 2.23 and Lemma 2.54 we obtain the following: If $\delta \in \mathbb{P}_* \setminus \{0\}$ is non-primitive, then $[\delta]_{\approx^\delta} = \mathbb{T}\frac{\delta}{\mathbf{v}^\delta(1-e^\delta)}$ is contained in the orbit of δ, and $[\delta]_{\approx^\delta}$ belongs to at least two different simple closed curves in $\mathbb{T}/\approx^\delta$. Otherwise, the period of $[\delta]_{\approx^\delta}$ with respect to $h/\approx^\delta$ coincides with $PER(\delta)$. In the first case, $PER(\delta)$ is the product of the ramification order of the point $[\delta]_{\approx^\delta}$ and its period in the Abstract Julia set. From this $\hat{\alpha} = \hat{\gamma}$ is easy to see.

For $\alpha, \gamma \notin \mathbb{P}_\infty$, conjugacy of $(\mathbb{T}, h, ')/\approx^\alpha$ and $(\mathbb{T}, h, ')/\approx^\gamma$ is implicated by $\hat{\alpha} = \hat{\gamma}$, which is an immediate consequence of Theorem 4.38. Since the

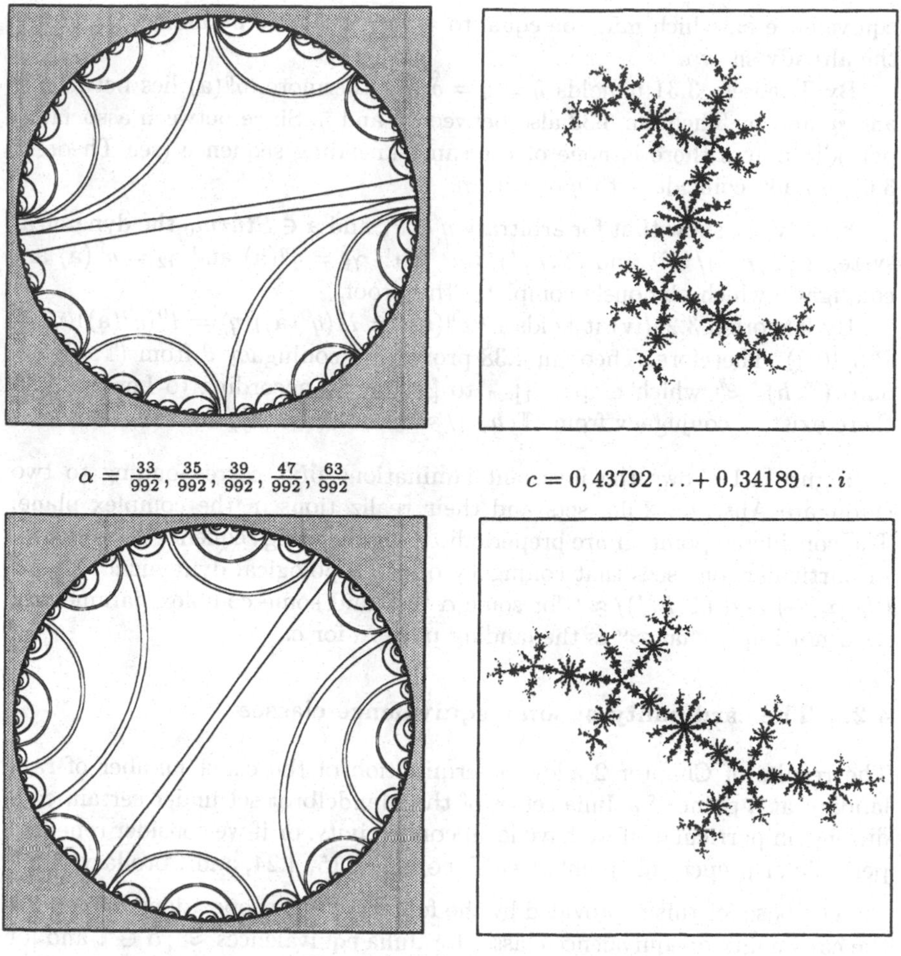

$\alpha = \frac{33}{992}, \frac{35}{992}, \frac{39}{992}, \frac{47}{992}, \frac{63}{992}$ $\qquad c = 0,43792\ldots + 0,34189\ldots i$

$\alpha = \frac{289}{992}, \frac{297}{992}, \frac{299}{992}, \frac{315}{992}, \frac{319}{992}$ $\qquad c = -0,56220\ldots + 0,64281\ldots i$

Fig. 4.1. conjugate Abstract Julia sets

1-frequencies of different non-periodic sequences in $Sturm_0$ are different (see Lemma 3.37), according to Theorem 3.36 the following remains to be shown: For $\alpha, \gamma \in \mathbb{P}_\infty$ the conditions $\hat{\alpha} = \hat{\gamma}$ and $\mathbf{a}(\alpha) = \mathbf{a}(\gamma)$ imply conjugacy of $(\mathbb{T}, h, ')/\approx^\alpha$ and $(\mathbb{T}, h, ')/\approx^\gamma$.

Let $\delta, \eta \in \mathbb{P}_*, \alpha, \gamma \in \mathbb{P}_\infty$, and let some $\mathbf{a} \in Sturm_0$ with $\alpha = q^\delta(\mathbf{a}), \gamma = q^\eta(\mathbf{a})$ and $\hat{\alpha} = \hat{\gamma}$ be given. According to the discussion below (3.10) we obtain $\hat{\delta} = \hat{\eta}$, and by the already shown there exists a conjugacy θ from $(\mathbb{T}, h, ')/\approx^\delta$ onto $(\mathbb{T}, h, ')/\approx^\eta$. We apply Lemma 4.39 with $\approx = \approx^\alpha$ and get the Julia equivalence $\tilde{\approx}$, which must be equal to $\approx^{q^{\tilde{\eta}}(\mathbf{a})}$ for some periodic $\tilde{\eta}$, again by the already shown.

By Theorem 3.31 it holds $\hat{\eta} = \hat{\bar{\eta}} = \hat{\delta}$. Furthermore, $q^{\tilde{\eta}}(\mathbf{a})$ lies between $\tilde{\eta}$ and $\overline{\tilde{\eta}}$, and by Theorem 3.33 also between η and $\overline{\eta}$. Since between associated periodic points there is none of the same kneading sequence (see Theorem 3.6), $\tilde{\eta}$ must coincide with η or with $\overline{\eta}$.

Finally, we show that for arbitrary $\eta \in \mathbb{P}_*$ and $\mathbf{a} \in Sturm_0$ the dynamical systems $(\mathbb{T}, h, ')/\approx^{\gamma_1}$ and $(\mathbb{T}, h, ')/\approx^{\gamma_2}$ with $\gamma_1 = q^{\eta}(\mathbf{a})$ and $\gamma_2 = q^{\overline{\eta}}(\mathbf{a})$ are conjugate, which obviously completes the proof.

By Theorem 3.26(iv) it holds $I^{\eta}(q^{\eta}(\mathbf{a})) = I^0(q^0(\mathbf{a}))\langle\eta\rangle = I^0(q^0(\mathbf{a}))\langle\overline{\eta}\rangle = I^{\overline{\eta}}(q^{\overline{\eta}}(\mathbf{a}))$. Therefore, Theorem 4.38 provides a conjugacy θ from $(\mathbb{T}, h)/\approx^{\eta}$ onto $(\mathbb{T}, h)/\approx^{\overline{\eta}}$ which maps $[\gamma_1]_{\approx^{\eta}}$ to $[\gamma_2]_{\approx^{\overline{\eta}}}$. So according to Lemma 4.39 there exists a conjugacy from $(\mathbb{T}, h, ')/\approx^{\gamma_1}$ onto $(\mathbb{T}, h, ')/\approx^{\gamma_2}$. ■

Figure 4.1 shows the invariant laminations $\eth\mathcal{B}^{\alpha}$ corresponding to two conjugate Abstract Julia sets and their realizations in the complex plane. The considered points α are preperiodic with kneading sequence $\hat{\alpha} = 00001\overline{0}$. In particular, one sees that conjugacy of the topological dynamical systems $(J_c, p_c, -)$ and $(\mathbb{T}, h, ')/\approx^{\alpha}$ for some $\alpha \in \mathbb{T}$ and some complex parameter c need not imply that $\approx^{\alpha}$ is the landing pattern for c.

4.2.3 The cardinality of some equivalence classes

The results of Chapter 2 allow determination of the exact number of rays landing at a point of a Julia set or of the Mandelbrot set under certain conditions; in particular, if we have local connectivity, or if we consider repelling periodic or preperiodic points (see Theorems 1.15, 4.24, and Corollary 4.28).

The base for this is provided by the following two propositions concerning the cardinality of equivalence classes for Julia equivalences $\approx^{\alpha}; \alpha \in \mathbb{T}$ and for the equivalence relation $\sim$. The first proposition is a consequence of Theorems 2.48, 2.51, 2.57, Corollary 2.23 and Proposition 2.65.

Proposition 4.41. (Cardinality of equivalence classes of $\approx^{\alpha}; \alpha \in \mathbb{T}$)
For $\alpha, \beta \in \mathbb{T}$ the following statements are mutually exclusive:

(i) $\sigma^n(I^{\alpha}(\beta)) \not\cong \hat{\alpha}$ *for all* $n \in \mathbb{N}$,
(ii) $\sigma^n(I^{\alpha}(\beta)) \cong \hat{\alpha}$ *for some* $n \in \mathbb{N}$ *and* $\alpha \notin \mathbb{P}$,
(iii) $h^n(\beta) = \alpha$ *for some* $n \in \mathbb{N}$ *and* $\alpha \in \mathbb{P}_{\infty}$,
(iv) $h^n(\beta) = \alpha$ *for some* $n \in \mathbb{N}$, $\alpha \in \mathbb{P}_*$, *and* $h^k(\alpha) \neq \overline{\alpha}$ *for all* $k \in \mathbb{N}$,
(v) $h^n(\beta) = \alpha$ *for all* $n \in \mathbb{N}$, $\alpha \in \mathbb{P}_*$, *and* $h^k(\alpha) = \overline{\alpha}$ *for some* $k \in \mathbb{N}$.

Moreover, each equivalence class $[\beta]_{\approx^{\alpha}}$ is finite and its cardinality is given as follows:

$$CARD([\beta]_{\approx^\alpha}) = \begin{cases} CARD^\alpha(I^\alpha(\beta)) & \text{if } (i) \\ 2CARD^\alpha(I^\alpha(\beta)) & \text{if } (ii) \\ 2 & \text{if } (iii) \text{ or } (iv) \\ PER(\alpha)/PER(\overline{\mathbf{v}^\alpha(1-e^\alpha)}) & \text{if } (v) \\ 1 & \text{else} \end{cases} .$$ ■

Now the second statement follows immediately from Theorem 3.19.

Proposition 4.42. (Cardinality of equivalence classes of $\sim$)
For $\alpha \in \mathbb{T}$ the following statements are valid

(i) If $\alpha \in \mathbb{P}_\infty$ or $\alpha = 0$, then $CARD([\alpha]_\sim) = 1$.
(ii) If $\alpha \in \mathbb{P}_* \setminus \{0\}$, then $CARD([\alpha]_\sim) = 2$.
(iii) If $\alpha \in \mathbb{T} \setminus \mathbb{P}$, then $CARD([\alpha]_\sim) = CARD^\alpha(\hat{\alpha})$, and $CARD([\alpha]_\sim) > 2$ implies that α is preperiodic. ■

In view of Proposition 4.42 the question whether there are equivalence classes of every given finite cardinality arises. The answer is positive, as the following example shows.

Example 4.43. For $n = 2, 3, 4, \dots$ let $\alpha_n = \frac{2^n+1}{(2^n-1)2^n}$ and $\gamma_n = \frac{1}{2^n-1}$. By use of formula (2.7), one easily checks that $\overline{\gamma_n} = \frac{2}{2^n-1}$ and that in each case α_n is equal to the point $Q^{\alpha_n}(\frac{1}{2}) = q^{\alpha_n}(0\overline{1})$ obtained from $\frac{1}{2}$ by tuning. (Note that the only point being $\sim$-equivalent to $\frac{1}{2} = q^0(0\overline{1})$ is $\frac{1}{2}$ itself.)

According to Proposition 3.14(iii), it holds $e^{\alpha_n} = 1$ for all $n = 2, 3, 4, \dots$, and by $\widehat{\gamma_n} = \overline{0^{n-1}*}$ and Theorem 3.31(i) it follows $\widehat{\alpha_n} = 0^{n-1}1\overline{0}$. So the α_n structure graph of $\widehat{\alpha_n}$ has n components, which by Theorem 2.19 means $CARD([\alpha]_\sim) = n$. ■

We want to conclude with two statements saying how large the sets $\mathbb{T}\langle\alpha\rangle$ for $\alpha \in \mathbb{P}_*$ and the set of points with some periodic or preperiodic itinerary are. We do not want to give the proofs here and refer to [78].

Let $\alpha \in \mathbb{T}$ be fixed. If the itinerary of a point $\beta \in \mathbb{T}$ is periodic or preperiodic, then by Corollary 2.23 either β is periodic or preperiodic itself, or $\alpha \in \mathbb{P}_\infty$ and $h^n(\gamma) \in \mathbb{P}^\alpha_\infty$ for some $n \in \mathbb{N}$. In the second case, $h^n(\gamma) \in \bigcup_{\alpha\in\mathbb{P}_*} q^\alpha(Sturm)$ by Theorem 3.36.

The union of all orbits containing elements of this set has Hausdorff dimension 0 (see [78], Theorem 3). Since the set of periodic or preperiodic points in $\mathbb{T}$, in other words, the set of rationals in $\mathbb{T}$, is countable, we can conclude as follows:

Proposition 4.44. *The set of all points $\beta \in \mathbb{T}$ with periodic or preperiodic itinerary $I^\alpha(\beta)$ for some $\alpha \in \mathbb{T}$ has Hausdorff dimension 0.* ■

Note that by Theorem 2.4 the set of points with kneading sequence $\overline{0}$ is uncountable, and so that Proposition 4.44 is non-trivial.

For $\alpha \in \mathbb{P}_* \setminus \{0\}$ and $m = PER(\alpha)$, the set $\mathbb{T}\langle\alpha\rangle$ is self-similar with respect to the two linear contractive maps given as follows: both maps have domain

$[\alpha, \overline{\alpha}]$ - we assume $\alpha < \overline{\alpha}$ - and contraction factor 2^{-m} and the fixed points are α and $\overline{\alpha}$. (The concept of a self-similar set is given in Appendix A.2.)

Therefore, by HUTCHINSON's results in ([68]), one gets the following statement (see [78], Proposition 3):

Proposition 4.45. *Let $\alpha \in \mathbb{P}_* \setminus \{0\}$ and let $m = PER(\alpha)$. Then the set $\mathbb{T}\langle\alpha\rangle$ has Hausdorff dimension $\frac{1}{m}$. Moreover, the $\frac{1}{m}$-dimensional Hausdorff measure of $\mathbb{T}\langle\alpha\rangle$ is finite and strictly positive.* ∎

A. Appendix: Invariant and completely invariant factors

In the following we discuss invariant and completely invariant equivalence relations, which we have introduced in Definition 1.2, from the conceptional viewpoint. We illustrate our discussion by a lot of examples.

A.1 Simple statements

I. Additional remarks to complete invariance. For an f-invariant equivalence relation $\equiv$ on some topological space X one easily shows equivalence of complete invariance to both of the following two statements:

(i) $(f/\equiv)^{-1}([y]_\equiv) = \{[x]_\equiv \mid x \in f^{-1}(y)\}$ for all $y \in X$,

(ii) for all $x, y \in X$ with $f(x) \equiv y$ there exists an $\tilde{x}$ satisfying $x \equiv \tilde{x}$ and $f(\tilde{x}) = y$.

Remark. The concept of an invariant equivalence relation is not new. In topological dynamics it is mostly used in connection with a topological group (of homeomorphisms) acting on a topological space (e.g., see [50, 57, 3, 168]).

It is rather easy to construct examples of invariant but not completely invariant equivalence relations. Here we give a simple one, where the map considered is a homeomorphism.

Example A.1. Consider the discrete topological dynamical system $(\mathbb{Z}, f)$ where $\mathbb{Z}$ is the set of integers and $f(z) = z - 1$ for all $z \in \mathbb{Z}$. Let $\equiv$ be the equivalence relation whose only non-trivial equivalence class consists of all negative numbers. Then $\equiv$ is invariant, but it fails to be completely invariant since the images of negative numbers do not form an equivalence class.

Remark. If f is a continuous map on a topological space X whose m-th iterate is equal to the identity for some m, then each f-invariant equivalence relation is completely invariant. This results as follows: If $f(E) \subseteq F$ for some equivalence classes E and F, then $E \subseteq f^{m-1}(F)$. In case that the equivalence relation is invariant, this implies $E = f^{m-1}(F)$, which is only possible if $F = f(E)$.

In particular, one sees that for finite X and injective f the concepts 'f-invariant' and 'completely f-invariant' coincide.

II. Closeness of an equivalence relation. Our definition of an invariant equivalence relation $\equiv$ includes that $\equiv$ is closed as a subset of the product space $X \times X$. This is equivalent to continuity of the underlying identification: if $(x_i)_{i\in I}$ and $(y_i)_{i\in I}$ are two Moore-Smith sequences in X converging to x and y and if $x_i \equiv y_i$ for all $i \in I$, then $x \equiv y$.

Moreover, $\equiv$ is closed as a subset of $X \times X$ if the quotient topology on $X/\equiv$ is a Hausdorff one (see [51], 2.4.C.).

The statement that $\equiv$ is closed as a subset of $X \times X$ is strongly related to the usual concept of a closed equivalence relation: $\equiv$ is said to be *closed* if the projection $\phi_\equiv$ is closed, i.e. if it maps closed sets to closed ones. (This is valid if and only if $Hull_\equiv(A)$ is closed for every closed subset A of X.)

If X is normal and $\equiv$ is closed in this sense, then the quotient topology on $X/\equiv$ becomes Hausdorff (see [51]), hence $\equiv$ is closed in $X \times X$. Note that in the case of a compact Hausdorff space closeness of $\equiv$, closeness of $\equiv$ as a subset of $X \times X$ and the statement that the quotient topology is a Hausdorff one coincide (see [142], 22.15.).

III. Semi-conjugacies and invariant equivalence relations. By definition, the projection $\phi_\equiv$ for some invariant equivalence relation $\equiv$ is a semi-conjugacy from a given topological dynamical system onto its factor.

Conversely, a semi-conjugacy ϕ from a topological dynamical system $(X,(f_i)_{i\in I})$ onto a topological dynamical system $(Y,(g_i)_{i\in I})$ induces an equivalence relation $\equiv$ on X by $x_1 \equiv x_2 :\Longleftrightarrow \phi(x_1) = \phi(x_2)$ with the property that all f_i map equivalence classes into equivalence classes.

Clearly, $(Y,(g_i)_{i\in I})$ is semi-conjugate to $(X,(f_i)_{i\in I})/\equiv$ by virtue of the map $y \in Y \mapsto \phi^{-1}(y) \in X/\equiv$. Although this map is bijective, it need not be a conjugacy since the quotient topology with respect to $\equiv$ is possibly stronger than the original topology on Y.

A conjugacy is given iff ϕ forms a quotient map (i.e. iff in Y the sets with open preimages are open), in particular if ϕ is closed. By this and the considerations in II. we have the following important fact:

If X is a compact Hausdorff space and Y is a (compact) Hausdorff one, then the induced equivalence relation $\equiv$ is invariant on $(X,(f_i)_{i\in I})$, and $(Y,(g_i)_{i\in I})$ can be identified with $(X,(f_i)_{i\in I})/\equiv$.

IV. Invariant equivalence relations with finite equivalence classes. Now let $\equiv$ be an invariant equivalence relation on a topological dynamical system (X, f), and assume that all equivalence classes are finite. If $\equiv$ identifies a point $x \in X$ with $f^n(x)$ for some $n \in \mathbb{N}$, it holds $x \equiv f^{kn}(x)$ for all $k \in \mathbb{N}$. So the orbit of x must be periodic or preperiodic.

In a similar way one sees that a periodic or preperiodic point can only be equivalent to a periodic or preperiodic one. This can be strengthened in the case of a completely invariant equivalence relation.

Proposition A.2. (Completely invariant equivalence relations with finite equivalence classes)
On a topological dynamical system (X, f) let $\equiv$ be a completely invariant equivalence relation with only finite equivalence classes. Then $x_1 \equiv x_2$ for $x_1, x_2 \in X$ implies exactly one of the following statements:

(i) x_1 and x_2 are periodic.

(ii) x_1 and x_2 are preperiodic.

(iii) Both x_1 and x_2 are neither periodic nor preperiodic.

In particular, $x \in X$ is periodic if there exists an $m \in \mathbb{N}$ with $x \equiv f^m(x)$.

Proof: Assume that x_0 is a point with $f^m(x_0) = x_0$ for some $m \in \mathbb{N}$, and that x_1 is a non-periodic point satisfying $x_0 \equiv x_1$. Then for each $k = 2, 3, 4, \ldots$ there exists an $x_k \in X$ with $f^{(k-1)m}(x_k) = x_1$ and $x_k \equiv x_0$ (see I., condition (ii)). Since x_1 is non-periodic, the points x_k for $k \in \mathbb{N}$ are mutually different. Consequently, $[x_0]_\equiv$ must be an infinite set, in contradiction to the assumption of our proposition. Therefore, a point equivalent to a periodic one must be periodic itself.

The corresponding statement for the preperiodic case can be obtained by 'iterating' the preperiodic case to the periodic one. ■

Note that invariant equivalence relations with finite equivalence classes (being not completely invariant) allow the identification of periodic points with preperiodic ones:

Example A.3. On $\mathbb{N}_0$ with the discrete topology consider the map f given by $f(0) = 0$ and $f(x) = x - 1$ for $x \neq 0$. The equivalence relation which only identifies 0 and 1, has finite equivalence classes but 0 is a fixed point and 1 is preperiodic.

V. Operations. Finally, we come to some operations with invariant equivalence relations. The statements we give are self-explanatory, and their verification does not need special ideas. Similar statements are used by different authors and in various forms (see [168], pp. 674-690, also [50, 57, 3]).

If $\equiv_1$ is an equivalence relation on a set X and $\equiv_2$ an equivalence relation on $X/\equiv_1$, then by $x_1 \equiv x_2 :\Longleftrightarrow [x_1]_{\equiv_1} \equiv_2 [x_2]_{\equiv_1}$ there is defined an equivalence relation $\equiv$ on X, which we want to denote by $\equiv_2 \circ \equiv_1$.

In case that X is a topological space and $\equiv_2$ is closed in $X/\equiv_1 \times X/\equiv_1$, the equivalence relation $\equiv$ is closed in $X \times X$ since it is the preimage of $\equiv_2$ with respect to the product map $\phi_{\equiv_1} \times \phi_{\equiv_1}$. This yields the following statement:

Proposition A.4. (Composition of invariant equivalence relations)
Let $\equiv_1$ *be an invariant (completely invariant) equivalence relation on a topological dynamical system* $(X,(f_i)_{i\in I})$ *and* $\equiv_2$ *an invariant (completely invariant) equivalence relation on* $(X,(f_i)_{i\in I})/\equiv_1$.

Then $\equiv_2 \circ \equiv_1$ *forms an invariant (completely invariant) equivalence relation on the topological dynamical system* $(X,(f_i)_{i\in I})$. ■

If $\equiv_1$ and $\equiv_2$ are equivalence relations on a set X and $\equiv_1$ is contained in $\equiv_2$ (as a subset of $X \times X$), then $[x_1]_{\equiv_1} \equiv [x_2]_{\equiv_1} :\Longleftrightarrow x_1 \equiv_2 x_2$ defines an equivalence relation $\equiv$ on $X/\equiv_1$. We want to denote the latter by $\equiv_2 / \equiv_1$. Obviously $\equiv_2$ and $(\equiv_2 / \equiv_1) \circ \equiv_1$ coincide.

Considering II., one easily obtains the following

Proposition A.5. (The 'quotient' of two invariant equivalence relations)
Let $\equiv_1$ *and* $\equiv_2$ *be invariant equivalence relations on a topological dynamical system* $(X,(f_i)_{i\in I})$. *Further, assume that* $\equiv_1$ *is contained in* $\equiv_2$ *and that* $X/\equiv_2$ *is a Hausdorff space.*

Then $\equiv_2 / \equiv_1$ *is invariant on* $(X,(f_i)_{i\in I})/\equiv_1$. *Moreover,* $\equiv_2 / \equiv_1$ *is completely invariant if* $\equiv_2$ *is.* ■

Finally, for a given family $(\equiv_j)_{j\in J}$ of invariant equivalence relations on a topological dynamical system $(X,(f_i)_{i\in I})$, the equivalence relation $\bigcap_{j\in J} \equiv_j$ is also invariant. Therefore, if $E \subset X \times X$, there exists a minimal invariant equivalence relation $\equiv$ containing E. We call it *generated* by E and the elements of E *generation rules* for $\equiv$.

In general the intersection of completely invariant equivalence relations can fail to be completely invariant. Let us give an example for that.

Example A.6. Consider the set $X = \{x_1, x_2, x_3, x_4, x_5, x_6, x_7\}$ with the discrete topology, the map f defined by $f(x_1) = f(x_2) = f(x_3) = x_1, f(x_4) = f(x_5) = x_2$ and $f(x_6) = f(x_7) = x_3$, and the two equivalence relations $\equiv_1, \equiv_2$ on X with $X/\equiv_1 = \{\{x_1\},\{x_2,x_3\},\{x_4,x_6\},\{x_5,x_7\}\}$ and $X/\equiv_2 = \{\{x_1\},\{x_2,x_3\},\{x_4,x_7\},\{x_5,x_6\}\}$. They are completely f-invariant but their intersection is not.

A.2 Shift-invariant factors

On the set $\mathbb{S}^{\mathbb{N}}$ of all sequences over a finite alphabet $\mathbb{S} = \{0, 1, \ldots, q\}$ consider the shift σ and its inverse branches, the *backward-shifts* $\tau_i; i = 0, 1, \ldots, q$ defined by $\tau_i(s_1 s_2 s_3 \ldots) = i s_1 s_2 \ldots$ for $s_1 s_2 s_3 \ldots \in \mathbb{S}^{\mathbb{N}}$. For further notions see Definition 2.1.

Now we want to discuss shift-invariance of equivalence relations, i.e. invariance with respect to some of the maps $\tau_0, \tau_1, \ldots, \tau_q, \sigma$. We consider $\mathbb{S}^{\mathbb{N}}$ as being equipped with the product topology, in other words, as a Cantor discontinuum (see [142]). To motivate our discussion let us start with a simple example.

Example A.7. (Simple topological description of the Tent map)
On the interval $[0,1]$ consider the *Tent map* ρ defined by $\rho(x) = 2x$ for $x \le \frac{1}{2}$ and $\rho(x) = 2 - 2x$ for $x \ge \frac{1}{2}$ (compare [138, 36]).

For studying the dynamics of ρ it is helpful to divide $[0,1]$ into the intervals $A_0 = [\frac{1}{2}, 1]$ and $A_1 = [0, \frac{1}{2}]$, and to code each point $x \in [0,1]$ by a sequence of symbols from $\{0,1\}$ whose n-th member is 0 if $\rho^{n-1}(x) \in A_0$ and 1 if $\rho^{n-1}(x) \in A_1$. Exactly the points with an iterate equal to $\frac{1}{2}$ possess two representations. In particular, $\frac{1}{2}$ provides the sequences $00\overline{1}$ and $10\overline{1}$.

It is not hard to see that each sequence in $\{0,1\}^{\mathbb{N}}$ appears as the coding of a point in $[0,1]$ and that different points yield different codings. Moreover, the coding is continuous at each point with no iterate equal to $\frac{1}{2}$. Therefore, $[0,1]$ is obtained from $\{0,1\}^{\mathbb{N}}$ by identifying the sequence $\mathbf{w}00\overline{1}$ with $\mathbf{w}10\overline{1}$ for each word $\mathbf{w}$ of symbols 0,1.

$\{0,1\}^{\mathbb{N}}$ and $[0,1]$ are compact Hausdorff spaces. So by the considerations in A.1, III. the identification defines an invariant equivalence relation $\equiv$ on $(\{0,1\}^{\mathbb{N}}, \sigma, \tau_0, \tau_1)$ with the following property: From the topological viewpoint, factorization of $\{0,1\}^{\mathbb{N}}, \sigma, \tau_0$ and τ_1 by $\equiv$ yields $[0,1], \rho$, and the two inverse branches of ρ with images A_0 and A_1.

Many statements which are obviously valid for the shift map remain true for the Tent map. For example, the number of periodic points for a given period is not changed, and both topological entropies are equal to $\ln 2$.

The proceeding in Example A.7 is typical in attempting to understand a complicated (chaotic) map, in particular in ergodic theory (see [32]): The domain of the map is covered by finitely many sets. Each of the sets is represented by a symbol, and all points are coded by looking in which of the sets their iterates are contained. One investigates the given map by use of the system of symbol sequences obtained.

Note that in one-dimensional real dynamics the coding of a point by a symbol sequence was called itinerary and developed into a very powerful tool by MILNOR and THURSTON [118] and other authors. The adaption of this concept to the angle-doubling map is central in our book.

Self-similar sets as shift-invariant factors. One class of examples for shift-invariant equivalence relations is originated in the investigation of the topology of self-similar sets (compare [73, 5, 74]).

Let (X,d) be a complete metric space and consider contractions $f_i; i = 0,1,\ldots,q$ acting on X. (A map f on a metric space X is said to be a *contraction* if $d(f(x_1), f(x_2)) \le r d(x_1, x_2)$ for some fixed positive number and all $x_1, x_2 \in X$.) Then there exists a unique non-empty compact set A satisfying $A = \bigcup_{i=0}^{q} f_i(A)$ and called the *self-similar set* for $f_0, f_1, \ldots, f_q$ (see [68]).

The set sequence $(f_{s_1} f_{s_2} \ldots f_{s_n}(A))_{n=1}^{\infty}$ assigned to a given symbol sequence $\mathbf{s} = s_1 s_2 s_3 \ldots \in \{0,1,\ldots,q\}^{\mathbb{N}}$ is monotonically decreasing. Since all sets in the sequence are compact, their intersection is not empty, and since the f_i are contractions, this intersection is a single set which depends continuously on $\mathbf{s}$. Conversely, to each $x \in A$ there exists a sequence $\mathbf{s} = s_1 s_2 s_3 \ldots \in \{0, 1, \ldots, q\}^{\mathbb{N}}$ with $\{x\} = \bigcap_{n=1}^{\infty} f_{s_1} f_{s_2} \cdots f_{s_n}(A)$. Such a sequence $\mathbf{s}$ is called an *address* of x (compare [10]).

Precisely those points which belong to more than one of the pieces $f_i(A); i = 0, 1, \ldots, q$ have more than one address. The map assigning to each sequence the point coded forms a semi-conjugacy from the topological dynamical system $(\{0, 1, \ldots, q\}^{\mathbb{N}}, \tau_0, \tau_1, \ldots, \tau_q)$ onto $(A, f_0, f_1, \ldots, f_q)$. According to the discussion in A.1, III., $(A, f_0, f_1, \ldots, f_q)$ can be considered as an invariant factor of $(\{0, 1, \ldots, q\}^{\mathbb{N}}, \tau_0, \tau_1, \ldots, \tau_q)$.

The corresponding invariant equivalence relation identifies sequences being the address of the same point in A. Let us illustrate this by a well-known example.

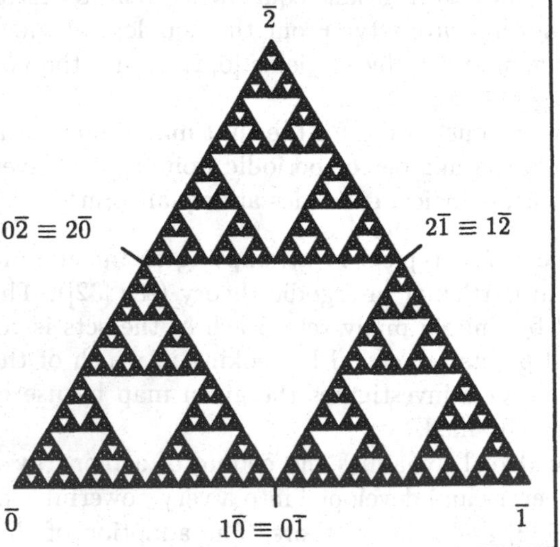

Fig. A.1. Sierpiński triangle as shift-invariant factor

Example A.8. (Sierpiński triangle as shift-invariant factor of $\{0, 1, 2\}^{\mathbb{N}}$)
For the linear contractive maps $f_0, f_1, f_2 : \mathbb{R}^2 \mapsto \mathbb{R}^2$ given by $f_0((x, y)) = (\frac{x}{2}, \frac{y}{2})$, $f_1((x, y)) = (\frac{x+1}{2}, \frac{y}{2})$, $f_2((x, y)) = (\frac{x}{2} + \frac{1}{4}, \frac{y}{2} + \frac{\sqrt{3}}{4})$, the self-similar set A is no more than the Sierpiński triangle (see Figure A.1).

A is 'spanned' by the fixed points $z_0 = (0, 0)$ of f_0, $z_1 = (1, 0)$ of f_1 and $z_2 = (\frac{1}{2}, \frac{\sqrt{3}}{2})$ of f_2 having the addresses $\overline{0}, \overline{1}$ and $\overline{2}$, respectively. Two different pieces $f_i(A), f_j(A); i, j \in \{0, 1, 2\}$ of A intersect in the point $f_j(z_i) = f_i(z_j) = \frac{1}{2}(z_i + z_j)$, which is mapped to z_i by f_j and to z_j by f_i. Consequently, this point has the two addresses $j\overline{i}$ and $i\overline{j}$ (see Figure A.1).

Further, it is not hard to check that two different sequences $\mathbf{s}, \mathbf{t} \in \{0,1,2\}^{\mathbb{N}}$ are addresses of one and the same point iff $\{\mathbf{s}, \mathbf{t}\} = \{\mathbf{w}j\bar{i}, \mathbf{w}i\bar{j}\}$ for some different $i, j \in \{0,1,2\}$ and some word $\mathbf{w} \in \{0,1,2\}^*$.

Therefore, the topological dynamical system (A, f_0, f_1, f_2) is conjugate to $(\{0,1,2\}^{\mathbb{N}}, \tau_0, \tau_1, \tau_2)/\equiv$, where $\equiv$ is the invariant equivalence relation generated by $E = \{(1\bar{0}, 0\bar{1}), (2\bar{1}, 1\bar{2}), (0\bar{2}, 2\bar{0})\}$ (compare [5], Theorem 8).

Supplementarily, we want to refer to the papers [31, 8] which contain deeper information on the topological structure of the Sierpiński-triangle.

The invariant equivalence relation in Example A.8 is already the least one containing E and identifying images of equivalent points. Nevertheless, it is closed. We call this kind of generation of an invariant equivalence relation *algebraic.*

Invariant factors of simplest type. One easily sees that the interval $[0,1]$ is the self-similar set for $f_0(x) = 1 - \frac{x}{2}$ and $f_1(x) = \frac{x}{2}$, and that the restrictions of the two maps f_0, f_1 provide the two branches of the Tent map ρ. So the description of $[0,1]$ as an invariant factor of $(\{0,1\}^{\mathbb{N}}, \tau_0, \tau_1)$ given above is the same as in Example A.7.

The invariant equivalence relation is algebraically generated by the rule $(00\bar{1}, 10\bar{1})$. In contrast to the equivalence relation in Example A.8, we have σ-invariance and beyond this complete σ-invariance.

We finish our discussion of shift-invariant factors by describing a generalization of such simple identification as that obtained for the Tent map.

Proposition A.9. (Simplest invariant factors of $(\{0,1\}^{\mathbb{N}}, \tau_0, \tau_1, \sigma)$)
For an equivalence relation $\equiv$ on $\{0,1\}^{\mathbb{N}}$ the following statements are equivalent:

(i) $\equiv$ is invariant on $(\{0,1\}^{\mathbb{N}}, \tau_0, \tau_1, \sigma)$ and has only finite equivalence classes, and the set $\phi_{\equiv}(\tau_0(\{0,1\}^{\mathbb{N}})) \cap \phi_{\equiv}(\tau_0(\{0,1\}^{\mathbb{N}}))$ consists of exactly one point.

(ii) $\equiv$ is the invariant equivalence relation on $(\{0,1\}^{\mathbb{N}}, \tau_0, \tau_1)$ which is algebraically generated by $\{(0\mathbf{s}, 1\mathbf{s})\}$ for some non-periodic $\mathbf{s} \in \{0,1\}^{\mathbb{N}}$.

Proof: If (i) is valid, then there exists exactly one equivalence class E containing sequences with first symbols 0 and 1, say $0\mathbf{s}, 1\mathbf{t}$ for some $\mathbf{s}, \mathbf{t} \in \{0,1\}^{\mathbb{N}}$. We show that $\mathbf{s} = \mathbf{t}$.

Assume that $\mathbf{s} \neq \mathbf{t}$. Then $\sigma^n(0\mathbf{s}), \sigma^n(1\mathbf{t}) \in E$ for some $n \in \mathbb{N}$, hence $\sigma^{in}(0\mathbf{s}), \sigma^{in}(1\mathbf{t}) \in E$ for that n and all $i \in \mathbb{N}$. Therefore, all sequences in E are periodic, which follows from Proposition A.2 and the simple fact that $\equiv$ must be completely σ-invariant. Moreover, the σ- and τ_1-invariance of $\equiv$ imply that $1\mathbf{s} \in E$.

The sequences $0\mathbf{s}$ and $1\mathbf{s}$ cannot be periodic at the same time. So $\mathbf{s} = \mathbf{t}$, and Proposition A.2 and the last argument also show that $\mathbf{s}$ is non-periodic.

We have $E = \{0\mathbf{s}, 1\mathbf{s}\}$ for some non-periodic $\mathbf{s} \in \{0,1\}^{\mathbb{N}}$, and (by use of [5]; Theorem 8) one easily completes the proof. ∎

A.3 Further interesting examples

We want to conclude with some further examples which will show that (completely) invariant factors appear in various topological situations. The first example stands for the consideration of (Riemann) manifolds as completely invariant factors.

Example A.10. (n-torus as completely invariant factor)
Consider the translations $f_i; i = 1,2,\ldots,n$ given by $f_i((x_1,\ldots,x_i,\ldots,x_n)) = (x_1,\ldots,x_i+1,\ldots,x_n); (x_1,x_2,\ldots,x_n) \in \mathbb{R}^n$ on $\mathbb{R}^n$ and the equivalence relation which identifies each point of $\mathbb{R}^n$ with its images under the maps $f_i; i = 1,2,\ldots,n$. Obviously, $\equiv$ is completely invariant, and the corresponding completely invariant factor is homeomorphic to the n-torus.

In Chapter 1.1.2 we have mentioned the fact of fruitful interaction between dynamics and continuum theory. The following two examples should underline this fact.

Example A.11. (Suspension and matchbox manifolds)
A topological dynamical system (Y,g) can be considered as a description of a physical system: a state y is transformed into the state $g^n(y)$ after n steps, and if g is a homeomorphism from Y onto Y, then the past of each state is uniquely determined.

Systems which are continuous in time can be modelled by flows. By a *flow* a topological dynamical system $(X,(f_t)_{t\in\mathbb{R}})$ is understood, where $(f_t)_{t\in\mathbb{R}}$ is a one-parametric family of homeomorphisms satisfying the following properties: For all $t_1,t_2 \in \mathbb{R}$ it holds $f_{t_1+t_2} = f_{t_1} \circ f_{t_2}$, and for all $x \in X$ the map $t \mapsto f_t(x)$ is continuous. (In other words, the additive group of real numbers acts continuously on X.)

If g is a homeomorphism on a topological space Y, then there is a construction which embeds (Y,g) into a flow in such a way that g appears as the state transition after a time of length 1. The idea for this is already due to POINCARÈ (see [1, 54]).

Let I be the unit-interval, and on $X = Y \times I$ define a map f by $f((y,a)) = (g(y),a); y \in Y, a \in I$. Further, let $\equiv$ be the equivalence relation on X with proper identifications $(y,1) \equiv f(y,0); y \in Y$.

$\equiv$ is the union of the diagonal $\Delta = \{((y,a),(y,a)) \mid (y,a) \in X\}$ and the sets $M = \{((y,1),(g(y),0)) \mid y \in Y\}$ and $M' = \{((g(y),0),(y,1)) \mid y \in Y\}$, and M, M' can be considered as the product of $\{(0,1)\}$ and the graph of g.

If Y is Hausdorff, then Δ is closed in $X \times X$ and the graph of g is closed in Y (e.g., see [51], 2.3.C., 2.3.22.). This implies closeness of $\equiv$ in

$X \times X$. Moreover, one easily sees that $\equiv$ is completely invariant on (X, f). The corresponding completely invariant factor is called the ***suspension*** of the homeomorphism g.

$(X/\equiv, (f_t)_{t\in\mathbb{R}})$ with $f_t([(y,a)]_\equiv) = [(g^{[a+t]}(y), a+t-[a+t])]_\equiv$ for $t \in \mathbb{R}$ and $(y,a) \in X$ is a flow. (Here $[s]$ denotes the integer part of a real number s.) The factor map $f/\equiv$ coincides with f_1, and the map $y \in Y \mapsto [(y,0)]_\equiv$ embeds the topological dynamical system (Y, g) into the flow $(X/\equiv, (f_t)_{t\in\mathbb{R}})$.

Many attractors in differentiable dynamics are locally the topological product of a Cantor set and an interval. This was the reason for introducting the concept of a *matchbox manifold* by AARTS and MARTENS [1]. This concept is very instructive: Let X be a separable and metrizable topological space X. Then an open subset of X is said to be a *matchbox* if it is the product of the open interval $]0,1[$ and a zero-dimensional topological space, and X is called a matchbox manifold if it is the union of matchbox manifolds. Let us add the notion *match* for a connectedness component of a matchbox.

AARTS and MARTENS gave the following remarkable characterization of one-dimensional flows (see [1]): A matchbox manifold is said to be orientable if there exists a continuous orientation on the set of all matches, and this is equivalent to the statement that it can be represented as the suspension of a homeomorphism on a zero-dimensional space. A one-dimensional separable and metrizable space supports a flow without rest points if it is an orientable matchbox manifold.

Example A.12. (A simple topological description of the bucket handle and the solution of an old problem)
Starting from the Cantor set $A = \{\sum_{i=1}^{\infty} \frac{m_i}{3^i} | m_i \in \{0,2\}, i \in \mathbb{N}\}$, a planar continuum K can be constructed as follows (see [89, 69]):
As illustrated in Figure A.2, each $x \in A$ is connected with $1-x \in A$ by a semi-circle lying above the real line, and for each $x \in A$ and each $n \in \mathbb{N}$ the points $\frac{2+x}{3^n} \in A$ and $\frac{3-x}{3^n} \in A$ are connected by a semi-circle lying below it.

The topological space K obtained is a special (non-orientable) matchbox manifold, the ***bucket handle***. It is not arcwise connected and decomposes into infinitely many arcwise components called the ***composants***, and it seems that the components different from that containing the endpoint $0 \in A$ are mutually homeomorphic.

During the fifties Knaster asked whether this was really true. The problem remained open for a long time, but recently BANDT [4] was able to give an affirmative answer.

The important step from the 'rigid' to the dynamic consideration of K was provided by SMALE [157]. He obtained K as the attractor of a horseshoe map, and in the meantime K is much more widely known as the ***horseshoe*** than the bucket handle.

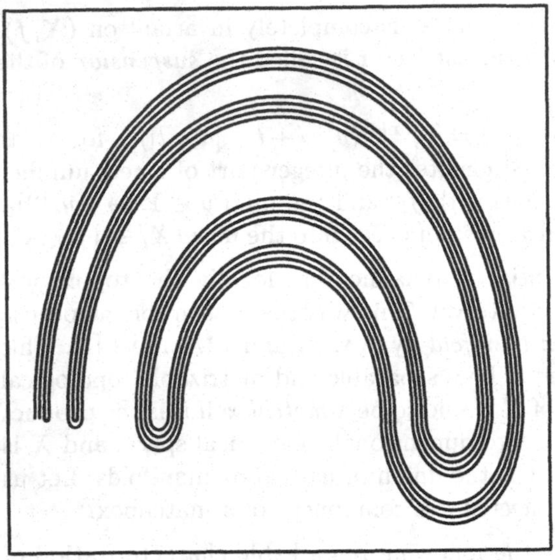

Fig. A.2. Bucket handle - the topological view

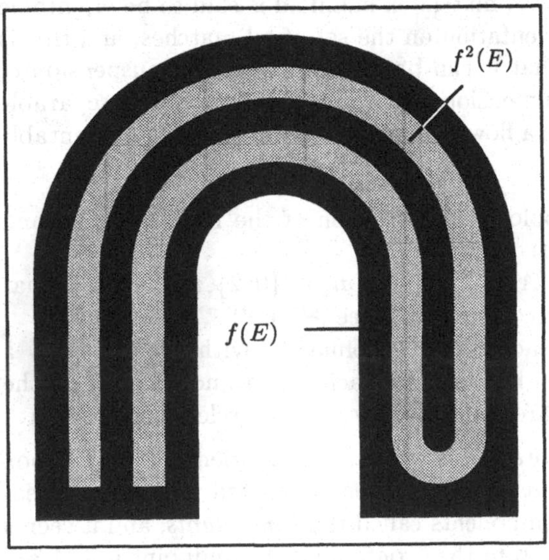

Fig. A.3. Horseshoe - the dynamical view

Roughly speaking, a horseshoe map f can be described as follows: First it contracts a square E horizontally and expands it vertically. Then it puts the result back inside E in a horseshoe-shaped figure A as illustrated by Figure A.3. K is homeomorphic to the intersection of the sets $A \supset f(A) \supset f^2(A) \ldots$.

Ignore the dimensioning in horizontal dimension. Then in the vertical direction f acts like the Tent map (see Example A.7). Therefore, K is homeomorphic to the inverse limit $\{\ldots x_{-2}x_{-1}x_0 \mid x_0, x_{-1}, x_{-2}, \ldots \in [0,1]; \rho(x_i) = x_{i+1}$ for $i = -1, -2, -3, \ldots\} \subset [0,1]^{\mathbb{N}}$. This is the starting point for a topological description of the bucket handle, which we want to sketch now.

It is useful to write the above inverse limit in an unusual form: appending to each left-side sequence $\ldots x_{-2}x_{-1}x_0$ the sequence $\rho(x_0)\rho^2(x_0)\rho^3(x_0)\ldots$, the space K can be considered as the subspace $\{\ldots x_{-2}x_{-1}x_0x_1x_2\ldots \mid \rho(x_i) = x_{i+1}$ for all $i \in \mathbb{Z}\}$ of $[0,1]^{\mathbb{Z}}$.

Now a look at Example A.7 suggests a description of K as an invariant factor of $\{0,1\}^{\mathbb{Z}}$ with respect to both the left- and the right-shift: the corresponding equivalence relation $\equiv$ identifies two sequences if they are equal on the left of some coordinate and coincide with $0\overline{1}$ on the right of this coordinate.

BANDT showed that $[\mathbf{s}_1]_{\equiv}$ and $[\mathbf{s}_2]_{\equiv}$ for $\mathbf{s}_1, \mathbf{s}_2 \in \{0,1\}^{\mathbb{Z}}$ belong to the same composant iff $\mathbf{s}_1$ and $\mathbf{s}_2$ are equal on the left of some coordinate. This was the key to solve the above problem.

Solenoids also belong to the class of matchbox manifolds, and in contrast to the Knaster continuum, they are orientable (e.g., compare [54]). Recently, DE MAN (see [102]) proved that any two composants of any two solenoids are homeomorphic. For this, he adapted the ideas from [4] to solenoids.

The main (completely) invariant equivalence relations considered in this book are the Julia equivalences. There is a direct generalization of them: for $d \in \mathbb{N} \setminus \{1\}$ replace the angle-doubling map on $\mathbb{T}$ by $\beta \mapsto d\beta \bmod 1$ and the symmetry $'$ by the rotation $\beta \mapsto (\beta + \frac{1}{d}) \bmod 1$.

It would be nice to have an extension of the theory of Julia equivalences to all completely invariant equivalence relations obtained for fixed d. The reader can foresee that one would get an abstract theory of connected Julia sets for the polynomials $z \mapsto z^d + c; c \in \mathbb{C}$ and of the corresponding connectedness locus consisting of all c with connected Julia sets. Note that LAU and SCHLEICHER [91, 92, 151, 152] have started a combinatorial investigation of such connectedness loci, and that by a recent result of MCMULLEN [110] they are universal for holomorphic families of rational maps in a certain sense.

The maps $z \mapsto z^d + c; c \in \mathbb{C}$ have only one (finite) critical point, namely 0. This is the reason for that their combinatorics can be described by only one kneading sequence. If a polynomial has more than one critical point, the situation is much more complicated.

Recently, KIWI [85] gave a substantial step to the understanding of combinatorics of such polynomials. Motivated by the work of BIELEFELD, FISHER and HUBBARD [13], he described the equivalence relations on the rationals in $\mathbb{T}$ which are induced by the rational dynamic ray landing for polynomials with all periodic orbits repelling.

KIWI's interesting work gives an idea to generalize the discussion in this book in view of an abstract theory of connected Julia sets of polynomials. However, already the understanding of the combinatorics of cubic maps is not nearly complete. In order to illustrate this, let us finish our discussion by stating an unsolved problem, which is due to THURSTON [167]. The statement of non-wandering triangles for the angle-doubling map (see Theorem 2.11) played an important role in the quadratic case, and it would be nice to have the same or a similar statement for $d = 3$ or higher.

Problem. Does for a given positive integer $d > 2$ a triangle with angles in $\mathbb{T}$ exist whose iterates under the map $\beta \in \mathbb{T} \mapsto d\beta \bmod 1$ are non-degenerate and have mutually disjoint interiors?

References

1. Aarts, J. M.; Martens, M., Flows on one-dimensional spaces, *Fund. Math.* 131 (1988), 53-67.
2. Atela, P., Bifurcations of dynamic rays in complex polynomials of degree two, *Ergod. Th. Dynam. Sys.* 12 (1991), 401-423.
3. Auslander, J., Minimal flows and their extensions, North-Holland 1988.
4. Bandt, C., Composants of the horseshoe, *Fund. Math.* 144 (1994), 231-241.
5. Bandt, C.; Keller, K., Self-similar sets 2: A simple approach to the topological structure of fractals, *Math. Nachr.* 145 (1991), 27-39.
6. Bandt, C.; Keller, K., Symbolic dynamics for angle-doubling on the circle, I. The topology of locally connected Julia sets, In: Ergodic Theory and Related Topics (U. Krengel, K. Richter, V. Warstat, eds.), *Lecture Notes in Math.* 1514, Springer-Verlag 1992, 1-23.
7. Bandt, C.; Keller, K., Symbolic dynamics for angle-doubling on the circle, II. Symbolic description of the abstract Mandelbrot set, *Nonlinearity* 6 (1993), 377-392.
8. Bandt, C.; Retta, T., Topological spaces admitting a unique fractal structure, *Fund. Math.* 141 (1992), 257-268.
9. Bandt, C.; Retta, T., Self-similar sets as inverse limits of finite topological spaces, In: Topology, Measures, and Fractals (C. Bandt, J. Flachsmeyer, H. Haase, eds.), Akademie-Verlag 1992, 41-46.
10. Barnsley, M. F., Fractals everywhere, Academic Press 1988.
11. Barnsley, M. F.; Devaney, R. L.; Mandelbrot, B.; Peitgen, H.-O.; Saupe, D.; Voss, R. F., The science of fractal images (With contributions by Y. Fisher and M. McGuire), Springer-Verlag 1988.
12. Beardon, A., Iteration of rational functions, Springer-Verlag 1991.
13. Bielefeld, B.; Fisher, Y.; Hubbard, J., The classification of critically preperiodic polynomials as dynamical systems, *Jour. AMS* 5 (1992), 721-762.
14. Bielefeld, B.; Fisher, Y.; Häseler, F. v., Computing the Laurent series of the map $\Psi: \mathbf{C} - \overline{D} \to \mathbf{C} - M$, *Adv. in Appl. Math.* 14 (1993), 25-38.
15. Blanchard, P., Complex analytic dynamics on the Riemann sphere, *Bull. Amer. Math. Soc.* 11 (1984), 85-141.
16. Blanchard, P.; Chiu, A., Complex dynamics: An informal discussion, In: Fractal Geometry and Analysis (J. Bélair, S. Dubuc, eds.), NATO ASI Series, Series C: Mathematical and Physical Sciences, Vol. 346, Kluwer 1989.
17. Bötkher, L. E., The principal laws of convergence of iterates and their application to analysis, *Izv. Kazan. Fiz.-Math. Obshch.* 14 (1904), 155-234.
18. Branner, B., The Mandelbrot set, In: Chaos and Fractals: The mathematics behind the computer graphics, R. L. Devaney and L. Keen (editors), Proceedings of symposia in applied mathematics, Vol. 39, Amer. Math. Soc., 1989, 75-105.
19. Branner, B.; Hubbard, J. H., The iteration of cubic polynomials, Part 1: The global topology of parameter space, *Acta Math.* 160 (1988), 143-206.

20. Branner, B.; Fagella, N., Homeomorphisms between limbs of the Mandelbrot set, to appear in *J. Geom. Anal.*
21. Brjuno, A. D., Convergence of transformations of differential equations to normal forms, *Dokl. Akad. Nauk USSR* 165 (1965), 987-989.
22. Brooks, R.; Matelski, J. P., The dynamics of 2-generator subgroups of $PSL(2, C)$, In: Proceedings of the 1978 Stony Brook Conference: Riemann Surfaces and Related Topics, Ann. of Math. Stud. 97, Princeton University Press, Princeton 1980.
23. Bruin, H., Combinatorics of the kneading map, *Int. J. Bifurcation Chaos* 5 (1995), 1339-1349.
24. Bullett, S., Sentenac, P., Ordered orbits of the shift, square roots, and the devil's staircase, *Proc. Camb. Phil. Soc.* 115 (1994), 451-481.
25. Carleson, L.; Gamelin, T., Complex dynamics, Springer-Verlag 1993.
26. Cayley, A., The Newton-Fourier imaginary problem, *Amer. J. Math.* 2 (1879), 97.
27. Collet, P.; Eckmann, J.-P., Iterated maps on the interval as dynamical systems, Birkhäuser 1980.
28. Collingwood, E. F.; Lohwater, A. J., The theory of cluster sets, Cambridge University Press 1966.
29. Coven, E. M.; Hedlund, G. H., Sequences with minimal block growth, *Math. Systems Theory* 7 (1973), 138-153.
30. Cremer, H., Über die Häufigkeit der Nichtzentren, *Math. Ann.* 115 (1938), 573-580.
31. Debski, W.; Mioduszewski, J., Simple plane images of the Sierpiński triangular curve are nowhere dense, *Colloq. Math.* 59 (1990), 125-140.
32. Denker, M.; Grillenberger, C.; Sigmund, K., Ergodic theory on compact spaces, *Lecture Notes in Mathematics* 527, Springer-Verlag 1976.
33. Denker, M.; Urbański, M., On the existence of conformal measures, *Trans. Amer. Math. Soc.* 328 (1991), 563–587.
34. Denker, M.; Urbański, M., Absolutely continuous invariant measures for expansive rational maps with rationally indifferent periodic points, *Forum Math.* 3 (1991), 561–579.
35. Denker, M.; Urbański, M., The dichotomy of Hausdorff measures and equilibrium states for parabolic rational maps, In: Ergodic Theory and Related Topics (U. Krengel, K. Richter, V. Warstat, eds.), *Lecture Notes in Math.* 1514, Springer-Verlag 1992, 90–113.
36. Devaney, R. L., An introduction to chaotic dynamical systems, 2nd ed., Addison Wesley 1989.
37. Devaney, R. L., Knaster-like continua and complex dynamics, *Ergod. Th. and Dynam. Sys.* 13 (1993), 627-634.
38. Devaney, R. L., The Mandelbrot set and the Farey tree, *Amer. Math. Monthly* 106 (1999), 289-302.
39. Douady, A., Systems dynamiques holomorphes, Séminaire Bourbaki, No. 599 (1982); *Astérisque* 105-106 (1983), 39-63.
40. Douady, A., Computing angles in the Mandelbrot set, In: Chaotic Dynamics and Fractals, Academic Press 1986, 155-168.
41. Douady, A., Descriptions of compact sets in C, In: Topological Methods in Modern Mathematics, Publish or Perish 1993, 429-465.
42. Douady, A.; Hubbard, J., A proof of Thurston's characterization of rational functions, *Acta Math.* 171 (1993), 263-297.
43. Douady, A.; Hubbard, J., Étude dynamique des polynômes complexes, *Publications Mathématiques d'Orsay* 84-02 (1984) (première partie) and 85-02 (1985) (deuxième partie).

44. Douady, A.; Hubbard, J., On the dynamics of polynomial-like mappings, *Ann. Sci. École Norm. Sup. (4)*, 18 (1985), 287-343.
45. Fagella, N., Surgery on the limbs of the Mandelbrot set, In: Progress in Complex Dynamics, H. Kriete (editor), Pitman Research Notes in Mathematical Sciences 387, 1998, 139-158.
46. Fatou, P., Sur les solutions uniformes de certaines équations fonctionnelles, *C.R. Acad. Sci. Paris* 143 (1906), 546-548.
47. Fatou, P., Sur les équations fonctionnelles, *Bull. Soc. Math. France* 47 (1919), 161-271.
48. Fatou, P., Sur les équations fonctionnelles, *Bull. Soc. Math. France* 48 (1920), 33-94.
49. Fatou, P., Sur les équations fonctionnelles, *Bull. Soc. Math. France* 48 (1920), 208-314.
50. Ellis, R., Lecture notes in topological dynamics, Benjamin 1969.
51. Engelking, R., General Topology, PWN, Warzawa 1977.
52. Falconer, K., The geometry of fractal sets, Cambridge University Press 1985.
53. Falconer, K., Fractal Geometry, John Wiley & Sons 1990.
54. Fokking, R., The structure of trajectories, Dissertation, Technical University Delft 1991.
55. Fisher, Y.; Giarrusso, D., A parameterization of the period 3 hyperbolic components of the Mandelbrot set, *Proc. Amer. Math. Soc.* 123 (1995), 3731-3737.
56. Ghys, E., Transformation holomorphe au voisinage d'une courbe de Jordan, *C. R. Acad. Sc. Paris* 289 (1984), 385-388.
57. Glasner, S., Proximal flows, *Lecture Notes in Mathematics* 517, Springer-Verlag 1988.
58. Goldberg, L., Milnor, J., Fixed points of polynomial maps. II: Fixed point portraits, *Ann. Scient. Ec. Norm. Sup.*, 4e *série*, t. 26 (1993).
59. Graczyk, J.; Świątek, G., Generic hyperbolicity in the logistic family, *Ann. of Math.* 146 (1997), 1-56.
60. Grispolakis, J.; Mayer, J. C.; Oversteegen, L. O., Building blocks for quadratic Julia sets, *Trans. A.M.S.* 351 (1999), 1171-1201.
61. Haïssinsky, P., Applications de la chirurgie holomorphe, nottament aux points paraboliques, Thèse, Université de Paris-Sud 1998.
62. Herman, M. R., Are there critical points on the boundary of singular domains, *Comm. Math. Phys.* 99 (1985), 593-612.
63. Herman, M. R., Recent results and some open questions on Siegel's linearization theorem of germs of complex analytic diffeomorphisms of $\mathbf{C}^n$ near a fixed point, VIIIth International Congress on Mathematical Physics (Marseille, 1986), World Sci. Publishing 1987, 138–184.
64. Hofbauer, F., The topological entropy of a transformation $x \mapsto ax(1-x)$, *Monatsh. Math.* 90 (1980), 117-141.
65. Hofbauer, F.; Keller, G., Quadratic maps without assymptotic measure, *Commun. Math. Phys.* 127 (1990), 319-337.
66. Hubbard, J. H., Local connectivity of Julia sets and bifurcation loci: Three Theorems of J.-C. Yoccoz, In: Topological Methods in Modern Mathematics, Publish or Perish 1993, 467-511.
67. Hubbard, J. H.; Schleicher, D., The spider algorithm, In: Complex Dynamics: The mathematics behind the Mandelbrot and Julia sets, *AMS Lecture Notes* (1994).
68. Hutchinson, J. E., Fractals and self-similarity, *Indiana Univ. Math. J.* 30 (1981), 713-747.
69. Janiszewski, Z., Oeuvres Choisies, PWN, Warzawa 1962.

70. Julia, G., Mémoire sur l'itération des fonctions rationnelles, *J. Math. Pure Appl.* 8 (1918), 47-245.
71. Jung, W., Families of homeomorphic subsets of the Mandelbrot set, Preprint, Aachen 1999.
72. Jungreis, I., The uniformization of the complement of the Mandelbrot set, *Duke Math. J.* 52 (1985), 935–938.
73. Kameyama, A., On the self-similar sets with frames, In: The study of dynamical systems (N. Aoki, edit.), World Scientific 1989, 1-9.
74. Kameyama, A., Julia sets and self-similar sets, *Topology Appl.* 54 (1993), 241-251.
75. Kauko, V., Trees of visible components in the Mandelbrot set, Preprint, Jyväskylä 1999.
76. Keller, K., The abstract Mandelbrot set - an atlas of abstract Julia sets, In: Topology, Measures, and Fractals (C. Bandt, J. Flachsmeyer, H. Haase, eds.), Akademie-Verlag 1992, 76-81.
77. Keller, K., Symbolic dynamics for angle-doubling on the circle, III. Sturmian sequences and the quadratic map, *Ergod. Th. and Dynam. Sys.* 14 (1994), 787-805.
78. Keller, K., Symbolic dynamics for angle-doubling on the circle, IV. Equivalence of abstract Julia sets, *Atti del Seminario dell'Universita de Modenà* XLII (1994), 301-321.
79. Keller, K., Invariante Faktoren, Juliaäquivalenzen und die abstrakte Mandelbrotmenge, Habilitationsschrift, Universität Greifswald 1995.
80. Keller, K., A note on the structure of quadratic Julia sets, *Comment. Math. Univ. Carolinae* 38 (1997), 395-406.
81. Keller, K., Correspondence and Translation Principles for the Mandelbrot set, *Stony Brook IMS Preprint 1997/4.*
82. Keller, K., Julia equivalences and abstract Siegel disks, In: Progress in Complex Dynamics, H. Kriete (editor), Pitman Research Notes in Mathematical Sciences 387, 1998, 86-101.
83. Khintchine, A., Kettenbrüche, Mathematisch-Naturwissenschaftliche Bibliothek, Teubner-Verlag Leipzig 1956.
84. Kiwi, J., Non-accessible critical points of Cremer polynomials, *Stony Brook IMS Preprint 1995/2.*
85. Kiwi, J., Rational rays and critical portraits of complex polynomials, *Stony Brook IMS Preprint 1997/15.*
86. Klein, B. G., Homomorphisms of symbolic dynamical systems, *Math. Systems Theory* 6 (1972), 107-122.
87. Koenigs, M. G., Recherches sur les integrals de certain équations fonctionnelles, *Ann. Sci. École Norm. Sup. (3),* 1 (1884), supplément, 1-44.
88. Kriete, H., Herman's proof of the existence of critical points on the boundary of singular domains, In: Progress in Complex Dynamics, H. Kriete (editor), Pitman Research Notes in Mathematical Sciences 387, 1998, 31-40.
89. Kuratowski, K., Théorie des continues irréductibles entre deux points I, *Fund. Math.* 3 (1922), 200-231.
90. Kuratowski, K., Topology, Vol. 1, 2, PWN, Warszawa 1966/1968.
91. Lau, E.; Schleicher, D., Symmetries of fractals revisited, *Math. Intell.* 18 (1996), 45-51.
92. Lau, E.; Schleicher, D., Internal addresses in the Mandelbrot set and irreducibility of polynomials, *Stony Brook IMS Preprint 1994/19.*
93. Lavaurs, P., Une déscription combinatoire de l'involution définie par M sur les rationnels à dénominateur impair, *C. R. Acad. Sc. Paris Série I, t. 303* (1986), 143-146.

94. Leau, L., Étude sur les équations fonctionelles à une ou plusièrs variables, *Ann. Faculté Sci. Toulouse* 11 (1897), E.1-E.110.
95. Levin, G. Disconnected Julia sets and rotation sets, *Ann. Sci. École Norm. Sup. (4),* 29 (1996), 1-22.
96. Levin, G.; van Strien, S., Local connectivity of the Julia sets of real polynomials, *Ann. of Math.* 147 (1998), 471-541.
97. Lyubich, M. Yu., Dynamics of rational transformations: topological picture, *Uspekhi Mat. Nauk* 41 (1986) no. 4 (250), 35-95.
98. Lyubich, M. Yu., Renormalization Ideas in conformal dynamics, In: Cambridge Seminar 'Current Developments in Mathematics', International Press 1995, 155-184.
99. Lyubich, M. Yu., Dynamics of quadratic polynomials III: parapuzzle and SBR measures, *Stony Brook IMS Preprint 1996/5.*
100. Lyubich, M. Yu., Feigenbaum-Coullet-Tresser Universality and Milnor's Hairiness Conjecture, to appear in *Ann. of Math.*
101. Lyubich, M. Yu., Dynamics of quadratic polynomials, I-II, *Acta Math.* 178 (1997), 185-297.
102. de Man, R., On composants of solenoids, *Fund. Math.* 147 (1995), 181-188.
103. Mañé, R.; Sad, P.; Sullivan, D., On the dynamics of rational maps, *Ann. Sci. École Norm. Sup. (4),* 16 (1983), 193-217.
104. Mandelbrot, B., The Fractal Geometry of Nature, Freeman, New York 1983.
105. Mayer, J. C., Complex Dynamics and Continuum Theory, In: Continua (H. Cook, W. T. Ingram, K. T. Kuperberg, A. Lelek, P. Minc, eds.), *Lect. Notes Pure Appl. Math.* 170, Dekker 1995, 133-157.
106. Mayer, J. C.; Oversteegen, L. G., Continuum theory, In: Recent Progress in general topology (M. Hušek, J. van Mill, eds.), North Holland 1992, 453-492.
107. McMullen, C., Frontiers in complex dynamics, *Bull. Amer. Math. Soc. (N.S.)* 31 (1994), 155-172.
108. McMullen, C., Complex Dynamics and Renormalization, Ann. of Math. Stud. 135, Princeton University Press, Princeton 1994.
109. McMullen, C., Renormalization and 3-Manifolds which fiber over the circle, Ann. of Math. Stud. 142, Princeton University Press, Princeton 1996.
110. McMullen, C., The Mandelbrot set is universal, Preprint, Harvard University 1997.
111. Metzler, W., The 'mystery' of the quadratic Mandelbrot set, *Amer. J. Phys.* 62 (1994), 813–814.
112. van Mill, J.; Reed, G. M. (eds.), Open problems in topology, North-Holland 1990.
113. Milnor, J., Self-similarity and hairiness in the Mandelbrot set, In: Computers in Geometry and Topology (Tangora, edit.), *Lect. Notes Pure Appl. Math.* 114, Dekker 1989, 211-257.
114. Milnor, J., Dynamics in one complex variable: Introductory Lectures, *Stony Brook IMS Preprint 1990/5.*
115. Milnor, J., Local Connectivity of Julia sets: Expository Lectures, *Stony Brook IMS Preprint 1992/11.*
116. Milnor, J., Errata for 'Local Connectivity of Julia sets: Expository Lectures', *Stony Brook IMS Preprint 1992/11.*
117. Milnor, J., Periodic orbits, external rays and the Mandelbrot set: An expository account, *Stony Brook IMS Preprint 1999/3.*
118. Milnor, J.; Thurston, W. P., On iterated maps of the interval, *Lecture Notes in Mathematics* 1342 (1988), 465-563.
119. Morse, M.; Hedlund, G. A., Symbolic dynamics II: Sturmian trajectories, *Amer. J. Math.* 62 (1940), 1-42.

120. Nadler, S. B., Continuum Theory, Marcel Dekker 1992.
121. Peitgen, H.-O.; Richter, P., The beauty of fractals, Springer-Verlag 1986.
122. Peitgen, H.-O.; Jürgens, H.; Saupe, D., Chaos and fractals, New frontiers of science, Springer-Verlag 1992.
123. Penrose, C. S., On quotients of the shift associated with dendrite Julia sets of quadratic polynomials, PhD thesis, University of Warwick 1990.
124. Penrose, C. S., Quotients of the shift associated with dendrite Julia sets, Preprint, London 1994.
125. Perez-Marco, R., Solution complete au probleme de Siegel de linearisation d'une application au voisinage d'un point fixe (d'apres J.-C. Yoccoz), Séminaire Bourbaki, Vol. 1991/92; *Astérisque* 206 (1992), 273-310.
126. Perez-Marco, R., Sur les dynamiques holomorphes non linearisables et une conjecture de V.I. Arnold, *Ann. Sci. École Norm. Sup. (4),* 26 (1993), 565-644.
127. Perez-Marco, R., Topology of Julia sets and hedgehogs, Preprint 48, Université Paris-Sud 1994.
128. Perez-Marco, R., Fixed points and circle maps, *Acta Math.* 179 (1997), 243-294.
129. Petersen, C. L., On the Pommerenke-Levin-Yoccoz inequality, *Ergod. Th. and Dynam. Sys.* 12 (1993), 785–806.
130. Petersen, C. L., Local connectivity of some Julia sets containing a circle with an irrational rotation, *Acta Math.* 177 (1997), 163-224.
131. Petersen, C. L., Puzzles and Siegel disks, In: Progress in Complex Dynamics, H. Kriete (editor), Pitman Research Notes in Mathematical Sciences 387, 1998, 50-85.
132. Poirier, A., On the realization of fixed point portraits, *Stony Brook IMS Preprint 1991/20.*
133. Poirier, A., Hubbard Forests, *Stony Brook IMS Preprint 1992/12.*
134. Poirier, A., On Postcritically Finite Polynomials, Part 1: Critical Portraits. *Stony Brook IMS Preprint 1993/5.*
135. Poirier, A., On Postcritically Finite Polynomials, Part 2: Hubbard Trees, *Stony Brook IMS Preprint 1993/7.*
136. Poirier, A., Coexistence of Critical Orbit Types in Sub-Hyperbolic Polynomial Maps, *Stony Brook IMS Preprint 1994/10.*
137. Pommerenke, Ch., On conformal mapping and iteration of rational functions, *Complex Variables Theory Appl.* 5 (1986), 117–126.
138. Preston, C., Iterates of piecewise monoton mappings on an interval, *Lecture Notes in Mathematics* 1347 (1988).
139. Rees, M., A partial description of parameter space of rational maps of degree two: Part 1, *Acta Math.* 168 (1992), 11-87.
140. Remmert, R., Funktionentheorie 1./2., Springer-Verlag 1992/1995.
141. Riedl, J.; Schleicher, D., On the locus of crossed renormalization, In: S. Morosawa (ed.), Proceedings of the Kyoto workshop on complex dynamics, Kyoto 1997.
142. Rinow, W., Lehrbuch der Topologie, Deutscher Verlag der Wissenschaften 1975.
143. Rogers, J. T., Is the boundary of a Siegel disk a Jordan curve?, *Bull. Amer. Math. Soc. (N.S.)* 27 (1992), 284-287.
144. Rogers, J. T., Singularities in the boundaries of local Siegel disks, *Ergod. Th. and Dynam. Sys.* 12 (1992), 803-821.
145. Rogers, J. T., Critical points on the boundaries of Siegel disks, *Bull. Amer. Math. Soc. (N.S.)* 32 (1995), 317-321.

146. Rogers, J. T., Recent results on the boundaries of Siegel disks, In: Progress in Complex Dynamics, H. Kriete (editor), Pitman Research Notes in Mathematical Sciences 387, 1998, 41-49.
147. Rogers, J. T.; Mayer, J. C., Indecomposible continua and the Julia sets of polynomials, *Proc. Amer. Math. Soc.* 117 (1993), 795-802.
148. Schleicher, D., Internal Adresses in the Mandelbrot set and irreducibility of polynomials, PhD thesis, Cornell University 1994.
149. Schleicher, D., Wann kann man ein nichtlineares dynamisches System geradebiegen?, *Spektrum der Wissenschaft*, Dezember 1994, 96-99.
150. Schleicher, D., Rational External Rays of the Mandelbrot Set, *Stony Brook IMS Preprint 1997/13.*
151. Schleicher, D., On Fibers and Local Connectivity of Mandelbrot and Multibrot Sets, *Stony Brook IMS Preprint 1998/13a.*
152. Schleicher, D., On Fibers and Renormalization of Julia Sets and Multibrot Sets, *Stony Brook IMS Preprint 1998/13b.*
153. Schröder, M., Über unendlichviele Algorithmen zur Auflösung der Gleichungen, *Math. Ann.* 2 (1870), 317-365.
154. Schröder, M. R., Fractals, chaos, power laws, W. H. Freeman and Company 1991.
155. Shishikura, M., The boundary of the Mandelbrot set has Hausdorff dimension two, Complex analytic methods in dynamical systems (Rio de Janeiro, 1992), *Astérisque* 222 (1994), 389–405.
156. Siegel, C. L., Iteration of analytic functions, *Ann. of Math. (2)* 43 (1942), 607-612.
157. Smale, S., Differentiable dynamical systems, *Bull. Amer. Math. Soc.* 73 (1967), 747-817.
158. Sørensen, D. E. K., Complex dynamical systems, PhD thesis, Technical University of Denmark 1995.
159. Sullivan, D., Conformal dynamical systems, In: Geometric Dynamics (J. Palis ed.), *Lecture Notes in Math.* 1007, Springer-Verlag 1983, 725-752.
160. Steinmetz, N., Rational iteration: complex analytic dynamical systems, de Gruyter 1993.
161. Steinmetz, N., On Sullivan's classification of periodic stable domains, *Complex Variables, Theory Appl.* 14 (1990), 211-214.
162. Sullivan, D., Quasiconformal homeomorphisms and dynamics III: Topological conjugacy classes of analytic endomorphisms, IHES Preprint, 1984.
163. Sullivan, D., Quasiconformal homeomorphisms and dynamics I: Solution of the Fatou-Julia problem on wandering domains, *Annals Math.* 122 (1985), 401-418.
164. Świątek, G., Hyperbolicity is dense in the real quadratic family, *Stony Brook IMS Preprint 1992/10.*
165. Tan Lei, Similarity between Mandelbrot set and Julia sets, *Commun. Math. Phys.* 134 (1990), 587-617.
166. Tan Lei; Yin Yongcheng, Local connectivity of the Julia set for geometrically finite rational maps, Preprint, École Normale Superieure de Lyon 1994.
167. Thurston, W. P., On the combinatorics and dynamics of iterated rational maps, Preprint, Princeton 1985.
168. de Vries, J., Elements of topological dynamics, Kluwer 1993.
169. Walters, P., An introduction to ergodic theory, Springer-Verlag 1981.
170. Weitkämper, J., Konjugation quadratischer Polynome, Dissertation, Marburg 1988.
171. Yoccoz, J. C., An introduction to small divisors problems, From number theory to physics (Les Houches, 1989), Springer 1992, 659–679.

172. Yoccoz, J. C., Théorème de Siegel, nombres de Brjuno et polynômes quadratique, *Astérisque* 231 (1995), 3-88.
173. Yunping Jiang, Renormalization and geometry in one-dimensional and complex dynamics, World Scientific 1996.

Index

Further symbols used

$CARD(A)$	cardinality of a set A
$\mathrm{cl}\,A$	closure of a subset A of a topological space
∂A	boundary of a subset A of a topological space
$\max(A)$	maximimum of a partially ordered set A
$\min(A)$	minimum of a partially ordered set A
$\inf(A)$	infimum of a partially ordered set A
$\sup(A)$	supremum of a partially ordered set A
$\mathrm{conv}\,A$	convex hull of a set A
$\log$	natural logarithm
$\mathbb{C}$	set of complex numbers
$\mathbb{N}$	set of natural numbers
$\mathbb{N}_0$	set of natural numbers including 0
$\mathbb{Q}$	set of rational numbers
$\mathbb{R}$	set of real numbers
$\mathbb{R}^+$	set of strictly positive real numbers
$\mathbb{Z}$	set of integers
$\|z\|$	absolute value of a complex number z

Printing: Weihert-Druck GmbH, Darmstadt
Binding: Buchbinderei Schäffer, Grünstadt

4. Lecture Notes are printed by photo-offset from the master-copy delivered in camera-ready form by the authors. Springer-Verlag provides technical instructions for the preparation of manuscripts. Macro packages in TEX, LATEX2e, LATEX2.09 are available from Springer's web-pages at

http://www.springer.de/math/authors/b-tex.html.

Careful preparation of the manuscripts will help keep production time short and ensure satisfactory appearance of the finished book.

The actual production of a Lecture Notes volume takes approximately 12 weeks.

5. Authors receive a total of 50 free copies of their volume, but no royalties. They are entitled to a discount of 33.3% on the price of Springer books purchase for their personal use, if ordering directly from Springer-Verlag.

Commitment to publish is made by letter of intent rather than by signing a formal contract. Springer-Verlag secures the copyright for each volume. Authors are free to reuse material contained in their LNM volumes in later publications: A brief written (or e-mail) request for formal permission is sufficient.

Addresses:

Professor F. Takens, Mathematisch Instituut,
Rijksuniversiteit Groningen, Postbus 800,
9700 AV Groningen, The Netherlands
E-mail: F.Takens@math.rug.nl

Professor B. Teissier
Université Paris 7
UFR de Mathématiques
Equipe Géométrie et Dynamique
Case 7012
2 place Jussieu
75251 Paris Cedex 05
E-mail: Teissier@ens.fr

Springer-Verlag, Mathematics Editorial, Tiergartenstr. 17,
D-69121 Heidelberg, Germany,
Tel.: *49 (6221) 487-701
Fax: *49 (6221) 487-355
E-mail: lnm@Springer.de

